Chemistry and Toxicology of Pyrrolizidine Alkaloids

A. R. MATTOCKS

Toxicology Unit
Medical Research Council Laboratories
Carshalton, Surrey, United Kingdom

1986

ACADEMIC PRESS

Harcourt Brace Jovanovich, Publishers
London   Orlando   San Diego   New York
Austin   Montreal   Sydney   Tokyo   Toronto

0537474/

CHEMISTRY

rpl B001122658

COPYRIGHT © 1986 BY ACADEMIC PRESS INC. (LONDON) LTD.
ALL RIGHTS RESERVED.
NO PART OF THIS PUBLICATION MAY BE REPRODUCED OR
TRANSMITTED IN ANY FORM OR BY ANY MEANS, ELECTRONIC
OR MECHANICAL, INCLUDING PHOTOCOPY, RECORDING, OR
ANY INFORMATION STORAGE AND RETRIEVAL SYSTEM, WITHOUT
PERMISSION IN WRITING FROM THE PUBLISHER.

ACADEMIC PRESS INC. (LONDON) LTD.
24–28 Oval Road
LONDON NW1 7DX

United States Edition published by
ACADEMIC PRESS, INC.
Orlando, Florida 32887

British Library Cataloguing in Publication Data

Mattocks, A. R.
        Chemistry and toxicology of pyrrolizidine alkaloids.
        1. Pyrrolizidine alkaloids
        I. Title
        581.19'242   QK898.P9

Library of Congress Cataloging-in-Publication Data

Mattocks, A. R.
        Chemistry and toxicology of pyrrolizidine alkaloids.

        Bibliography: p.
        Includes index.
        1. Pyrrolizidines.   2. Pyrrolizidines–Toxicology.
I. Title.
QP801.P855M38   1986   615.9'5   85-19937
ISBN   0–12–480570–1 (U.S.: alk. paper)

PRINTED IN THE UNITED STATES OF AMERICA

86 87 88 89   9 8 7 6 5 4 3 2 1

Contents

v

Preface

Pyrrolizidine alkaloids (PAs) are found in a large number of plant species occurring throughout the world. Many of them are cytotoxic and these are often responsible for poisoning livestock and people; they include some of the relatively few substances occurring in higher plants which are known to be carcinogens. That there is a growing interest in these compounds is shown by an increase of some 50% in the number of annual literature references relating to their chemistry and biology during the past decade.

Three categories of information are included in this book. Firstly, there is data, much of it tabulated, including structures and plant sources of PAs, physical properties and up-to-date information on animal toxicity. Secondly, there is a general coverage of the chemical and biological properties of PAs. A historical survey is not included, but the book is sufficiently comprehensive to obviate frequent consultation of early literature which may not be easily accessible. Such chemistry has been included as is necessary for understanding metabolic reactions, for the isolation and analysis of PAs and for the preparation of putative metabolites and some synthetic analogues. However, much of the chemistry concerned with the elucidation of structures and stereochemistry of PAs has been adequately reviewed elsewhere and is not repeated here. Syntheses of PAs have received increasing attention in recent years and a survey is included of synthetic methods which could be of practical use for preparing these alkaloids, analogues or radioactive derivatives. Thirdly, recent research on PAs is reviewed critically, principally covering the period from 1968 to mid-1984, during which time many new PAs and plant sources have been discovered and major advances have been made in understanding mechanisms of toxic actions.

This book is intended to be useful in the laboratory, particularly to chemists, biochemists, cell biologists, pathologists and others working with biologically active PAs and derivatives, and I have included such data as I myself would wish to consult frequently. It will also be of interest to veterinary and public health workers concerned with the poisoning of livestock and people by plants and plant products, including foods and herbal medicines. Those engaged in cancer research will find up-to-date surveys of the carcinogenic, mutagenic and antitumor activities of PAs and their metabolites. The uses of PAs and pyrrolic derivatives as research tools, especially in producing cardiopulmonary disease and for their striking antimitotic activity, will interest experimental pathologists. Furthermore, because the book embraces a variety of disciplines, it is expected to be a useful source of background information about fields outside the readers' own specializations; for example, for chemists who wish to learn of the diverse biological actions of PAs and for biologists who require an understanding of the chemistry of PAs and the chemical basis for the reactions of their toxic metabolites.

The book is organized as follows. Chapter 1 includes the structures and plant sources of most of the PAs referred to in later chapters; structure numbers allocated here are used throughout the book. Basic information on chemical and physical properties, isolation, analysis and synthesis are covered in Chapters 2–5. The remainder of the book deals with the metabolism and various biological actions of PAs in animals, people and isolated cell systems. Alongside the effects of PAs, I have included discussions of metabolites believed to be responsible for these actions and of synthetic analogues which have similar effects. The final chapter attempts to summarize current understanding of structure–activity relationships, drawing on information from previous chapters.

I would like to express my appreciation to colleagues who have collaborated in various aspects of our own work, mentioned in this book, including the late Dr. J. M. Barnes, Dr. W. H. Butler, Dr. J. R. P. Cabral, Dr. H. Elizabeth Driver, Mr. R. F. Legg, Dr. Muriel Ord, Dr. R. Plestina and Dr. I. N. H. White; also to Dr. T. A. Connors for his encouragement. I would also like to acknowledge the valuable technical assistance of Alan Alexander, Rodney Jones, Ian Bird and Miss Janet Brown. Finally, my particular thanks is due to Mrs. Margaret Bateman for much patient work typing the manuscript.

A. R. Mattocks

Chemistry and Toxicology
of Pyrrolizidine Alkaloids

1

Structures and Plant Sources of Pyrrolizidine Alkaloids

I. INTRODUCTION

Most of the pyrrolizidine alkaloids (PAs) discussed in this book are esters of hydroxylated 1-methylpyrrolizidines: the amino alcohols are called necines, and the acid moieties necic acids (Manske, 1931). Hepatotoxic PAs are esters of unsaturated necines (having a 1,2-double bond); in this book these alkaloids are referred to as 'unsaturated' PAs, whereas PAs with a saturated necine moiety are called 'saturated' PAs, even if there is unsaturation in the acid moiety.

The main purpose of this chapter is to provide a source of information on the potentially hepatotoxic unsaturated PAs, with their plant sources and chemical structures; these are numbered so that reference can be made to them throughout the book. Most of the known saturated ester PAs, which often coexist with unsaturated PAs in plants, are separately listed. The structures of the necine bases and many of the acid components of PAs are also brought together here.

II. PLANT SOURCES AND STRUCTURES

Useful lists of known PAs with their plant sources and chemical structures have been given by Leonard (1950, 1960), Warren (1955, 1966), Bull *et al.* (1968) and Robins (1982a). Atal and Sawhney (1973) have reviewed PAs from Indian *Crotalarias,* and Pilbeam *et al.* (1983) have screened the seeds of 80 *Crotalaria*

1

species for alkaloids (unidentified, except for monocrotaline). Smith and Culvenor (1981) have given a comprehensive list of plants containing PAs which are known, or expected to be, hepatotoxic.

Table 1.1 lists plants known to contain ester PAs with an unsaturated necine moiety; these alkaloids are listed in Table 1.3. Saturated PAs (prefixed by 'S' in Table 1.1) are listed in Table 1.4. Some additional plants, which have (so far) been found to contain only non-hepatotoxic PAs, are shown in Table 1.2.

Table 1.1 does not distinguish between 'major' and 'minor' alkaloids. For any plant, all the listed alkaloids are not necessarily found together at the same time; the same species growing at different locations or in different seasons may contain different alkaloids. The reader is strongly advised to consult the original literature.

PAs are frequently accompanied in the plant by variable proportions of their corresponding N-oxides. In some instances they may be wholly in the N-oxide form. The N-oxides are not listed separately here; they are usually reduced to the basic alkaloids when PAs are extracted from plants (see Chapter 2). A few PA N-oxides have been isolated directly from plants. These include isatidine (retrorsine N-oxide) (Christie et al. 1949), anadoline (Culvenor et al., 1975a), and the N-oxides of heliotrine and lasiocarpine (Culvenor, 1954), indicine (Kugelman et al., 1976), europine (Zalkow et al., 1978), and curassavine (crude) (Subramanian et al., 1980). Some pyrrolizidine N-oxides, with their plant sources, have been listed by Phillipson (1971) and by Phillipson and Handa (1978).

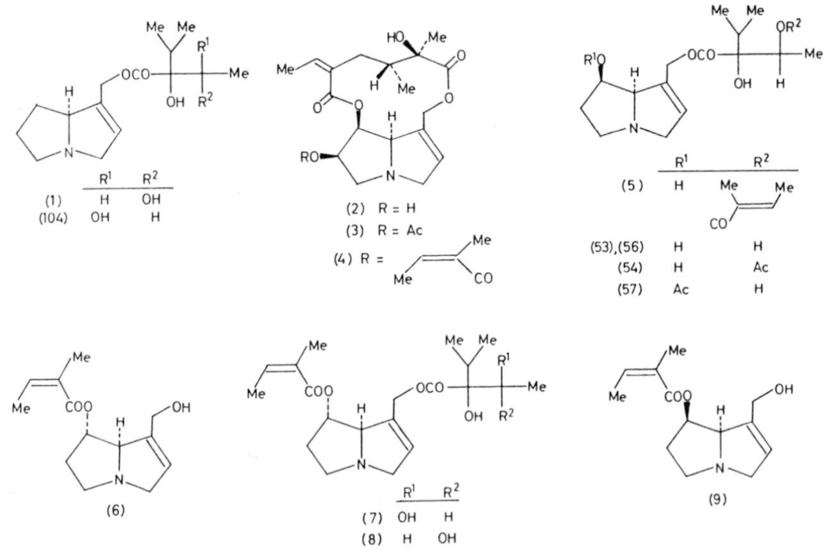

Scheme 1.1.

(10)

(11), (47) R = H
(48) R = Ac

(12) R = H
(13) R = OH

	R^1	R^2	R^3
(14)	OH	Et	H
(59)	OH	Et	Ac
(113)	Et	H	H

(15)

	R^1	R^2
(16)	Me	OH
(18),(58) - diastereomers of (16)		
(42)	OH	Me

Scheme 1.2.

	R^1	R^2	R^3
(17)	H	Me	OH
(29)	OH	H	H

(19)

(20)

	R	R
(21)	H	OH
(72)	H	H
(73)	Ac	H

	R^1	R^2	R^3
(22)	H	H	OH
(69)	OAc	OAc	Me
(127)	H	OAc	Me

(23)

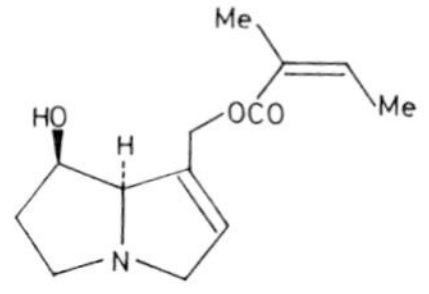

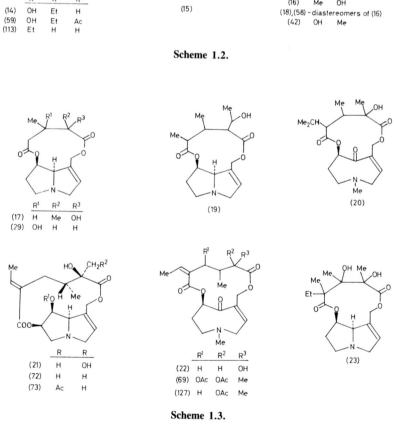

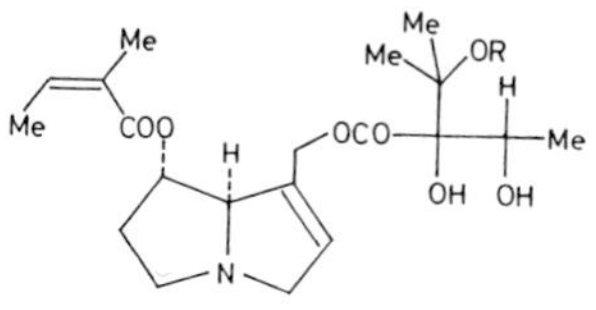

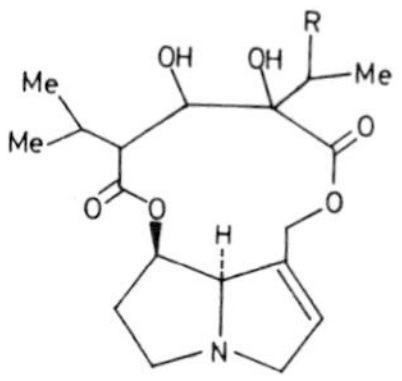

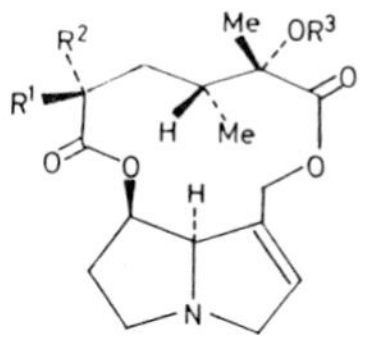

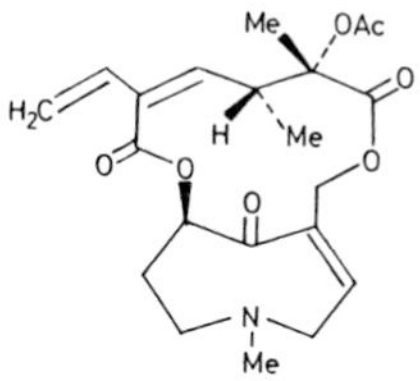

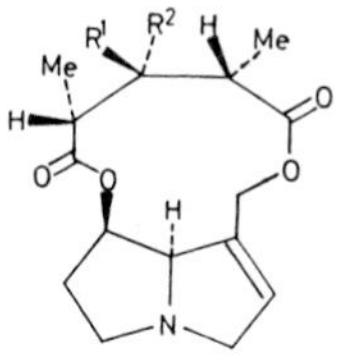

Scheme 1.3.

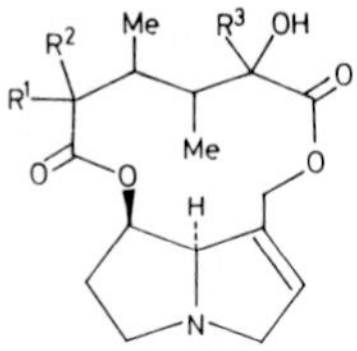

	R^1	R^2	R^3
(24)	Me	H	H
(89)	Me	OH	CH$_2$OH
(90)	Me	OH	CH$_2$Cl
(94)	—CH$_2$—		Me

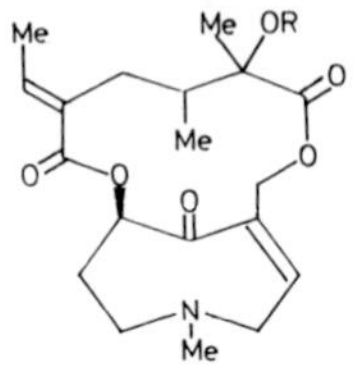

(26)	R = H
(27)	R = Ac

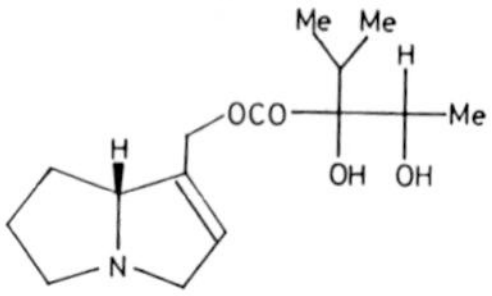

(28)

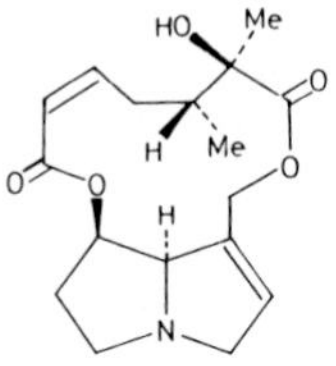

(30)

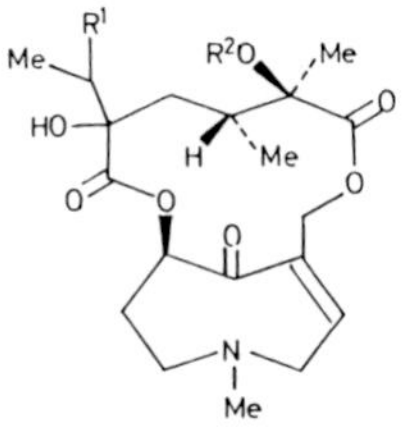

	R^1	R^2
(31)	Cl	Ac
(39)	OAc	Ac
(40)	OH	Ac
(78)	OH	H

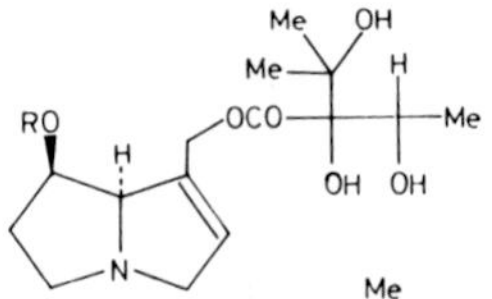

(32)	R =
(111)	R = Ac

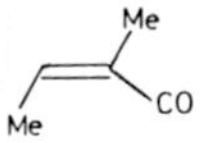

Scheme 1.4.

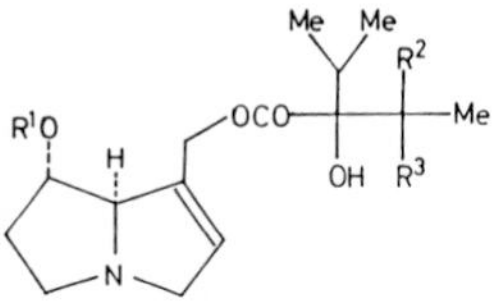

	R^1	R^2	R^3
(33)	H	H	OH
(34)	Ac	H	OH
(88)	H	OH	H

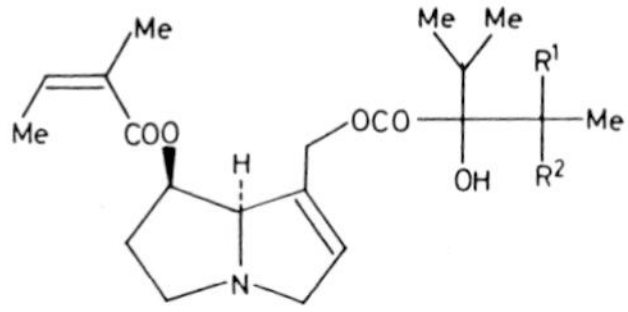

	R^1	R^2
(35)	OH	H
(106)	H	OH

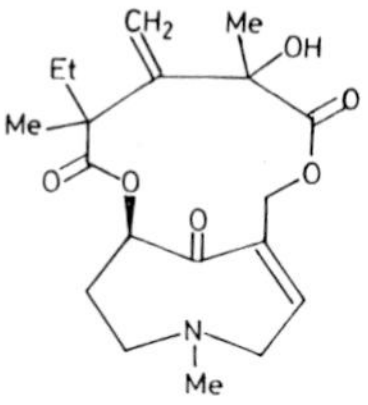

(36)

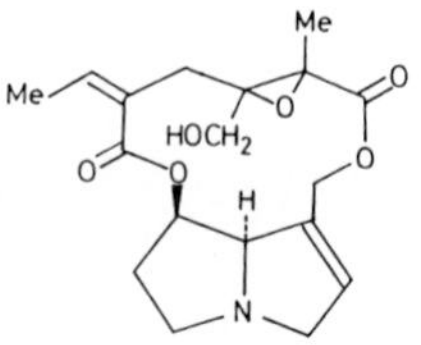

(37)

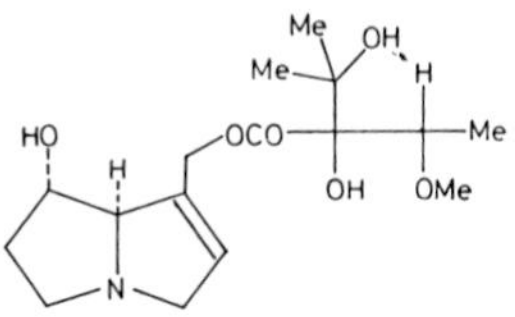

(38)

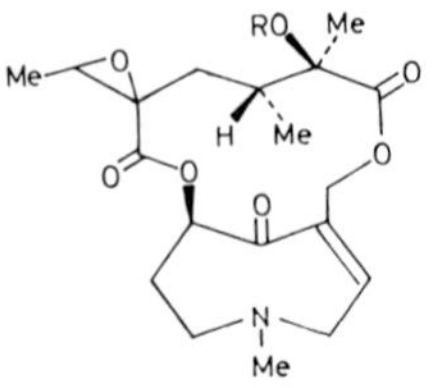

(41)	R = Ac	(α-epoxide)
(79)	R = H	(α-epoxide)
(81)	R = H	(β-epoxide)
(82)	R = Ac	(β-epoxide)

Scheme 1.5.

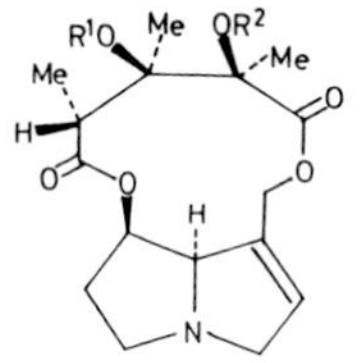

	R^1	R^2
(43)	Me>CHCO Et-	H
(74)	H	H
(75)	—CHMe—	
(100)	Ac	H

(44)

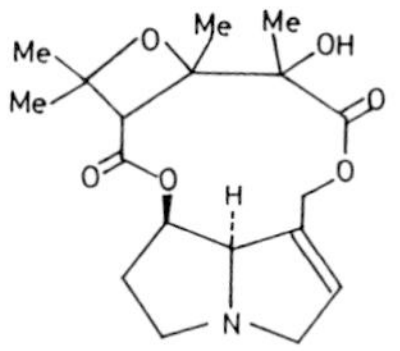

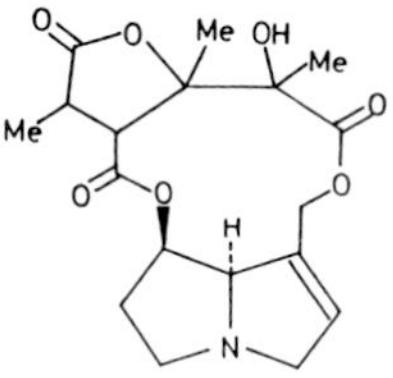

(45)

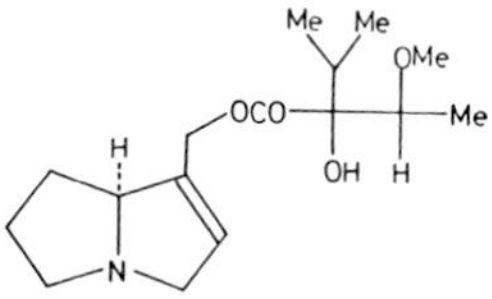

(46)

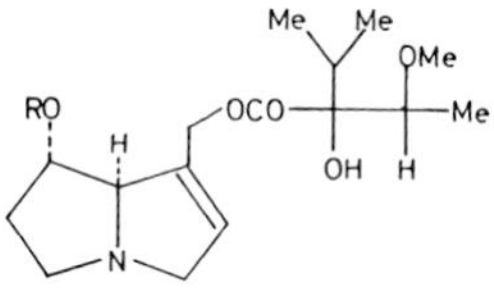

(49) R = H
(50) R =

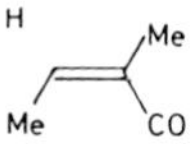

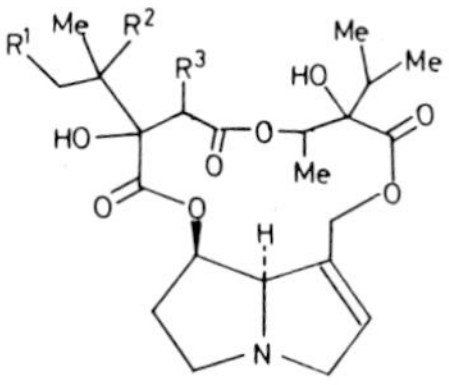

Me Me CO

	R^1	R^2	R^3
(51)	Me	H	H
(80)	H	H	H
(101)	Me	OH	OH
(102)	H	H	OH
(103)	Me	H	OH

Scheme 1.6.

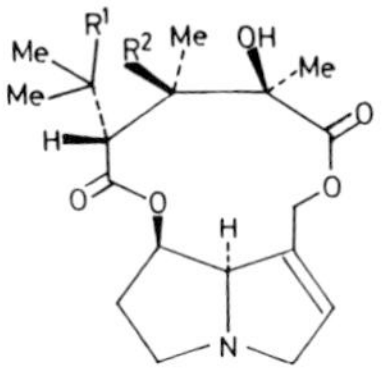

	R^1	R^2
(52)	H	H
(110)	H	OH
(120)	OH	OH

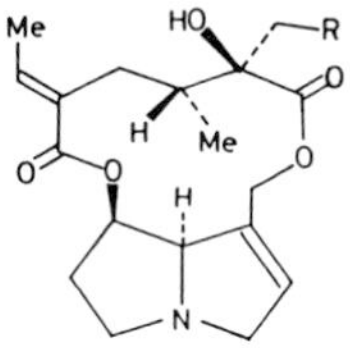

(55) R = H
(112) R = OH

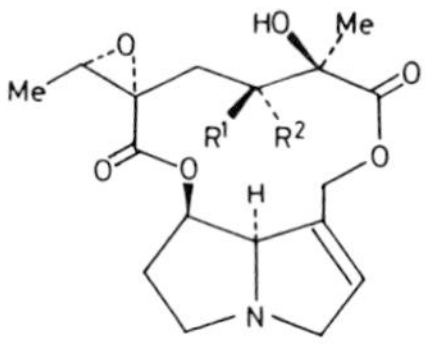

	R^1	R^2
(60)	H	Me
(63)	—CH$_2$—	

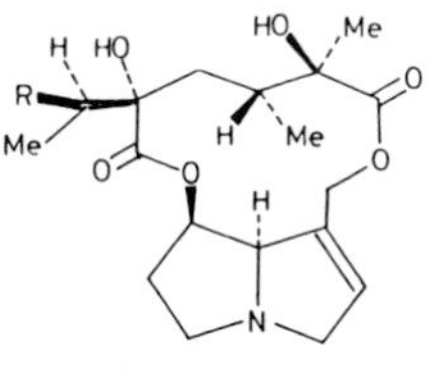

(61) R = OH
(62) R = Cl

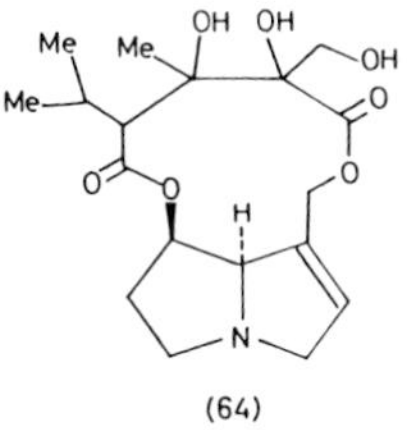

(64)

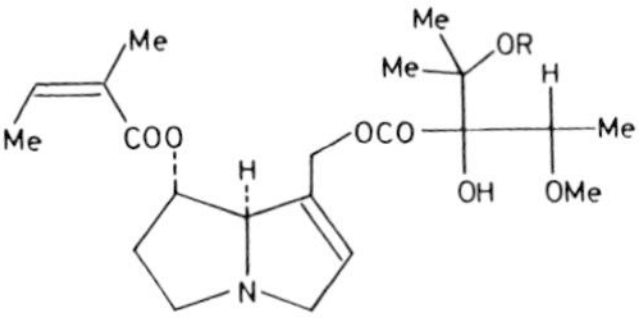

(65) R = H
(66) R = Ac

Scheme 1.7.

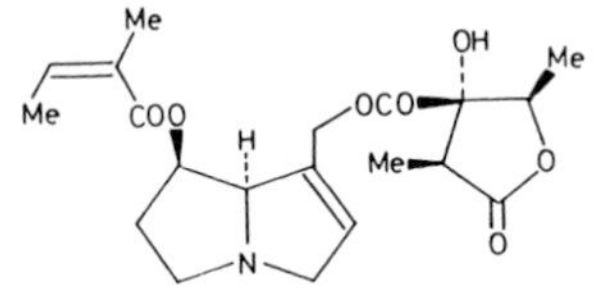

(67)

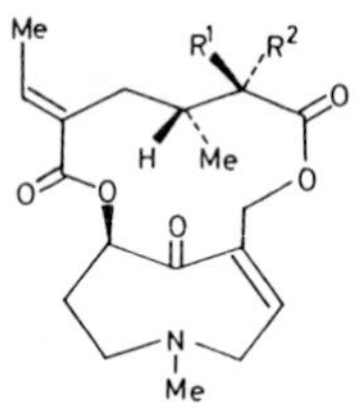

	R¹	R²
(68)	Me	OAc
(76)	OH	Me

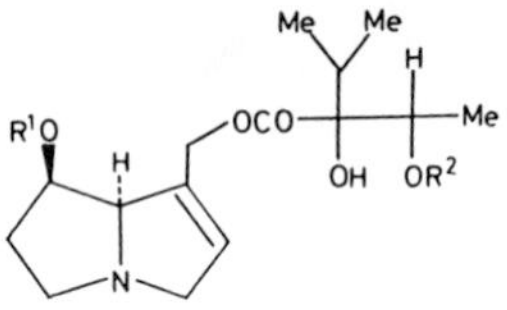

	R¹	R²
(70)	H	H
(71)	Ac	H
(118)	H	Ac
(119)	Ac	Ac

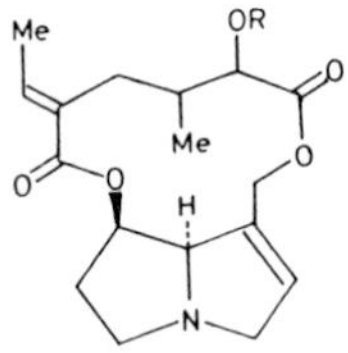

(77) R = H
(25) R = Ac

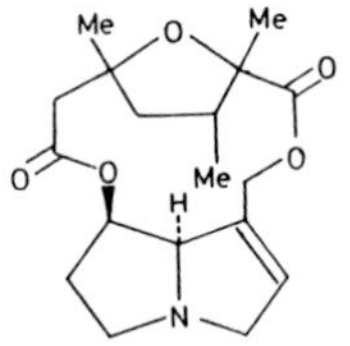

(83)

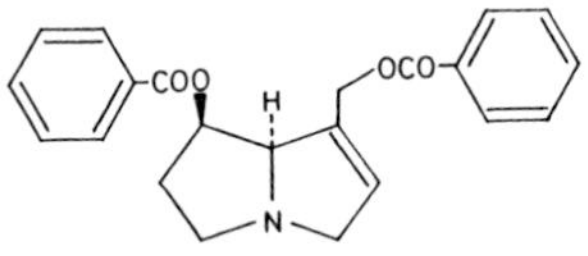

(84)

Scheme 1.8.

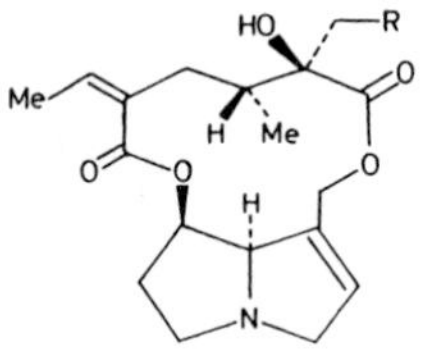

(85) R = OH
(92) R = H

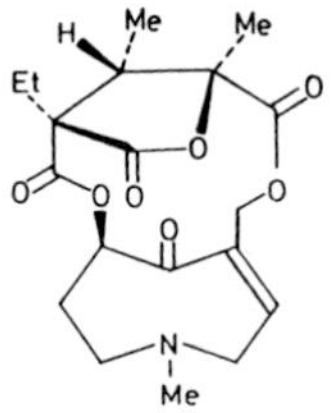

(86)

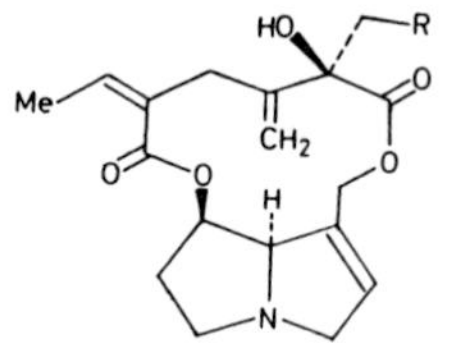

(87) R = OH
(93) R = H

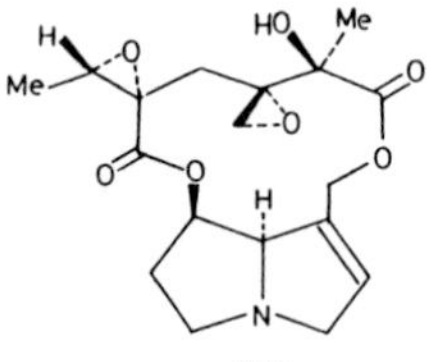

(91)

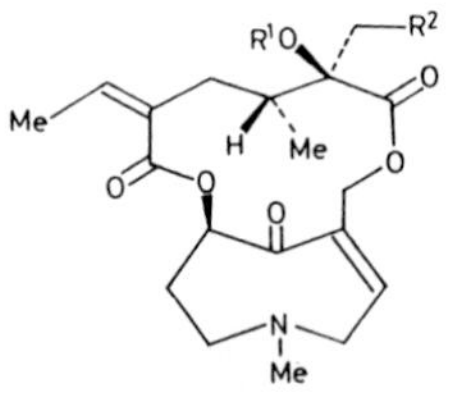

	R¹	R²
(95)	H	H
(96)	Ac	H
(97)	H	OH

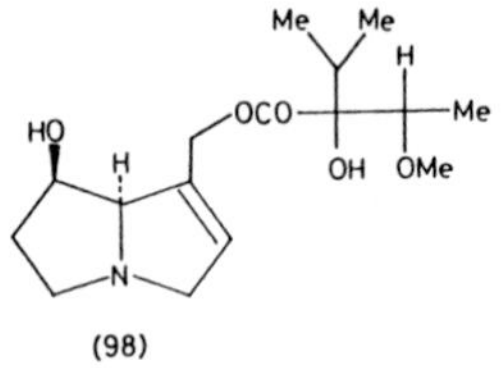

(98)

Scheme 1.9.

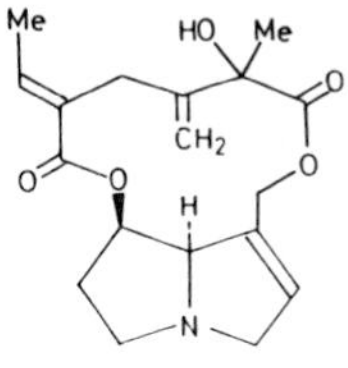

(99)

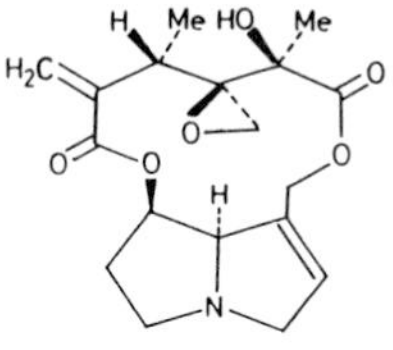

(105)

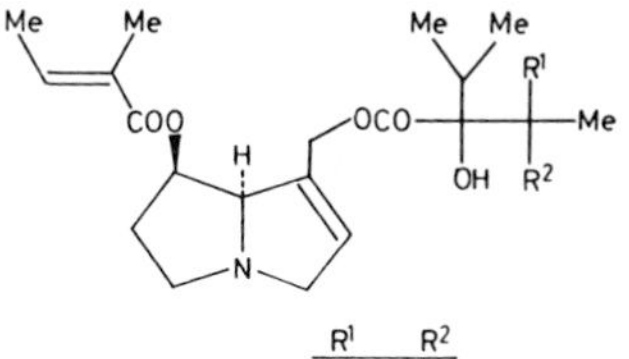

	R^1	R^2
(107)	H	OH
(123)	OH	H

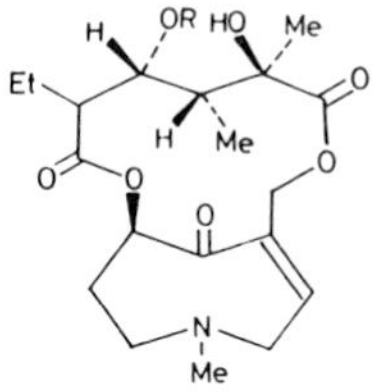

(108) R = H
(109) R = Ac

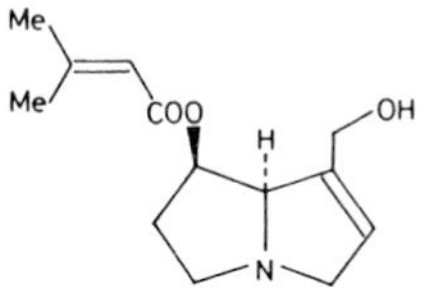

(114)

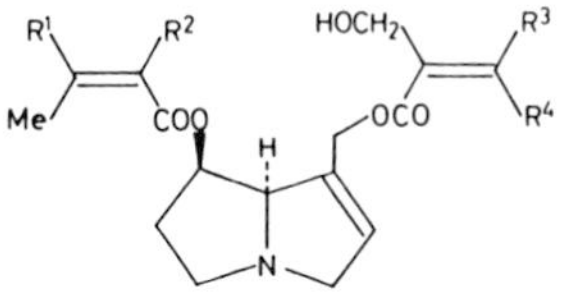

	R^1	R^2	R^3	R^4
(115)	Me	H	H	Me
(116)	H	Me	H	Me
(117)	H	Me	Me	H

Scheme 1.10.

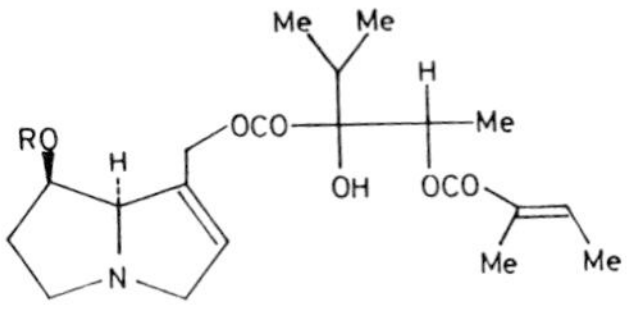

(121) R = H
(122) R = Ac

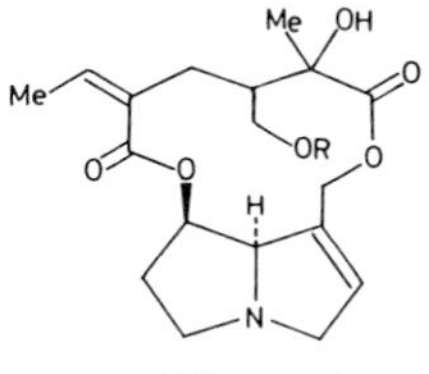

(124) R = H
(125) R = Ac

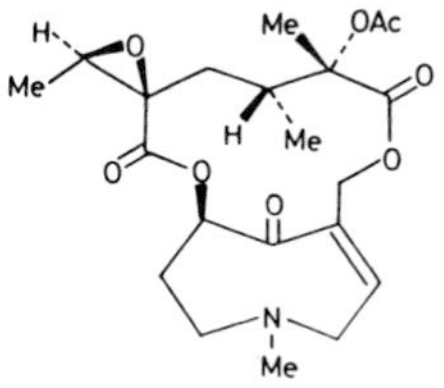

(126)

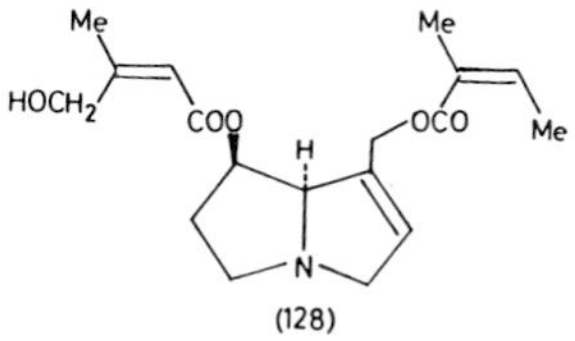

(128)

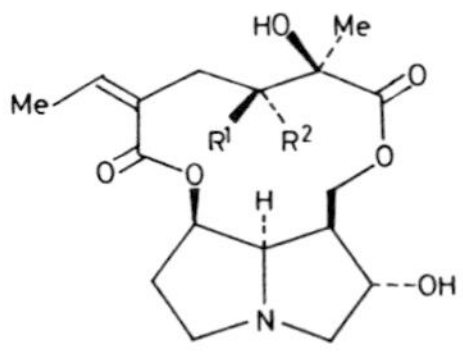

	R^1	R^2
(130)	$-CH_2-$	
(155)	H	Me

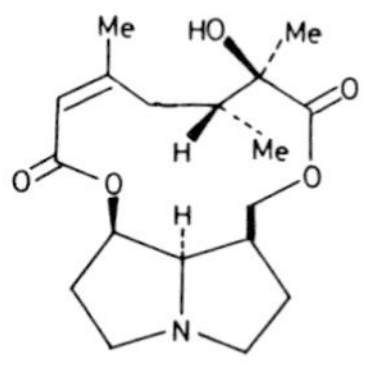

(131)

Scheme 1.11.

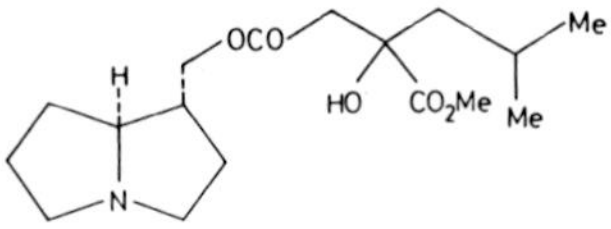

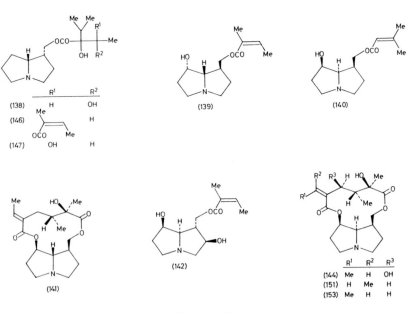

(132)

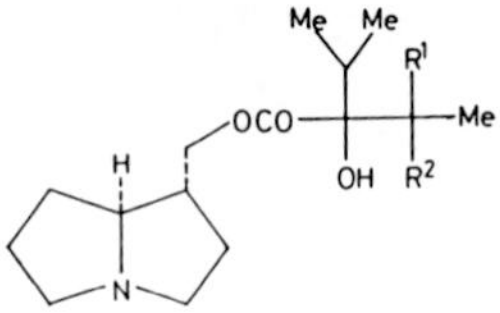

R¹   R²
(133), (160)   H   OH
(143), (158)   OH   H

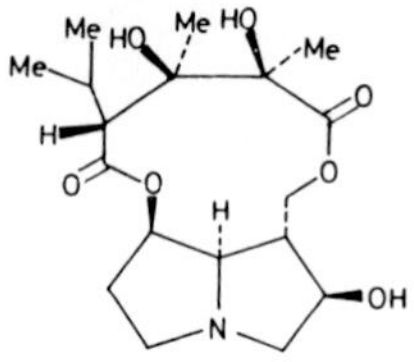

(134)

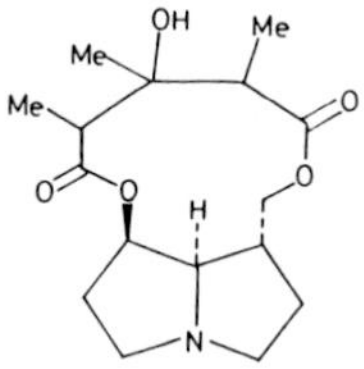

(135), (145)

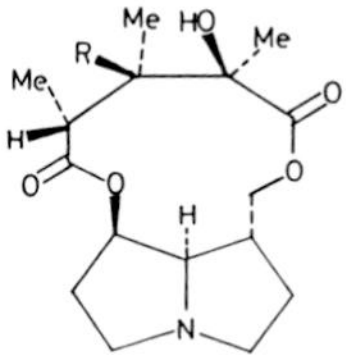

(136)   R = OH
(154)   R = H

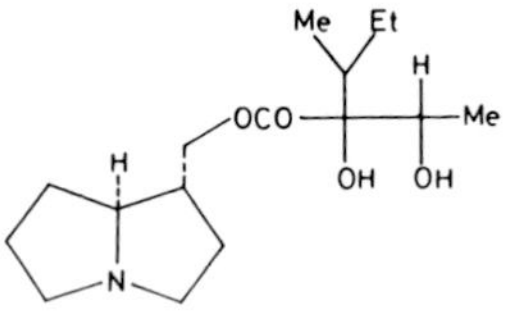

(137)

Scheme 1.12.

R¹   R²
(138)   H   OH
(146)
(147)   OH   H

(139)

(140)

(141)

(142)

R¹   R²   R³
(144)   Me   H   OH
(151)   H   Me   H
(153)   Me   H   H

Scheme 1.13.

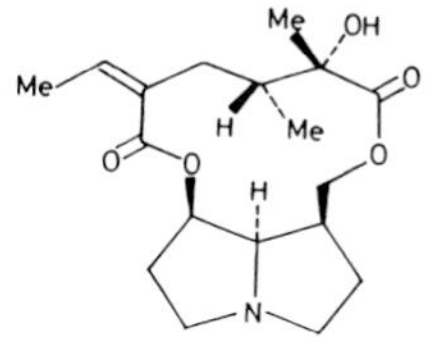

(145a)

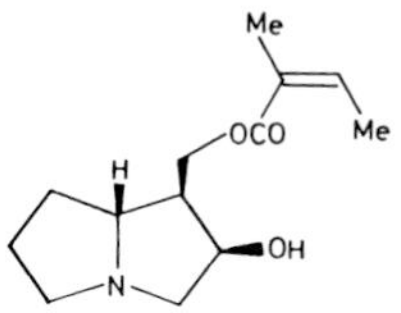

(148)

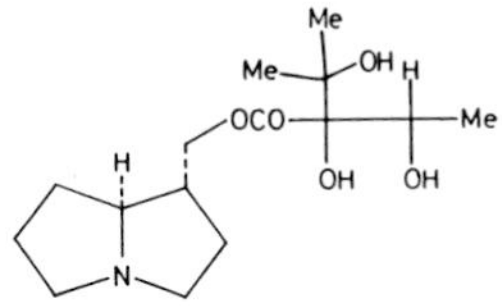

(149)

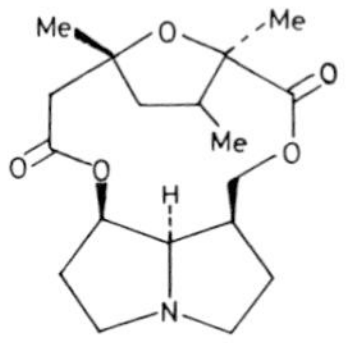

(150)

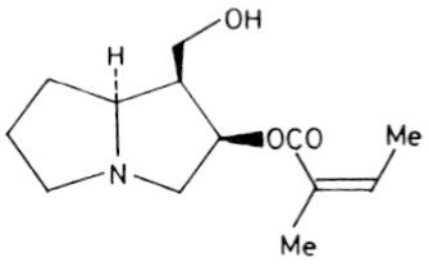

(152)

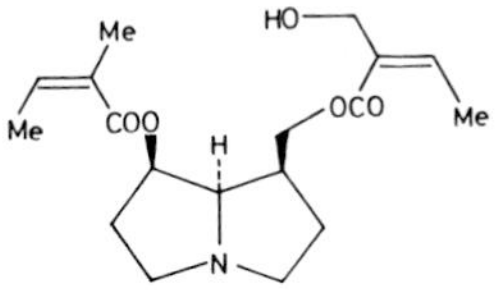

(156)

Scheme 1.14.

A group of saturated amino pyrrolizidines occurs in some grasses (family Graminae). These are not hepatotoxic, but they may play a part in other toxic actions of these plants. Thus *Festuca arundinacea* contains loline (festucine) (161) (Yates and Tookey, 1965). This alkaloid is also found in *Lolium cuneatum,* along with norloline (162), lolinine (163) and others: for a list, see Robins (1982a). An association has recently been demonstrated between the presence of PAs and an endophytic fungus infection [*Epichloe typhina* (Fr.) Tul.] in *Festuca arundinacea* (Jones *et al.*, 1983). This raises the possibility, as yet unconfirmed, that loline alkaloids might be phytoalexins produced only in plants stressed by such infections.

A number of PAs contain a dihydropyrrolizine nucleus. These include the dihydropyrrolizinone senaetnine (164) and related esters, and acetoxy derivatives,

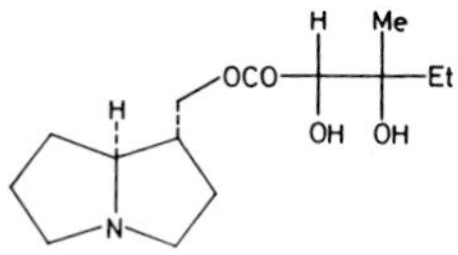

(157)

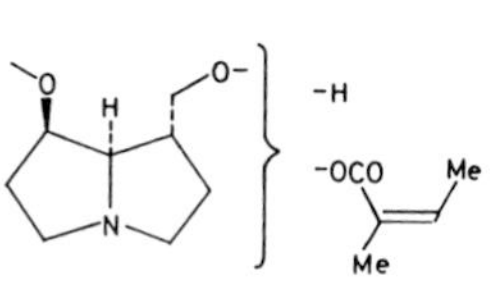

(159)

Scheme 1.15.

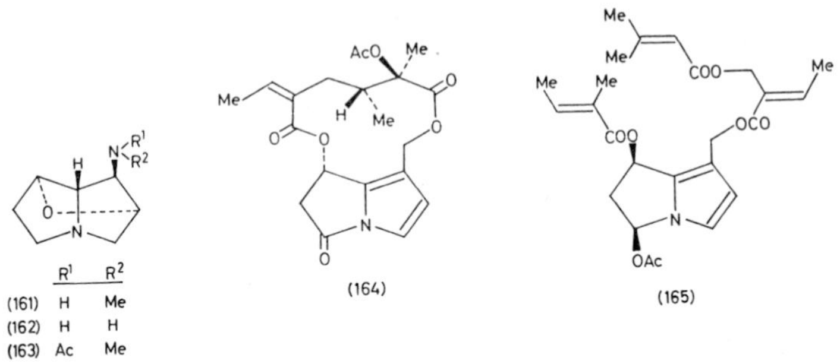

	R¹	R²
(161)	H	Me
(162)	H	H
(163)	Ac	Me

(164)

(165)

Scheme 1.16.

e.g. senampeline A (165), which are found in some South African *Senecio* species (*S. aetnensis; S. aucheri; S. pterophorus*) (Bohlmann *et al.*, 1977a,b). These alkaloids have not yet been tested for hepatotoxicity, a surprising omission in view of their similarity to the dihydropyrrolizine esters formed by metabolism of hepatotoxic PAs.

III. PYRROLIZIDINE BASES

Most PAs are derivatives of 1-methylpyrrolizidine, exemplified by heliotridane (167), rather than of pyrrolizidine (166) itself (Fig. 1.1). 1-Methylenepyr-rolizidine (168) and related compounds occur naturally in some *Crotalaria* species (see Smith and Culvenor, 1981). Hydroxylated pyrrolizidines (169–186) constitute the amino alcohol moiety of ester PAs; they are known as necines and can be obtained by hydrolysing the esters. Unesterified necines are rarely extracted from plants; Culvenor and Smith (1957a) obtained retronecine (183), in the form of its *N*-oxide (isatinecine), from *Crotalaria retusa* seed; trachelanthamidine (172) has been extracted from *Heliotropium strigosum* (Mattocks, 1964a), and turneforcidine (174) from *Crotalaria candidans* (Suri *et al.*, 1982). Necines, including retronecine, trachelanthamidine, supinidine (182) and lindelofidine (170), have also been found in various *Heliotropium* species by Birecka *et al.* (1983), and free necines are probably present in many other plant species along with their esters.

The distribution of free and esterified necine bases in *Heliotropium* species has been studied by Catalfamo *et al.* (1982a,b) and Birecka *et al.* (1983). Heliotridine was the dominant necine in Australian and Asian species investigated, whereas retronecine or trachelanthamidine were dominant in American species. In some plants, four or five different necines were present.

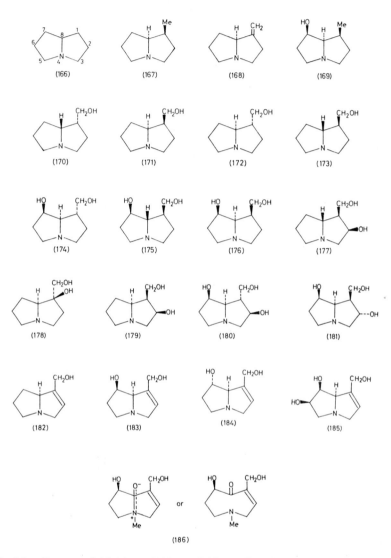

Fig. 1.1. Some pyrrolizidine bases: (166) pyrrolizidine; (167) heliotridane; (168) 1-methylene-8α-pyrrolizidine; (169) retronecanol; (170) (+)-isoretronecanol, (lindelofidine); (171) (−)-iso-retronecanol; (172) (−)-trachelanthamidine; (173) (+)-trachelanthamidine (laburnine); (174) (−)-turneforcidine; (175) (−)-hastanecine; (176) (−)-platynecine; (177) (+)-macronecine; (178) curas-sanecine; (179) petasinecine; (180) croalbinecine; (181) (−)-rosmarinecine; (182) (−)-supinidine; (183) (+)-retronecine; (184) (+)-heliotridine; (185) (+)-crotanecine; (186) otonecine.

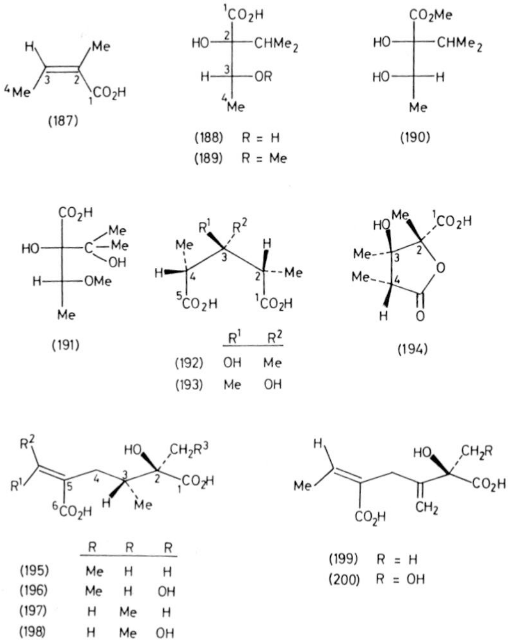

Fig. 1.2. Some necic acids: (187) angelic acid (*cis*-2-methyl-2-butenoic acid); (188) (+)-tra-chelanthic acid (2,3-dihydroxy-2-isopropylbutanoic acid); (189) heliotric (heliotrinic) acid; (190) (−)-viridofloric acid; (191) lasiocarpic acid; (192) fulvinic acid; (193) crispatic acid; (194) mono-crotalic lactone; (195) senecic acid; (196) isatinecic acid; (197) integerrinecic acid; (198) retronecic acid; (199) seneciphyllic acid; (200) riddelliic acid.

IV. NECIC ACIDS

Many different necic acids form the acid moiety of pyrrolizidine ester alkaloids. They are often closely related, e.g. as stereoisomers. The acid is usually obtained when the PA is hydrolysed or hydrogenolysed, but sometimes a decomposition product or the lactone of a hydroxy acid is formed (see Chapter 5). The structures of some necic acids are shown here (187–200: Fig. 1.2): these, or their isomers, occur in over half of the known ester PAs.

V. NOMENCLATURE

For practical purposes, PAs are generally known by their trivial names (usually derived from the names of their source plants). Systematic chemical names are unwieldy and are infrequently used. A useful note on the systematic naming of

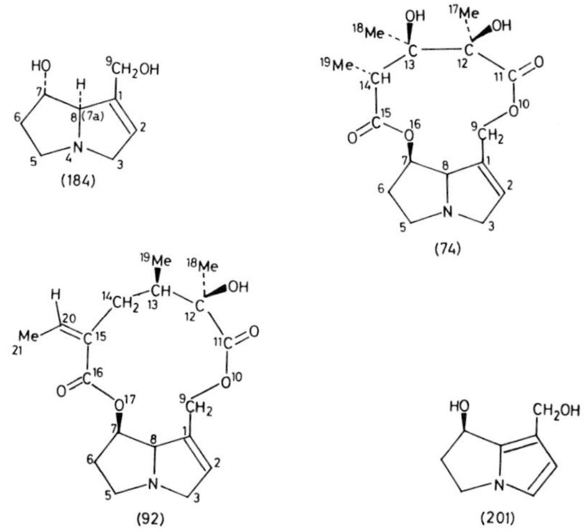

Scheme 1.17.

PAs is given by Culvenor *et al.* (1971b). The necine bases are named either as pyrrolizidine (if saturated) or pyrrolizine (if unsaturated) derivatives; thus heliotridine (184) is 7α-hydroxy-1-hydroxymethyl-5,6,7,8α-tetrahydro-3*H*-pyrrolizine. The 1-CH$_2$ position is usually designated C-9, and the 8-position is sometimes 7a. (An alternative numbering used by *Chemical Abstracts* is based on 1*H*-pyrrolizine, making heliotridine be 2,4,5,7α-tetrahydro-1-hydroxy-1*H*-pyrrolizine-7-methanol; this system is never used by pyrrolizidine alkaloid chemists.) Systematic names can accordingly be derived for non-macrocyclic esters of necines: e.g. supinine (104) is (5,6,7,8α-tetrahydro-3*H*-pyrrolizine-1-yl)-methyl 2,3-dihydroxy-2-isopropylbutanoate.

For macrocyclic diesters a different system is used: the stem-names 'crotalanine' and 'senecanine' are given to 11- and 12- membered macrocyclic structures respectively (Culvenor *et al.*, 1971b). Thus, monocrotaline (74) is 12β,13β-dihydroxy-12α,13α,14α-trimethylcrotal-1-enine (or -trimethyl-1,2,-didehydro crotalanine); senecionine (92) is 15-ethylidene-12β-hydroxy-12α, 13β-dimethylsenec-1-enine, and senkirkine (95) is 15-ethylidene-12β-hydroxy-4,12α,13β-trimethyl-8-oxo-4,8-secosenec-1-enine. *Chemical Abstracts* uses the above numbering but a different naming system. Thus, senecionine becomes 12-hydroxysenecionan-11,16-dione.

Dehydropyrrolizidines, which are important metabolites of many PAs, are best related to 5*H*-pyrrolizine, so as to keep the same ring numbering as the parent alkaloids. Dehydroretronecine (201) is thus 7β-hydroxy-1-hydroxymethyl-6,7-dihydro-5*H*-pyrrolizine.

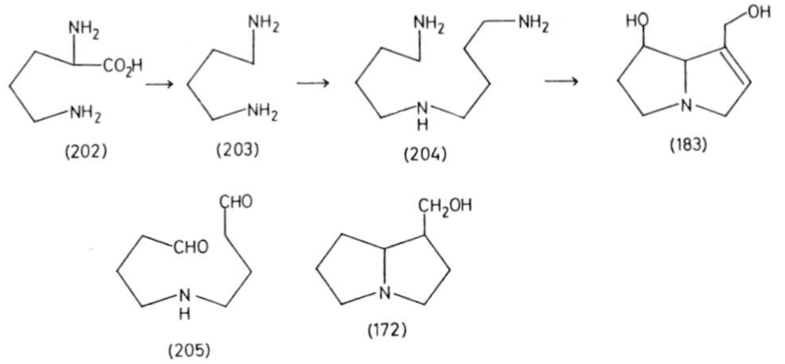

Scheme 1.18.

VI. BIOSYNTHESIS

It is not intended to give a detailed account here of the work on the biosynthesis of PAs in plants. The earlier work has been reviewed by Bull *et al.* (1968), and more recent work is discussed by Robins (1982a) and by Culvenor (1978), who also gives a useful survey of the occurrence and distribution of PAs among families and genera of angiosperm plants.

The only necine base whose biosynthesis has been studied in detail is retronecine (183). To summarize a great deal of work, it has been established that retronecine is formed from two units of ornithine (202) (Nowacki and Byerrum, 1962; Bottomley and Geissman, 1964; Bale and Crout, 1975), via putrescine (203) (Robins and Sweeney, 1981; Kahn and Robins, 1981a; Rana and Robins, 1983) and the symmetrical intermediate homospermidine (204) (Khan and Robins, 1981b), or the dialdehyde (205) (Grue-Sørensen and Spenser, 1981, 1982). Plants used in these studies included *Crotalaria spectabilis, Senecio douglasii, S. magnificus, S. isatideus* and *S. vulgaris,* containing the retronecine-based alkaloids monocrotaline, senecionine, retrorsine, and others. The mode of incorporation into retronecine of putrescine, labelled with deuterium in various positions, has recently been studied by Rana and Robins (1983; 1984) and by Grue-Sørensen and Spenser (1983).

The biosyntheses of a number of necic acids have also been investigated. For details, see the reviews cited above.

The biological synthesis of trachelanthamidine (172) from homospermidine (204) has been simulated *in vitro* (Robins, 1982b).

TABLE 1.1 Plant Sources of Unsaturated Pyrrolizidine Alkaloids

Plant	Alkaloids[a]	References[b]
Apocynaceae		
1. *Parsonsia eucalyptophylla* (F. Muel.)	53 or 56; 70	1
2. *P. heterophylla* A. Cunn.	51, 80	2
3. *P. spiralis* Wall.	101, 102, 103	2
4. *P. straminea* [(R. Br.) F. Muel.]	53 or 56; 70	1
Boraginaceae		
4a. *Alkanna tinctoria* Tausch	9, 116	297
5. *Amsinckia hispida* (Ruis & Pav.) I. M. Johnston	35, 56, 70	3
6. *A. intermedia* Fisch. & Mey.	35, 56, 70, 98	3
7. *A. lycopsoides* Lehm.	35, 56, 70	3
7a. *A. menziesii*	56, 70, 71, 118, 119	276
8. *Anchusa arvensis* (L.) Bieb. (*Lycopsis arvensis* L.)	33 or isomer	4
9. *A. officinalis* L.	70, 71	4, 5
10. *Asperugo procumbens* L.	1 or 104, 33	4
10a. *Borago officinalis* L.	1, 56, 57, 70, 71, 104	298, 299
11. *Caccinia glauca* Savi.	84	6
12. *Cynoglossum amabile* Stapf. & Drummond	1, 33	7
13. *C. australe* R. Br.	28, S138	7
14. *C. creticum*	33, 47	8
15. *C. glochidiatum* Wall. ex Lindl.	1	9
16. *C. lanceolatum* Forsk.	28, S138	9
17. *C. latifolium* R. Br.	9, 67	10
18. *C. officinale* L.	6, 33, 47, 48	11–15
19. *C. pictum* Ait.	33, 47	16, 17
20. *C. viridoflorum* Pallas ex Lehm.	47, S158	18, 19
21. *Echium diffusum*	[c]	20
22. *E. italicum*	32	20, 21
23. *E. lycopsis* L. (*E. plantagineum* L.)	32, 35	22
24. *E. vulgare* L.	11, 47, 48	4, 23, 24
25. *Heliotropium acutiflorum*	49	25
26. *H. amplexicaule* Vahl.	53	21
27. *H. arbainense*	38, 49, 65	8

(*continued*)

TABLE 1.1 (*Continued*)

Plant	Alkaloids[a]	References[b]
28. *H. arborescens* L. (*H. peruvi-anum* L.)	65	26
29. *H. arguzioides* Kar. & Kir.	49, 110	27, 28
30. *H. curassavicum* L.	6, 49, 65, S133, S137, S143[d]	29, 30, 268, 269
31. *H. dasycarpum* Ledeb.	49	28
32. *H. eichwaldii* Steud.	6, 49, 50, 65	31–33
33. *H. europaeum* L.	38, 46, 49, 65, 66	34–37
34. *H. indicum* L.	33, 46, 49, 53, 54, 65, 104	38, 39, 261, 262
35. *H. lasiocarpum* Fisch. & Mey.	49, 65	40
36. *H. maris-mortui*	38, 65	8, 41
37. *H. olgae* Bunge	49, 52	43, 51
38. *H. popovii* subsp. *gillianum* H. Riedl.	49	44
39. *H. ramosissimum* (*H. persicum*)	49	45
40. *H. rotundifolium*	38	41
40a. *H. spathulatum*	e	279
41. *H. steudneri* Vatke	70	46
42. *H. supinum* L.	6, 7, 8, 33, 47, 49, 65, 104	47, 289
43. *H. transoxanum*	49	25
44. *Lappula glochidiata*	33	48
45. *L. intermedia*	65	49
46. *Lindelofia angustifolia* (Schrenk) Brand	1, 33	9, 32
47. *L. spectabilis* Lehm.	33, 34, 74	32, 33
48. *L. stylosa* (Kar. & Kir.) Brand	33, S147, S158	50, 51
49. *L. tschimganica*	33, S158	52
50. *Lithospermum officinale* L.		4
51. *Messerschmidia sibirica*	10, 70	53
51a. *Myosotis scorpioides* L. (*M. palustris* L.)	107, 121, 122, 123	277
52. *M. sylvatica* Hoffm.	10, 47, 48, S158	54
53. *Paracynoglossum imeretinum* (Kusnez.) Pop.	33, 47	55, 56
54. *Rindera austroechinata* M. Pop.	33	52

TABLE 1.1 (*Continued*)

Plant	Alkaloids[a]	References[b]
55. *R. baldschuanica* Kusnezov	33, 88, S158	52
56. *R. cyclodonta* Bge.	33	57
57. *R. echinata* Regel	33, S158	57
58. *R. oblongifolia* M. Pop.	33	52
59. *Solenanthus circinatus* Ledeb.	33	58
60. *S. coronatus* Regel	33	51
61. *S. karateginus* Lipsky	33	58
62. *S. turkestanicus* (Regel & Smirnov) Kusnezov	88	51
63. *Symphytum asperum* Lepech.	11, 33, 47; 34, 48 (or isomers)	4, 59–61
64. *S. caucasicum*	11, 32, 33, 65	62, 254
65. *S. officinale* L.	32, 33, 47, 65, 107; 48 (or isomer); S160	4, 63–66
66. *S. orientale*	5, 32, 107	67, 68
67. *S. tuberosum*	5, 32	69
68. *S.* × *uplandicum* Nyman	32, 56, 57, 70, 71, 106, 107, 111	4, 70
69. *Tournefortia sarmentosa* Lam.	104	71
70. *Trichodesma africana*	38, 56, 110	8, 293
71. *T. incanum* Alph. DC.	52, 110	72, 73
72. *T. zeylanicum* R. Br.	104	74
Compositae		
73. *Adenostyles alliariae*	93, S153	75, 292
73a. *A. glabra*	93	292
74. *A. rhombifolia* (Willd.) Pim.	93, S153, S156	76
75. *Brachyglottis repanda* Forst. & Forst.	92, 95	77
76. *Cacalia floridana*	39, 40, 41, 79	78
77. *C. hastata* L. subsp. *orientalis* Kitamura	55	79
78. *C. yatabei* Maxim.	113	80
79. *Conoclinium coelestinum* (L.) DC	33, 56	81
80. *Doronicum macrophyllum*	31, 40, 79	82
81. *Emilia flammea* Cass.	36, 79	83
82. *E. sonchifolia* DC.	92	54
83. *Erechtites heiracifolia* (L.) Raf. ex DC.	92, 93	84
84. *Eupatorium altissimum* L.	6, 88	81
85. *E. cannabinum* L.	33, 104	85

(*continued*)

TABLE 1.1 (*Continued*)

Plant	Alkaloids[a]	References[b]
86. *E. compositifolium* Walt.	56, 70	81
87. *E. maculatum* L.	33, S158	86
88. *E. serotinum* Michx.	88, 104	87
89. *E. stoechadosmum* Hance	104, S147	88
90. *Farfugium japonicum* Kit.	95	89
90a. *Gynura scandens*	124, 125	278
90b. *G. segetum* (Lour.) Merr.	92	290, 296
91. *Ligularia brachyphylla* Hand.-Mazz.	15, 69	90
92. *L. clivorum*	15	91, 252
93. *L. dentata* (A. Gray) Hara	15, 68, 126, 127, S145a	92, 291
94. *L. elegans* Cass. (*L. macrophylla* (Ledeb) DC.)	15, 69	90
94a. *Petasites albus* L.	95	280
95. *P. hybridus* (L.) P. Gaertn et al.	55, 92, 95, etc.	21, 280
96. *P. japonicus* Maxim.	81, 82, 95	93–96
97. *P. laevigatus* (Willd.) Reichenb. (*Nardosmia laevigata*)	92, 95, S153	97
97a. *Senecio adonidifolius*	92	132, 285
98. *S. aegyptius* L.	79, 87, 92	98, 99
99. *S. alpinus* L. Scop.	55, 60, 62, 63, 92, 93	100, 101
100. *S. ambrosioides*	85, 87, 92, 93	102
101. *S. ampullaceus* Hook	85, 92, 93	103–105
102. *S. antieuphorbium* (L.) Sch. Bip.	55, 95	106, 107
103. *S. aquaticus* Hill	93	108, 131
104. *S. aureus* L.	40, 41, 79, 92	109, 110, 266, 273
105. *S. auricola* Bourg.	76	111
106. *S. bipinnatisectus* Belcher	85	112
107. *S. borysthenicus*	93	113, 169
108. *S. brasiliensis* DC.	55, 60, 85, 92, 93	21, 102
109. *S. bupleuroides* DC	85	115
109a. *S. cacaliaster*	114, 115, 128, S131	295
110. *S. cannabifolius* Less	89, 93	116, 251
111. *S. carthemoides* Greene	92, 93	103, 105
112. *S. chrysanthemoides*	93	117
113. *S. cineraria* DC.	60, 79, 85, 92, 93	118–121
114. *S. congestus*	92, S153	122

TABLE 1.1 (*Continued*)

Plant	Alkaloids[a]	References[b]
114a. *S. cruentus*	85, 87, 92, 93	281
115. *S. cymbalarioides*	85, 87, 92, 93	123
116. *S. desfontainei* Druce	79, 87, 92, 93	98, 99
117. *S. discolor* DC.	85, 92	124, 125
118. *S. doria* L. (*S. paucifolius*)	93	126
119. *S. doronicum* L.	30, S131	127
120. *S. douglasii* DC. (*S. longi-lobus* Benth.)	85, 87, 92, 93	103–105
121. *S. durieui* Gay	55	111
122. *S. eremophilus* Richards	85, 87, 92, 93	103–105
123. *S. erraticus* Berthol.	37, 55, 79, 92, 93	128–130
124. *S. erucifolius* L.	37, 85, 92, 93	130–132
124a. *S. faberi* Hemsl.	55	282
125. *S. floridanus* Sch. Bip.	39, 40, 41, 79	78
126. *S. fluviatilis* Wallr.	41, 79, 93	133
127. *S. formosus*	55, 85	134
128. *S. fremonti* Torr. & A. Gray	92, 93	102
129. *S. fuchsii* Gmel (*S. nemorensis* subsp. *fuchsii*)	92, S140	135, 136
130. *S. glabellus* (Turcz.) DC.	92	137
131. *S. glaberrimus* DC.	85	131
132. *S. graminifolius* Jacq.	85	138
133. *S. grandifolia*	93, S153	163
134. *S. griesbachii* Baker	85	140
135. *S. ilicifolius* Thunb.	92	84, 138
136. *S. incanus* L. subsp. *carniolicus* (Willd.) Br.	55, 93	100
137. *S. integerrimus* Nutt.	55, 92, S153	110
138. *S. isatideus* DC.	85	114, 131
139. *S. jacobaea* L.	60, 61, 62, 63, 79, 85, 92, 93, 95	25, 118, 131, 132, 141–144
140. *S. kirkii* Hook f. ex Kirk	95, 96	145
141. *S. kleinia* Sch. Bip.	55, 95	265, 288
142. *S. krylovii*	93	146
143. *S. kubensis* Grossh.	93	147
144. *S. latifolius* DC.	85	148, 149
145. *S. lautus* Forst. f. ex Willd.	92	54
146. *S. longiflorus* Sch. Bip.	92, 93	150
S. longilobus Benth—see *S. douglasii*		
147. *S. magnificus* F. Muell.	55, 92	151, 152, 286

(*continued*)

TABLE 1.1 (*Continued*)

Plant	Alkaloids[a]	References[b]
148. *S. minimus* Poir	93	112
149. *S. morrisonensis*	55	54
150. *S. nebrodensis* L. var. *sicula*	55, 92	153
151. *S. nemorensis* L. var. *bulgaricus*	83, S131, S150	154
152. *S. nemorensis* L. var. *subdecurrens*	83, S131, S150	155, 156
S. nemorensis L. subsp. *fuchsii*—see *S. fuchsii*		
153. *S. othonnae* Bieb.	31, 40, 78, 79, 93	54
154. *S. othonniformis* Fourcade	14, 59	157, 158
154a. *S. otites* Kunze ex DC.	c	284
155. *S. ovirensis*	9	159
S. palmatus Pall.—see *S. cannabifolius*		
156. *S. paludosus* L.	60, 93	131, 160
157. *S. pampaenus* Cabrera	92	161
158. *S. pancicii* Degen	92, 93	162
159. *S. paniculatus* Berg. (*S. grandifolius*)	92, S153	163
160. *S. paucicalyculatus* Klatt	85	164
161. *S. paucifolius* S. G. Gmel	93	165
162. *S. pellucidus* (*S. ruderalis*)	85	166
163. *S. petasitis* DC.	92	99
164. *S. pierotii*	76, 95	167
165. *S. platyphylloides* Somm. & Lev.	93, S153	54
S. platyphyllus (Bieb.) DC.—see *S. rhombifolius*		
166. *S. pojarkovae*	93, S156	168
167. *S. praealtus* Bertol. (*S. borysthenicus*)	93	169
168. *S. propinquus*	93	170
169. *S. procerus* L.	95	171
170. *S. pseudo-arnica* Less.	92	110
171. *S. pterophorus* DC.	85, 92, 93, S155	84, 138, 172
172. *S. quadridentatus* Labill. (*Erechtites quadridentata*)	85, 92, 93	173
173. *S. racemosus*	93	174
174. *S. renardii* Winkl.	79, 93, 95	21
175. *S. retrorsus* DC.	85	114, 175

TABLE 1.1 (*Continued*)

Plant	Alkaloids[a]	References[b]
176. *S. rhombifolius* (Willd.) Sch. Bip. (*S. platyphyllus*)	93, S153, S156	148
177. *S. riddellii* Torr. & A. Gray	85, 87	103, 110, 180
178. *S. rivularis* DC.	6	91, 129
179. *S. ruderalis* Harvey	85	166
180. *S. scandens*	92, 93	176
181. *S. sceleratus* Schweickerdt	85, 89, 90	177–179
182. *S. spartioides* Torr. & A. Gray	85, 87, 92, 93, 99	110, 181
183. *S. spathulatus* A. Rich.	55, 92, 93	112, 182
184. *S. squalidus* L.	55, 92	131, 183, 184
185. *S. stenocephalus* Maxim	93	185
186. *S. subalpinus* Koch	92, 93	100, 186
187. *S. swaziensis* Compton	85, 105	187, 188
187a. *S. tenuifolius* Burm.	55, 92, 95, 96	283
188. *S. tomentosus*	79, 92	189, 190
189. *S. triangularis* Hook	92, 114, 115, 116, 117, etc.	191, 274, 275
190. *S. uintahensis*	92, 95	123
191. *S. vernosus* Harvey	95	131
192. *S. vernalis* Waldst. & Kit.	85, 92, 94, 95	192
193. *S. viminalis* Bremek.	85, 92	150
194. *S. viscosus* L.	55, 92	129, 131, 183
195. *S. vulgaris* L.	85, 87, 92, 93	109, 131, 183, 193
196. *S. werneriaefolius*	85, 92	123
197. *Syneilesis palmata* Maxim	92, 108, 109	194, 195
198. *Tussilago farfara* L.	95 etc.	196, 271, 272
Leguminosae		
199. *Crotalaria aegyptiaca* Benth.	20, 74	8, 222, 287
200. *C. agatiflora* Schweinf.	2, 3, 4, 21, 72, 73	197, 198
201. *C. anagyroides* H. B. & K.	2, 92	197, 199
202. *C. assamica* Benth.	74	200, 287
203. *C. axillaris* Ait.	12, 13	201, 202
204. *C. barbata* R. Graham	17	203
205. *C. brevifolia*	55, 112	204, 205
206. *C. burhia* Buch.-Ham.	23, 74	206
207. *C. candidans* Pericarp.	16, 58, etc.	245, 270
207a. *C. cephalotes* Steud. ex A. Rich.	74, etc.	287
207b. *C. crassipes*	86	226

(*continued*)

TABLE 1.1 (*Continued*)

Plant	Alkaloids[a]	References[b]
208. *C. crispata* F. Muell. ex Benth.	16, 42, 74	207
208a. *C. cunninghamii* R. Br.	74, etc.	287
209. *C. dura* Wood & Evans	29	208, 264
210. *C. fulva* Roxb.	42	209
211. *C. globifera* E. Mey	29, 44, 45, 110, 120	208, 267
212. *C. grahamiana* R. Wight & Walk. Arn.	43, 74, 75	210–212
213. *C. grantiana* Harvey	44, 45	213, 214, 263, 294
214. *C. incana* L.	2, 55, 112	137, 215, 216
215. *C. intermedia* Kotschy	55, 112	217
216. *C. juncea* L.	64, 87, 92, 93, 110	218
217. *C. laburnifolia* L.	2	205
218. *C. laburnifolia* subsp. *eldomae*	2, 22, 72, 95, 97	219
219. *C. leiloba* Bart. (*C. ferruginea* Wall.)	74	220
220. *C. leschenaultii* DC.	16, 74	221
221. *C. madurensis* R. Wight	16, 18, 42, 58, 72	32, 197, 222–225
222. *C. mitchellii* Benth.	74, 86	226
223. *C. mucronata* Desv. (*C. striata*; *C. pallida*)	25, 55, 77, 112	205, 227–230
224. *C. mysorensis* Roth.	74	231
225. *C. nana* Burm.	19, 24	232, 233
225a. *C. nitens* Kunth.	74	287
226. *C. novae-hollandae* DC.	74, 86	226
C. pallida—see *C. mucronata*		
227. *C. paniculata* Willd.	42	234
227a. *C. paulina* Schrank	74, etc.	287
228. *C. quinquefolia* L.	74	21, 287
229. *C. recta* Steud. ex Rich.	74, 110	235, 287
230. *C. retusa* L.	74, 86	236, 237, 253
231. *C. rubiginosa* Willd. (*C. wightiana*)	64, 110	238
232. *C. sagittalis* L.	74	239
233. *C. semperflorens* Vent.	20	240
C. serica Retz.—see *C. spectabilis*		
234. *C. sessiliflora*	74	241
235. *C. spartioides* DC.	85	242
236. *C. spectabilis* Roth (*C. retzii* Hichc.; *C. sericea* Retz)	74, 100	236, 243

TABLE 1.1 (*Continued*)

Plant	Alkaloids[a]	References[b]
237. *C. stipularia* Desv. *C. striata* DC.—see *C. mucronata*	74	220
238. *C. tetragona* Roxb.	55, 110	220
239. *C. usaramoensis* E. G. Baker	55, 85, 92, 112	137, 244
240. *C. verrucosa* L. *C. virgulata*—see *C. grantiana*	2, 26, 27	246, 247
241. *C. walkeri* Arn. *C. wightiana*—see *C. rubiginosa*	26, 27	248
Ranunculaceae		
242. *Caltha biflora*	92	249
243. *C. leptosepala*	92	249
Scrophulariaceae		
244. *Castillega rhexifolia* Rydb.	92	250, 300

[a] The unsaturated PAs are listed in Table 1.3; saturated PAs (prefixed 'S') are listed in Table 1.4; etc. means that at least one additional alkaloid (known or unknown) was found.

[b] References: 1, Edgar and Culvenor (1975); 2, Edgar *et al.* (1980); 3, Culvenor and Smith (1966a); 4, Pedersen (1975a); 5, Broch-Due and Aasen (1980); 6, Siddiqi *et al.* (1978a); 7, Culvenor and Smith (1967); 8, Zalkow *et al.* (1979); 9, Suri *et al.* (1975b); 10, Crowley and Culvenor (1962); 11, Man'ko and Borisyuk (1957); 12, Man'ko (1959); 13, Sykulska (1962); 14, Jerzmanowska and Sykulska (1964); 15, Pedersen (1970); 16, Man'ko and Marchenko (1972a); 17, Man'ko and Marchenko (1972b); 18, Man'ko (1972); 19, Men'shikov (1948); 20, Amil and Ates (1971); 21, Bull *et al.* (1968); 22, Culvenor (1956); 23, Man'ko (1964); 24, Karimov *et al.* (1975); 25, Akramov *et al.* (1968); 26, Marquez (1961); 27, Medvedeva and Zolotavina (1971); 28, Akramov *et al.* (1961a); 29, Rajagopalan and Batra (1977b); 30, Mohanraj *et al.* (1978); 31, Gandhi *et al.* (1966a); 32, Rao *et al.* (1974); 33, Suri *et al.* (1975); 34, Culvenor *et al.* (1954); 35, Culvenor (1954); 36, Crowley and Culvenor (1956); 37, Culvenor *et al.* (1975b); 38, Mattocks *et al.* (1961); 39, Mattocks (1967a); 40, Men'shikov (1932); 41, Zalkow *et al.* (1978); 42, Kropman and Warren (1949); 43, Sheveleva *et al.* (1969); 44, Mohabbat *et al.* (1976); 45, Habib (1975); 46, Schneider *et al.* (1975); 47, Crowley and Culvenor (1959); 48, Suri *et al.* (1978); 49, Man'ko and Vasil'kov (1968); 50, Akramov *et al.* (1961b); 51, Kiyamitdinova *et al.* (1967); 52, Akramov *et al.* (1965); 53, Hikichi *et al.* (1980); 54, Smith and Culvenor (1981); 55, Man'ko and Marchenko (1971); 56, Man'ko and Marchenko (1976); 57, Akramov *et al.* (1967); 58, Akramov *et al.* (1964); 59, Man'ko and Kotovskii (1970a); 60, Man'ko and Kotovskii (1970b); 61, Man'ko *et al.* (1970b); 62, Man'ko *et al.* (1972); 63, Furuya and Araki (1968); 64, Furuya and Hikichi (1971); 65, Man'ko *et al.* (1970a); 66, Huizing and Malingre (1979a); 67, Ulubelen and Doganca

(*continued*)

TABLE 1.1 (*Continued*)

(1971); 68, Culvenor *et al.* (1975a); 69, Ulubelen and Ocal (1977); 70, Culvenor *et al.* (1980b); 71, Crowley and Culvenor (1955); 72, Men'shikov and Rubinstein (1935); 73, Yunusov and Plekhanova (1953, 1959); 74, O'Kelly and Sargeant (1961); 75, Yakhontova *et al.* (1976); 76, Pimenov *et al.* (1975); 77, Mortimer and White (1967); 78, Cava *et al.* (1968); 79, Hayashi *et al.* (1972); 80, Hikichi *et al.* (1978); 81, Herz *et al.* (1981); 82, Alieva *et al.* (1976); 83, Kohlmuenzer *et al.* (1971); 84, Culvenor and Smith (1954); 85, Pedersen (1975b); 86, Tsuda and Marion (1963); 87, Locock *et al.* (1966); 88, Furuya and Hikichi (1973); 89, Furuya *et al.* (1971); 90, Klasek *et al.* (1971); 91, Klasek *et al.* (1967); 92, Hikichi *et al.* (1979); 93, Yamada *et al.* (1976a); 94, Yamada *et al.* (1976b); 95, Furuya *et al.* (1976); 96, Yamada *et al.* (1978); 97, Glizin and Senov (1965); 98, Klasek *et al.* (1968a); 99, Gharbo and Habib (1969); 100, Klasek *et al.* (1968b); 101, Luthy *et al.* (1981); 102, Adams and Gianturco (1956f); 103, Adams and Govindachari (1949c); 104, Warren *et al.* (1950); 105, Adams and Looker (1951); 106, Rodriguez and Gonzalez (1969); 107, Rodriguez and Gonzalez (1971); 108, Evans and Evans (1949); 109, Manske (1936); 110, Manske (1939); 111, Panizo and Rodriguez (1974); 112, White (1969); 113, Red'ko (1956); 114, Christie *et al.* (1949); 115, Sapiro (1949); 116, Alekseev (1964); 117, Wali and Handa (1964); 118, Barger and Blackie (1937); 119, Adams and Govindachari (1949b); 120, Habib (1974); 121, Klasek *et al.* (1975b); 122, Roeder *et al.* (1982b); 123, Roitman *et al.* (1979); 124, Schoental (1960); 125, Hennig (1961); 126, Constantinescu and Albulescu (1961); 127, Roeder *et al.* (1980a); 128, Kompis *et al.* (1960); 129, Santavy *et al.* (1962); 130, Sedmera *et al.* (1972); 131, Blackie (1937); 132, Ferry and Brazier (1976); 133, Klasek *et al.* (1973b); 134, Munoz Quevedo (1976); 135, Lemp (1973); 136, Wiedenfeld and Roeder (1979); 137, Adams and van Duuren (1953b); 138, de Waal (1941); 140, Motidome and Ferreira (1966); 141, Bradbury and Culvenor (1954); 142, Bradbury and Mosbauer (1956); 143, Bradbury and Masamune (1959); 144, Culvenor (1964); 145, Briggs *et al.* (1965); 146, Sapunova and Ban'kovskii (1968); 147, Khalilov and Telezhenetskaya (1973); 148, Bredenkamp *et al.* (1985); 149, Barger *et al.* (1935); 150, Warren (1966); 151, Culvenor (1962); 152, Gellert and Mate (1964); 153, Plescia *et al.* (1976); 154, Nghia *et al.* (1976); 155, Klasek *et al.* (1973a); 156, Klasek *et al.* (1980); 157, Coucourakis and Gordon-Gray (1970); 158, Coucourakis *et al.* (1972); 159, Roeder *et al.* (1980c); 160, Alekseev (1961a); 161, Novelli (1958); 162, Jizba *et al.* (1982); 163, Glonti (1956); 164, Pretorius (1949); 165, Alekseev and Ban'kovs'kii (1965); 166, Leisegang (1950); 167, Asada and Furuya (1982); 168, Chernova and Murav'eva (1974); 169, Alekseev (1961b); 170, Khalilov *et al.* (1972); 171, Jovceva *et al.* (1978); 172, Edgar *et al.* (1976); 173, Culvenor and Smith (1955); 174, Khmel (1961); 175, Manske (1931); 176, Batra and Rajagopalan (1977); 177, de Waal and Pretorius (1941); 178, de Waal *et al.* (1963); 179, Gordon-Gray (1967); 180, Adams and van Duuren (1953c); 181, Adams and Gianturco (1957); 182, Benn *et al.* (1979); 183, Barger and Blackie (1936); 184, Kropman and Warren (1950); 185, Konovalova and Orekhov (1937); 186, Trivedi and Santavy (1963); 187, Gordon-Gray *et al.* (1972); 188, Gordon-Gray and Wells (1974); 189, Adams *et al.* (1956); 190, Schroter and Santavy (1960); 191, Kupchan and Suffness (1967); 192, Roeder *et al.* (1979); 193, Segall (1979b); 194, Hikichi and Furuya (1974); 195, Hikichi and Furuya (1976); 196, Culvenor *et al.* (1976b); 197, Atal *et al.* (1966a); 198, Culvenor and Smith (1972); 199, Sethi and Atal (1964); 200, Group of *Crotalaria* Plant Research (1974); 201, Crout (1968a); 202, Crout (1969);

TABLE 1.1 *(Continued)*

203, Puri *et al.* (1973); 204, Sawhney and Atal (1966); 205, Sawhney *et al.* (1967); 206, Rao *et al.* (1975a); 207, Culvenor and Smith (1963); 208, Adams and van Duuren (1953a); 209, Schoental (1963); 210, Gandhi *et al.* (1966b); 211, Atal *et al.* (1969); 212, Rajagopalan and Batra (1977a); 213, Adams *et al.* (1942a); 214, Adams and Gianturco (1956e); 215, Mattocks (1968d); 216, Sawhney and Atal (1970); 217, Suri *et al.* (1975c); 218, Adams and Gianturco (1956b, 1956c,d); 219, Crout (1972); 220, Puri *et al.* (1974); 221, Suri and Atal (1967); 222, Mahran *et al.* (1979); 223, Habib *et al.* (1971); 224, Rao *et al.* (1975b); 225, Rao *et al.* (1975c); 226, Culvenor *et al.* (1967b); 227, Bhacca and Sharma (1968); 228, Atal *et al.* (1968); 229, Gandhi *et al.* (1968); 230, Batra *et al.* (1975); 231, Sawhney and Atal (1968); 232, Siddiqi *et al.* (1978b); 233, Siddiqi *et al.* (1978c); 234, Subramanian *et al.* (1968); 235, Crout (1968b); 236, Adams and Rogers (1939); 237, Culvenor and Smith (1957a); 238, Atal *et al.* (1966b); 239, Willette and Cammarato (1972); 240, Atal *et al.* (1967); 241, Huang *et al.* (1980); 242, Bruemmerhoff and de Waal (1961); 243, Culvenor and Smith (1957b, 1958); 244, Culvenor and Smith (1966b); 245, Suri *et al.* (1982); 246, Subramanian and Nagarajan (1967); 247, Suri *et al.* (1976b); 248, Suri *et al.* (1976a); 249, Stermitz and Adamovics (1977); 250, Stermitz and Suess (1978); 251, Asada *et al.* (1982a); 252, Birnbaum *et al.* (1971); 253, Wunderlich (1962); 254, Mel'kumova *et al.* (1974); 261, Hoque *et al.* (1976); 262, Kugelman *et al.* (1976); 263, Jones *et al.* (1982); 264, Brown *et al.* (1983); 265, Gonzalez and Calero (1958); 266, Resch *et al.* (1983); 267, Brown *et al.* (1984); 268, Subramanian *et al.* (1980); 269, Mohanraj *et al.* (1982); 270, Haksar *et al.* (1982); 271, Roeder *et al.* (1981); 272, Rosberger *et al.* (1981); 273, Roeder *et al.* (1983); 274, Rueger and Benn (1983a); 275, Roitman (1983b); 276, Roitman (1983a); 277, Resch *et al.* (1982); 278, Wiedenfeld (1982); 279, Catalfamo *et al.* (1982a); 280, Luthy *et al.* (1983a); 281, Asada *et al.* (1982b); 282, Wei *et al.* (1982); 283, Bhakuni and Gupta (1982); 284, Reyes *et al.* (1982); 285, Ferry (1972); 286, Crout *et al.* (1972); 287, Pilbeam *et al.* (1983); 288, Rodriguez *et al.* (1971); 289, Pandey *et al.* (1982, 1983); 290, Hua *et al.* (1983); 291, Asada and Furuya (1984); 292, Wiedenfeld *et al.* (1984); 293, Omar *et al.* (1983); 294, Smith and Culvenor (1984); 295, Roeder *et al.* (1984a); 296, Liang and Roeder (1984); 297, Roeder *et al.* (1984b); 298, Luthy *et al.* (1984); 299, Larsen *et al.* (1984); 300, Roby and Stermitz (1984).

[c] Contains unidentified retronecine or heliotridine esters.

[d] The more recent work shows only saturated PAs.

[e] Contains unidentified retronecine, supinidine and trachelanthamidine esters.

TABLE 1.2 Some Plants Found Only to Contain Non-hepatotoxic PAs[a]

Boraginaceae
  Heliotropium strigosum
  H. ovalifolium
  Lindelofia macrostyla
  L. olgae
  L. pterocarpa
  Macrotomia echioides
  Paracaryum himalayense
  Tournefortia sibirica
  Trachelanthus hissoricus
  T. korolkovii

Compositae
  Cacalia hastata
  C. robusta
  Kleinia kleinioides
  Notonia petraea
  Senecio adnatus (*S. hygrophyllus*)
  S. aetnensis
  S. amphibolus
  S. angulatus
  S. aronicoides
  S. aucheri
  S. barbertonicus
  S. brachypodus
  S. cissampelinus
  S. francheti

  S. hygrophyllus
  S. inaequidens
  S. macrophyllus
  S. mikanoides
  S. pauciligulatus
  S. pubigerus
  S. pulviniformis
  S. rosmarinifolius
  S. sarracenius
  S. schvetzovii
  S. sylvaticus
  S. taiwanensis
  S. tournefortis

Leguminosae
  Crotalaria albida
  C. aridicola
  C. damarensi
  C. goreensis
  C. grandistipulata
  C. lachnophera
  C. maypurensis
  C. medicaginea (*C. trifoliastrum*)
  C. natalita
  C. podocarpa
  C. rhodesiae
  C. stolzii

---

[a] The alkaloids are either non-ester PAs, esters with a saturated necine moiety or esters of the dihydropyrrolizinone type (e.g., senaetnine). Plants in a few other families, particularly the Orchidaceae, also contain non-hepatotoxic PAs. For most of the references, see Smith and Culvenor (1981) and Robins (1982a).

TABLE 1.3 Unsaturated Ester PAs (Esters of Unsaturated Necines)

Alkaloid[a]		Necine	Necic acids(s)	References[b]	Plant sources[c]
1	Amabiline	(−)-supinidine	(−)-viridofloric	Culvenor and Smith (1967)	10?, 12, 15, 46
2	Anacrotine (crotalaburnine)	crotanecine	senecic	Atal et al. (1966a); Culvenor and Smith (1972); Mattocks (1968a)	200, 201, 214, 217, 218, 240
3	[2] 6-Acetyl	crotanecine	acetic; senecic	Culvenor and Smith (1972)	200
4	[2] 6-Angelyl	crotanecine	angelic; senecic	Culvenor and Smith (1972)	200
5	Anadoline	retronecine	tigloyl trachelanthic	Culvenor et al. (1975a)	66, 67
6	7-Angelyl heliotridine (rivularine)	heliotridine	angelic	Crowley and Culvenor (1959)	18, 30, 42, 84, 178
7	trachelanthate	heliotridine	angelic; (+)-trachelanthic	Crowley and Culvenor (1959)	42
8	[7] viridoflorate	heliotridine	angelic; (−)-viridofloric	Crowley and Culvenor (1959)	42
9	7-Angelyl retronecine	retronecine	angelic	Crowley and Culvenor (1962)	17, 155, 189
10	9-Angelyl retronecine	retronecine	angelic	Hikichi et al. (1980)	51, 52
11	Asperumine	heliotridine	angelic; asperumic	Man'ko and Kotovskii (1970b); Mel'kumova et al. (1974)	24, 63, 64
12	Axillaridine	retronecine		Crout (1969)	203
13	[12] Axillarine	retronecine		Crout (1969)	203
14	Bisline	retronecine	isolinecic	Coucourakis and Gordon-Gray (1970); Coucourakis et al. (1972)	154

(continued)

TABLE 1.3 (Continued)

Alkaloid[a]		Necine	Necic acid(s)	References[b]	Plant sources[c]
15		Clivorine	otonecine	clivonecic	Klasek et al. (1967); Birnbaum et al. (1971)	9, 92, 93, 94
16		Crispatine	retronecine	crispatic	Culvenor and Smith (1963)	207, 208, 220, 221
17		Crobarbatine	retronecine		Puri et al. (1973)	204
		Croburhine: see crotalarine				
18	[16]	Cromadurine	retronecine	cromaduric	Rao et al. (1975b,c)	221
19		Cronaburmine	retronecine	cronaburmic	Siddiqi et al. (1978b)	225
20		Crosemperine	otonecine	incanic	Atal et al. (1967)	199, 233
21		Crotaflorine	crotanecine	retronecic	Culvenor and Smith (1972)	200
22		Crotafoline	otonecine	crotafolic	Crout (1972)	218
		Crotalaburnine: see anacrotine				
23		Crotalarine (croburhine)	retronecine		Rao et al. (1975a)	206
24		Crotananine	retronecine		Siddiqi et al. (1978c)	225
25	[77]	Crotastriatine (acetyl nilgirine)	retronecine	acetylnilgiric	Batra et al. (1975)	223
26		Crotaverrine	otonecine	integerrinecic (isomer)	K. A. Suri et al. (1976a); O. P. Suri et al. (1976)	240, 241
27	[26]	Acetyl	otonecine	acetylintegerrinecic (isomer)	K. A. Suri et al. (1976a); O. P. Suri et al. (1976)	240, 241
28		Cynaustine	(+)-supinidine	(−)-viridofloric	Culvenor and Smith (1967)	13, 16
29	[17]	Dicrotaline	retronecine	dicrotalic	Adams and van Duuren (1953a); Brown et al. (1983)	209, 211
30		Doronenine	retronecine	bulgarsenecic	Roeder et al. (1980a)	119
31		Doronine	otonecine (isomer)	acetyljaconinecic	Alieva et al. (1976)	80, 153

No.	[Ref]	Alkaloid	Necine base	Acid	Reference	Numbers
32		Echimidine	retronecine	angelic; echimidinic	Culvenor (1956)	22, 23, 64, 65, 66, 67, 68
33		Echinatine	heliotridine	(−)-viridofloric	Crowley and Culvenor (1959); Akramov et al. (1964)	8?, 10, 12, 14, 18, 19, 34, 42, 44, 46, 47, 48, 49, 53, 54, 55, 56, 57, 58, 59, 60, 61, 63, 64, 65, 79, 85, 87
34		7-Acetyl	heliotridine	acetic; (−)-viridofloric	O. P. Suri et al. (1975)	45, 63?
35		Echiumine	retronecine	angelic; (+)-trachelanthic	Culvenor (1956); Culvenor and Smith (1966a)	5, 6, 7, 23
36ᵃ		Emiline	otonecine		Kohlmeunzer et al. (1971)	81
37		Erucifoline	retronecine	erucifolinecic	Sedmera et al. (1972)	123, 124
38		Europine	heliotridine	lasiocarpic	Culvenor (1954); Culvenor et al. (1954)	27, 33, 36, 40, 70
39	[31]	Floricaline	otonecine	diacetyl jacolic	Cava et al. (1968)	76, 125
40	[31]	Flordanine	otonecine	acetyl jacolic	Cava et al. (1968); Roeder et al. (1983)	76, 80, 104, 125, 153
41		Florosenine (acetyl otosenine)	otonecine	acetyl jacobinecic	Cava et al. (1968); Roeder et al. (1983)	76, 104, 125, 126
42	[16]	Fulvine	retronecine	fulvinic	Culvenor and Smith (1963); Schoental (1963)	208, 210, 221, 227
43		Grahamine	retronecine	3-(2-methylbutyryl)-monocrotalic	Atal et al. (1969)	212
44		Grantaline	retronecine	grantalic	Jones et al. (1982); Brown et al. (1984)	211, 213
45		Grantianine	retronecine	grantianic	Adams et al. (1942a); Adams and Gianturco (1956e); Brown et al. (1984)	211, 213

(continued)

TABLE 1.3 (*Continued*)

Alkaloid[a]		Necine	Necic acid(s)	References[b]	Plant sources[c]
46		Heleurine	(−)-supinidine	heliotric	Culvenor (1954); Culvenor et al. (1954)	33, 34
47	[11]	Heliosupine	heliotridine	angelic; echimidinic	Crowley and Culvenor (1959)	13, 18, 19, 20, 24, 42, 52, 53, 63, 65
48	[11]	Acetyl	heliotridine	angelic; acetyl echimidinic	Pedersen (1970)	18, 24, 52, 63?, 65?
49		Heliotrine	heliotridine	heliotric	Culvenor (1954); Culvenor et al. (1954)	25, 27, 29, 30, 31, 32, 33, 34, 35, 37, 38, 39, 43
50		7-Angelyl Heterophylline	heliotridine	angelic; heliotric	O. P. Suri et al. (1975)	32
51		Incanine	retronecine		Edgar et al. (1980)	2
52			retronecine	incanic	Yunusov and Plekhanova (1953, 1959)	37, 71
53	[5]	Indicine	retronecine	(−)-trachelanthic	Mattocks et al. (1961); Kugelman et al. (1976)	1?, 4?, 26, 34
54	[5]	Acetyl	retronecine	acetyl trachelanthic	Mattocks (1967a)	34
55		Integerrimine (squalidine)	retronecine	integerrinecic	Klasek et al. (1968b); Culvenor and Smith (1966b)	77, 95, 99, 102, 108, 121, 123, 124a, 127, 136, 137, 141, 147, 149, 150, 183, 184, 187a, 194, 205, 214, 215, 223, 238, 239
56	[5]	Intermedine	retronecine	(+)-trachelanthic	Culvenor and Smith (1966a); Herz et al. (1981)	1?, 4?, 5, 6, 7, 68, 70, 79
57	[5]	7-Acetyl	retronecine	acetic; (+)-trachelanthic	Culvenor et al. (1980b)	86

No.	[No.]	Name	Base	Acid	Reference	Numbers
58	[16]	Isocromadurine	retronecine	isocromaduric	Rao et al. (1975c)	207, 221
59	[14]	Isoline	retronecine	acetyl isolinecic	Coucourakis and Gordon-Gray (1970); Coucourakis et al. (1972)	154
60		Jacobine	retronecine	jacobinecic	Bradbury and Culvenor (1954); Bradbury and Masamune (1959)	99, 108, 113, 139, 156
		Jacodine: see seneciphylline				
61		Jacoline	retronecine	jacolic	see jacobine	139
62	[61]	Jaconine	retronecine	jaconinecic	see jacobine	99, 139
63	[60]	Jacozine	retronecine	jacozinecic	Culvenor (1964)	99, 139
64		Junceine	retronecine	junceic	Adams and Gianturco (1956d); Atal et al. (1966b)	216, 231
65		Lasiocarpine	heliotridine	angelic; lasiocarpic	Culvenor (1954); Culvenor et al. (1954)	27, 28, 30, 33, 34, 35, 36, 45, 64, 65
66	[65]	Acetyl	heliotridine	angelic; acetyl lasiocarpic	Culvenor et al. (1975b)	33
67		Latifoline	retronecine	angelic; latifolic	Crowley and Culvenor (1962)	17
68		Ligularidine	otonecine		Hikichi et al. (1979); Asada and Furuya (1984)	93
69	[22]	Ligularine	otonecine	diacetyl hygrophyllinecic	Klasek et al. (1971)	91, 94
		α-Longilobine: see seneciphylline				
		β-Longilobine: see retrorsine				
70		Lycopsamine	retronecine	(−)-viridofloric	Culvenor and Smith (1966a); Herz et al. (1981)	1, 4, 5, 6, 7, 9, 41, 51, 68, 86

(continued)

31

TABLE 1.3 (Continued)

Alkaloid[a]	Necine	Necic acids(s)	References[b]	Plant sources[c]
71 7-Acetyl	retronecine	acetic; (−)-viridofloric	Culvenor et al. (1980b)	9, 68
72 [21] Madurensine	crotanecine	integerrinecic	Atal et al. (1966a); Culvenor and Smith (1972)	200, 218, 221
73 [21] 7-Acetyl	crotanecine	acetic; integerrinecic	Atal et al. (1966a)	200
74 [43] Monocrotaline	retronecine	monocrotalic	Adams and Rogers (1939)	47, 199, 202, 206, 207a, 207b, 208, 208a, 212, 219, 220, 222, 224, 225a, 226, 227a, 228, 229, 230, 232, 234, 236, 237
Acetyl: see spectabiline				
75 [43] Monocrotalinine	retronecine	ethylidenedioxy-monocrotalic acid	Rajagopalan and Batra (1977a)	212
Mucronatinine: see usaramine				
76 [68] Neosenkirkine	otonecine	integerrinecic	Panizo and Rodriguez (1974); Asada and Furuya (1982)	105, 164
77 Nilgirine	retronecine	nilgiric	Atal et al. (1968)	223
acetyl: see crotastriatine				
78 [31] Onetine	otonecine	jacolic	Wunderlich (1962)	153
79 [41] Otosenine (tomentosine)	otonecine	jacobinecic	Cava et al. (1968); Kompis et al. (1960)	76, 80, 81, 98, 104, 113, 116, 123, 125, 126, 139, 153, 174, 188
80 [51] Parsonsine	retronecine		Edgar et al. (1980)	2

No.	[Ref]	Alkaloid	Necine	Acid	Reference	Citations
81	[41]	Petasitenine (fukinotoxin)	otonecine	jacobinecic isomer (β-epoxide)	Yamada et al. (1976a,b); Furuya et al. (1976)	96
82	[41]	Acetyl (neopetasitenine)	otonecine	acetyl jacobinecic isomer	Yamada et al. (1976a,b)	96
		Renardine: see senkirkine				
83		Retroisosenine	retronecine		Nghia et al. (1976); Klasek et al. (1980)	151, 152
84		Retronecine dibenzoate	retronecine	benzoic	Siddiqi et al. (1978a)	11
85		Retrorsine (β-longilobine)	retronecine	isatinecic	Barger et al. (1935); Christie et al. (1949); Leisegang and Warren (1950)	100, 101, 106, 108, 109, 113, 113a, 115, 117, 120, 122, 124, 127, 131, 132, 134, 138, 139, 144, 160, 162, 171, 172, 175, 177, 179, 181, 182, 187, 191, 192, 193, 195, 196, 198, 235, 239
86		Retusamine	otonecine	retusaminic	Culvenor et al. (1967b); Wunderlich (1962)	207a, 222, 226, 230
87		Riddelliine	retronecine	riddellic	Adams and Govindachari (1949c); Adams and van Duuren (1953c)	98, 100, 114a, 115, 116, 120, 122, 177, 182, 195, 216
88	[33]	Rinderine	heliotridine	(+)-trachelanthic	Akramov et al. (1965); Locock et al. (1966)	55, 62, 84, 88
		Rivularine: see 7-angelyl heliotridine				
89	[24]	Sceleratine	retronecine	sceleranecic	de Waal et al. (1963); Gordon-Gray (1967)	181
90	[24]	Sceleratinyl chloride	retronecine	chlorodeoxysceleranecic	Gordon-Gray (1967)	181

(continued)

33

TABLE 1.3 (Continued)

Alkaloid[a]		Necine	Necic acids(s)	References[b]	Plant sources[c]
91		Senecicannabine	retronecine		Asada et al. (1982a)	110
92	[85]	Senecionine	retronecine	senecic	Barger and Blackie (1936); Kropman and Warren (1949, 1950)	75, 82, 83, 95, 97, 97a, 98, 99, 100, 101, 102, 104, 108, 111, 113, 114, 114a, 115, 116, 117, 120, 122, 123, 124, 128, 129, 130, 135, 137, 139, 145, 146, 147, 150, 157, 158, 159, 163, 170, 171, 172, 180, 182, 183, 184, 186, 187a, 188, 189, 190, 192, 193, 194, 195, 196, 197, 201, 216, 239, 242, 243, 244
93	[87]	Seneciphylline (jacodine; α-longilobine)	retronecine	seneciphyllic	Wali and Handa (1964); Klasek et al. (1968b)	73, 73a, 74, 83, 99, 100, 101, 103, 107, 108, 110, 111, 112, 113, 114a, 115, 116, 118, 120, 122, 123, 124, 126, 128, 133, 136, 139, 142, 143, 146, 148, 153, 156, 158, 161, 165, 166, 167, 168, 171, 173, 174, 176, 180, 182, 183, 185, 186, 216
94	[24]	Senecivernine	retronecine		Roeder et al. (1979)	192

No.	Ref	Name	Base	Acid	Reference	Numbers
95		Senkirkine (renardine)	otonecine	senecic	Briggs et al. (1965)	75, 90, 94a, 95, 96, 97, 102, 139, 140, 164, 169, 174, 187a, 190, 192, 198, 218
96	[95]	Acetyl	otonecine	acetyl senecic	Briggs et al. (1965)	140, 141, 187a
97	[95]	Hydroxy	otonecine	isatinecic	Crout (1972)	218
		Sericine: see spectabiline				
98		Sincamidine	retronecine	heliotric isomer	Culvenor and Smith (1966a)	6
99		Spartioidine	retronecine	seneciphyllic isomer	Manske (1939); Adams and Gianturco (1957)	182
100	[43]	Spectabiline (sericine)	retronecine	acetyl monocrotalic	Culvenor and Smith (1957b; 1958)	236
101	[51]	Spiracine	retronecine		Edgar et al. (1980)	3
102	[51]	Spiraline	retronecine		Edgar et al. (1980)	3
103	[51]	Spiranine	retronecine		Edgar et al. (1980)	3
		Squalidine: see integerrimine				
104	[1]	Supinine	(−)-supinidine	(+)-trachelanthic	Culvenor (1954); Crowley and Culvenor (1959)	10?, 34, 42, 69, 72, 85, 88, 89
105		Swazine	retronecine	swazinecic	Gordon-Gray and Wells (1974)	187
106	[35]	Symlandine	retronecine	(−)-viridofloric tiglic;	Culvenor et al. (1980b)	68
107		Symphytine	retronecine	(−)-viridofloric	Furuya and Araki (1968); Furuya and Hikichi (1971)	51a, 65, 66, 68
108		Syneilesine	otonecine		Hikichi and Furuya (1974; 1976)	197
109	[108]	Acetyl	otonecine		Hikichi and Furuya (1976)	197
110	[52]	Trichodesmine	retronecine	trichodesmic	Adams and Gianturco (1956c)	29, 70, 71, 211, 216, 229, 231, 238

(continued)

35

TABLE 1.3 (*Continued*)

Alkaloid[a]		Necine	Necic acids(s)	References[b]	Plant sources[c]
111	[32]	Uplandicine	retronecine	acetic; echimidinic	Culvenor et al. (1980b)	68
112	[55]	Usaramine (mucronatinine)	retronecine	retronecic	Sawhney et al. (1967); Culvenor and Smith (1966b)	205, 214, 215, 223, 239
113	[14]	Yamataimine	retronecine		Hikichi et al. (1978)	78
				Supplementary List		
114		7-Senecioyl retronecine	retronecine	senecioic	Rueger and Benn (1983a)	109a, 189
115		7-Senecioyl-9-sarracinyl retronecine	retronecine	senecioic; sarracinic	Rueger and Benn (1983a)	109a, 189
116	[115]	Triangularine	retronecine	sarracinic, angelic	Roitman (1983b)	189
117	[115]	Neotriangularine	retronecine	sarracinic (isomer), angelic	Roitman (1983b)	189
118	[71]	3'-Acetyl lycopsamine	retronecine	acetic; (−)-viridofloric	Roitman (1983a)	7a

119 [71]	3',7-Diacetyl lycopsamine	retronecine	acetic; (−)-viridofloric	Roitman (1983a)	7a
120 [52]	Globiferine	retronecine		Brown et al. (1984)	211
121	Scorpioidine (epimer of anadoline)	retronecine	tigloyl viridofloric	Resch et al. (1982)	51a
122 [121]	7-Acetyl	retronecine	acetic; tigloyl virido-floric	Resch et al. (1982)	51a
123 [107]	Myoscorpine (epimer of symphytine)	retronecine	tiglic; trachelanthic	Resch et al. (1982)	51a
124	Gynuramine	retronecine		Wiedenfeld (1982)	90a
125 [124]	Acetyl	retronecine		Wiedenfeld (1982)	90a
126	Ligularizine	otonecine		Asada and Furuya (1984)	93
127 [22]	Neoligularidine	otonecine		Asada and Furuya (1984)	93
128	Sencalenine	retronecine	angelic; hydroxy-senecic	Roeder et al. (1984a)	109a

a Bracketed numbers, where present, refer to structure diagrams.
b Principle references to isolation and structure.
c Numbers refer to plants listed in Table 1.1.
d The structure given for emiline (36) is incorrect. The revised structure is a 12-membered macrocyclic diester, the acid moiety being: —OCOCH(Et)CH$_2$C(=CH$_2$)C(Me)(OH)COO— (D. J. Robins, private communication).

TABLE 1.4 Ester PAs with a Saturated Necine Moiety

	Alkaloid	Necine	Necic acid(s)	References
130	Angularine	rosmarinecine	seneciphyllic	Porter and Geissman (1962)
131	Bulgarsenine	platynecine	bulgarsenecic	Nghia et al. (1976)
132	Cornucervine	trachelanthamidine	3-hydroxy-3-methoxycarbonyl-5-methylhexanoic	Brandange and Luning (1969)
133	Coromandalin	trachelanthamidine	(+)-viridofloric	Subramanian et al. (1980)
134	Croalbidine	croalbinecine	trichodesmic	Sawhney and Atal (1973)
135	Crocandine	(−)-turneforcidine	fulvinic	Siddiqi et al. (1979b)
136	Cropodine	(−)-turneforcidine	monocrotalic	Haksar et al. (1982)
137	Curassavine	trachelanthamidine	curassavic	Subramanian et al. (1980)
138	Cynaustraline	(+)-isoretronecanol	(−)-viridofloric	Culvenor and Smith (1967)
139	Farfugine	(+)-turneforcidine	angelic	Niwa et al. (1983b)
140	Fuchsisenecionine	platynecine	senecioic	Lemp (1973)
141	Hastacine	hastanecine	integerrinecic	Culvenor et al. (1968)
142	Helifoline	croalbinecine	angelic	Mohanraj et al. (1981)
143 [133]	Heliovicine	trachelanthamidine	(−)-trachelanthic	Subramanian et al. (1980)
144	Hygrophylline	platynecine	hygrophyllinecic	Schlosser and Warren (1965)
145 [135]	Isocrocandine	(−)-turneforcidine	cromaduric	Siddiqi et al. (1979b)
145a	Ligularinine	platynecine	senecic isomer	Asada and Furuya (1984)

No.	Ref.	Alkaloid	Necine	Acid	Reference
146	[138]	Lindelofamine	(+)-isoretronecanol	angelyl trachelanthic	Warren (1955)
147	[138]	Lindelofine	(+)-isoretronecanol	(+)-trachelanthic	Warren (1955)
148		Macrophylline	macronecine	angelic	Danilova and Utkin (1960); Warren (1966)
149		Macrotomine	trachelanthamidine	macrotomic	Warren (1955)
150		Nemorensine	platynecine	nemorensic	Klasek et al. (1973a)
151	[144]	Neoplatyphylline	platynecine	integerrinecic	Culvenor et al. (1968)
152		Petasinine	petasinecine	angelic	Yamada et al. (1978)
153	[144]	Playtphylline (senecifoline)	platynecine	senecic	Warren (1955)
154	[136]	Retusine	(−)-turneforcidine	2,3,4-trimethyl-4-hydroxyglutaric	Culvenor and Smith (1957a)
155	[130]	Rosmarinine	rosmarinecine	senecic	Richardson and Warren (1943); Koekemoer and Warren (1955)
156		Sarracine (mikanoidine)	platynecine	angelic, sarracinecic (sarracinic)	Culvenor and Geissman (1961b)
157		Strigosine	trachelanthamidine	2,3-dihydroxy-3-methylpentanoic	Mattocks (1964b)
158	[133]	Trachelanthamine	trachelanthamidine	(+)-trachelanthic	Men'shikov (1946); - Warren (1955)
159		Turneforcine	(−)-turneforcidine	angelic	Aasen et al. (1969)
160	[133]	Viridoflorine	trachelanthamidine	(−)-viridofloric	Men'shikov (1948); Warren (1955)

2

Extraction, Separation and Analysis of Pyrrolizidine Alkaloids

I. INTRODUCTION

It is often necessary to obtain pure PAs from plant material for chemical or other investigations; PAs are also extracted from plants and other biological sources for analytical purposes. It is unusual for a plant to contain a single PA. More often, several basic alkaloids are present, which may or may not be closely related. These are often accompanied by *N*-oxides which are usually, but not necessarily, those of alkaloids which are also present as free bases. Purification will entail separating individual alkaloids from such mixtures as well as from non-alkaloidal materials.

The PAs in a given plant species may vary considerably; for instance, *Senecio hygrophyllus* contains varying proportions of rosmarinine, platyphylline and hygrophylline depending on the stage of growth, season and district (Richardson and Warren, 1943; Schlosser and Warren, 1965). Ratios of alkaloid bases to *N*-oxides can vary between different parts of the plant and with the season of growth; thus in *Crotalaria retusa,* basic alkaloids accumulate in the seeds whereas *N*-oxides may predominate in the green parts (Culvenor and Smith, 1957a; Mattocks, 1971b).

For preparative purposes, to simplify separations, it is customary for pyrrolizidine *N*-oxides to be chemically reduced to bases during extraction from plants. One or more pure alkaloids can sometimes be separated from crude mixtures of bases by crystallization; more often, various techniques of column chromatography are necessary; occasionally, countercurrent separation or preparative high-performance liquid chromatography (HPLC) have been used.

For the analytical separation and identification of individual PAs, paper chromatography has largely been replaced by thin-layer chromatography (TLC); unsaturated PAs, their *N*-oxides, saturated alkaloids, and dehydro-alkaloids ('pyrrolic' metabolites) can all be located and distinguished using appropriate spray reagents. Early difficulties with the gas chromatography (GC) of PAs have largely been overcome, and recent methods of capillary GC coupled with mass spectrometry (MS) are the most sensitive and versatile means for identifying and determining PAs in complex mixtures. HPLC has also proved useful in this area, especially for the less volatile alkaloids and *N*-oxides.

The Ehrlich 'pyrrole' reaction has made possible a range of colourimetric methods for the detection and determination of unsaturated PAs and their *N*-oxides, as well as the identification of the major route of activation of these compounds to toxic metabolites in mammalian liver (Mattocks, 1968a). This colour reaction responds principally to the hepatotoxic group of PAs and their metabolites. For the non-hepatotoxic (saturated) alkaloids, colourimetric analyses using indicator dues can be used.

II. STABILITY OF THE ALKALOIDS

It is important to ensure that, as far as possible, alkaloids isolated from plants or other sources are qualitatively and quantitatively the same as those which were present in the original material. There are several ways in which alkaloids might be altered during isolation procedures.

A. Changes While in the Plant

Alkaloids may be changed by heat or by enzymic action when harvested plants are stored and dried. Bull *et al.* (1968) reported the loss of 50–80% of alkaloids during drying of some, but not all, *Crotalaria* and *Heliotropium* species. On the other hand, Pedersen (1975a) compared the alkaloid content of fresh leaves with that of dried leaves from several species of the Boraginaceae; allowing for the loss in weight on drying there were no changes in total alkaloid content, but changes were observed in the ratio of free bases to *N*-oxides, suggesting that enzymic oxidation or reduction of alkaloids could take place during the early stages of drying. Birecka *et al.* (1980) found no significant differences between the alkaloid contents of fresh and dried leaves of nine greenhouse-grown *Heliotropium* species. *N*-Oxides of unsaturated pyrrolizidine bases are slowly dehydrated by heat to pyrrolic products and thence decomposed to intractable polymers.

PAs (of *Senecio alpinus*) have been found to be stable when incorporated in hay but rapidly degraded when in silage, although low levels of alkaloids are more persistent (Candrian *et al.*, 1984b).

If fresh green plant material cannot be extracted straight away, it should be dried as quickly as possible to minimize enzyme action and fungal decomposition. A warm environment, with plenty of circulating air, is best. Direct sunlight should be avoided because of possible overheating; the alkaloids are not especially light-sensitive. After drying, the plant should be stored in a cool place.

B. Changes during Extraction Procedures

Pyrrolizidine esters may hydrolyse rapidly in aqueous solutions at high pH (>9). Solutions basified prior to extraction of alkaloids with organic solvents should be kept cool, and not stored. Mildly acid solutions are much more stable, but some alkaloids (e.g. the epoxide jacobine) are labile to halogen acids (Bull *et al.*, 1968). Ferrous ion in aqueous solution can catalyse decomposition of unsaturated pyrrolizidine *N*-oxides through conversion to pyrrolic derivatives (Mattocks, 1968c).

PAs should not be kept for long periods in contact with halogenated organic solvents. Impurities in commercial chloroform, especially chlorobromomethane, may form quaternary salts with tertiary bases (Williams, 1960); chloroform and dichloromethane may react slowly with PAs, especially at elevated temperatures (Zapesochnaya *et al.*, 1973). Halogenated solvents undergo photochemical reactions with amines, even in the absence of oxygen (Lautenberger *et al.*, 1968, and references therein), with formation of various products including the amine hydrochlorides. Hence, extractions with such solvents should not be conducted in bright light.

III. EXTRACTION OF ALKALOIDS

A. From Plants

1. Plant Sources of Pyrrolizidine Alkaloids

There are few commercial sources of PAs; it is often necessary to extract supplies of alkaloids from plants. The highest alkaloid levels generally occur in sub-tropical plants. *Crotalaria spectabilis* and *C. retusa* contain monocrotaline

(Adams and Rogers, 1939); an Indian *C. retusa* sample holds the record for the highest PA yield from seed, over 9% by weight (Kumari *et al.*, 1967). *C. spectabilis* seed in the United States also contains high levels of monocrotaline (Constantine *et al.*, 1967). In Australia it also yields spectabiline (acetyl-monocrotaline), and the two alkaloids can be separated by crystallization (Culnevor and Smith, 1957b). In southern Africa, *Senecio isatideus* leaves can hold up to 5% (dry weight) of PAs, mainly retrorsine *N*-oxide (isatidine) (Koekemoer and Warren, 1951); Johnson *et al.* (1985a) have reported PA levels varying from 0.18% to an incredible 17.99% in *Senecio riddellii* from different locations (see also Molyneux and Johnson, 1984). The best sources of heliotridine-based PAs are *Heliotropium* species: *H. europaeum* and *H. lasiocarpum* yield mixtures from which heliotrine and lasiocarpine can be separated by chromatography (Culvenor, 1954); *H. eichwaldii* contains mainly heliotrine (Gandhi *et al.*, 1966a).

In temperate climates, the greenhouse cultivation of high-yield plants may be possible. Plants native to such regions have relatively low alkaloid levels. *Senecio jacobaea* (ragwort; tansy ragwort) is widely available, but its alkaloid yield is poor—about 0.06% (Barger and Blackie, 1937)—and this may fall after flowering. It contains five or more alkaloids, difficult to separate (Bradbury and Culvenor, 1954), but since they are all esters of the amino alcohol retronecine (183), the latter may be obtained by hydrolysing the crude mixture.

2. Extraction Procedures

Plants often contain both basic PAs and their *N*-oxides. The latter are often incompletely extracted from aqueous solutions using organic solvents, but it is easy to chemically reduce them to the corresponding basic alkaloids before extraction. If the *N*-oxides are required, it is better to prepare these by oxidizing the purified alkaloids (Christie *et al.*, 1949; Koekemoer and Warren, 1951). Most procedures for extracting PAs from plants are based on that of Koekemoer and Warren (1951), and descriptions have been given in numerous publications. The dried plant, root or seed is milled, then extracted with hot or cold alcohol, either continuously (Soxhlet) or batchwise. The alcohol is evaporated, the alkaloids are dissolved in dilute aqueous acid, and chlorophyll and waxes are removed by extraction with ether or petroleum, or sometimes by filtration. *N*-Oxides are reduced by stirring zinc dust with the acidic solution, which is then filtered and made basic with ammonia, and the alkaloids are extracted using an organic solvent; chloroform is usual but ethyl acetate can sometimes be a convenient, less toxic alternative. Some highly water soluble alkaloids may be incompletely extracted; e.g. *Amsinckia intermedia* alkaloids are only fully extracted by chloroform if the aqueous liquor is saturated with salt (Culvenor and Smith, 1966a). Evaporation of the organic phase leaves a crude mixture of PAs which existed as basic alkaloids or *N*-oxides in the original plant. By omitting the reduction stage, usually only those alkaloids are obtained which were present as free bases; however, *N*-oxides of some alkaloids, e.g. lasiocarpine, may be extracted into chloroform. Subsequent reduction of the aqueous liquor, then re-extraction, gives the basic forms of those alkaloids which were *N*-oxides in the plant. Reduction may be unnecessary when extracting alkaloids from seeds, which may contain very little *N*-oxides (Mattocks, 1971b): e.g. the extraction of monocrotaline from *Crotalaria spectabilis* seed (Constantine *et al.*, 1967) and supinine from seeds of *Trichodesma zeylanicum* (O'Kelly and Sargeant, 1961).

PAs can also be extracted in similar ways from crushed or shredded fresh plant material. The use of fresh material is advisable when it is possible that sensitive alkaloids might not survive drying or storage, especially in a hot environment.

PAs, both as tertiary bases and *N*-oxides, can be isolated from alcoholic plant extracts using cation exchange resin (e.g. Dowex 50); a continuous extraction apparatus can be used for this, alcohol being circulated in turn through a bed of dried, ground plant material and then through the resin (Mattocks, 1961). The alkaloids are subsequently eluted from the resin using ammonia solution. In this way indicine, its *N*-oxide and minor alkaloids were isolated from *Heliotropium indicum* (Mattocks *et al.*, 1961; Mattocks, 1967a). This method avoids the tarry decomposition products often encountered after hot alcoholic extractions. It has recently been adapted for the large scale extraction of alkaloids from ragwort (*Senecio jacobaea*) (Deagen and Deinzer, 1977); the inclusion of a column of anion exchange resin removes acid from the solvent and enables the cation exchange resin to retain larger amounts of alkaloids. The method has also been used for the analytical isolation of ragwort alkaloids (Ramsdell and Buhler, 1979). Huizing and Malingre (1979a,b) used a redox polymer, 'Serdoxit', to reduce pyrrolizidine *N*-oxides to tertiary bases, and have claimed that a column of this can be used in conjunction with the ion exchange isolation procedure.

Prolonged soaking of plant material (fresh or dried) in dilute acid can be used to extract alkaloids such as senkirkine (Briggs *et al.*, 1965). This method has been adapted to the large scale extraction of PAs from *Senecio jacobaea* by Craig *et al.* (1984), who claim high recoveries. The dried plant material (23 kg) is soaked in dilute acetic acid (optimum: pH 4) for 2–3 days; the liquor is basified (pH 10.5) with ammonia and extracted with dichloromethane, which is then concentrated, and the alkaloids are extracted from the residue into dilute sulphuric acid.

Pyrrolizidine amino-alcohols such as retronecine and heliotridine are too water soluble to be extractable from aqueous solution by chloroform; however, extraction is possible if the solution is first saturated with potassium carbonate. Some bases and *N*-oxides are partially extracted by butan-1-ol, giving crude mixtures which may yield to chromatographic separation. The amino alcohols are better isolated by evaporating the aqueous liquor to dryness, preferably together with an absorbant material such as kieselguhr, then extracting the residue with chloroform.

A convenient method for recovering PAs from acidic aqueous solutions is to precipitate their reinecke salts (Brandange and Granelli, 1973). The bases can be recovered from the reineckates using ion-exchange resin (cf. Kum-Tatt, 1960).

An electrochemical method has been used for isolating alkaloids from *Senecio platyphylloides* (Vdoviko *et al.*, 1977).

B. From Other Materials

It is sometimes required to extract PAs from vegetable sources such as foodstuffs, or from animal tissues or fluids, for identification or quantitative analysis. Methods described above can be applied to herbal beverages, contaminated cereals and the like. PAs have been isolated for analytical purposes from honey (Deinzer *et al.*, 1977) and from milk (Dickinson *et al.*, 1976). Ames and Powis (1978) extracted indicine from plasma and urine: indicine *N*-oxide was reduced to the base *in situ* before extraction. Indicine *N*-oxide itself has been extracted with methanol from lyophilised human plasma (McComish *et al.*, 1980), and from urine samples using cation exchange resin (Evans *et al.*, 1979): the *N*-oxide was recovered from the resin using aqueous pyridine rather than ammonia as used previously (Mattocks *et al.*, 1961), to avoid hydrolysis. Recoveries of PAs from biological materials are often not quantitative. For instance, about 70% of retrorsine added to rat urine was recovered by extraction; retrorsine *N*-oxide reduced *in situ* with zinc and acid gave a similar recovery (Mattocks, 1967c). To facilitate accurate analyses, when the extraction is to be followed by separation procedures such as HPLC or GC, a similar known alkaloid can be added as internal standard; thus heliotrine or its *N*-oxide have been employed in the above indicine analyses.

The analytical isolation of alkaloids from urine, bile and blood of sheep given heliotrine was described by Jago *et al.* (1969). Neutral, deproteinated liquors were passed through florisil (a magnesium silicate adsorbant), which retained the alkaloids; the latter were subsequently eluted using methanol, with or without added ammonia.

In principle, the foregoing methods can be adapted for extracting PAs from solid animal tissues. However, many toxic PAs are metabolized very rapidly in animals (Mattocks, 1972b) to products which may bind strongly to tissues or be rapidly excreted. Hence the amounts of a PA ingested by an animal which are recoverable from tissues after a few hours may be very small.

IV. SEPARATION AND PURIFICATION METHODS

A. Preparative

1. Crystallization

Many PAs, and especially the macrocyclic diesters, are stable crystalline solids. Pure alkaloids are sometimes obtained from crude plant extracts by re-crystallization. Suitable solvents are methanol, ethanol, or combinations of these with ethyl acetate, acetone or diethyl ether.

The N-oxides usually crystallize from similar solvent systems, but they are sometimes hygroscopic, and often solvent of crystallization is present. The latter can be driven off by heat under reduced pressure, but the N-oxides of unsaturated PAs are liable to decompose under prolonged dehydrating conditions.

Some mixtures of alkaloids obtained from plants are difficult or impossible to separate by crystallization alone. Examples are the mixtures from *Senecio jacobaea* and *S. vulgaris*. In the past some 'pure' alkaloids have been described which later proved to be mixtures which co-crystallized; for example, 'pterophine' and 'hieracifoline' were both shown to be mixtures of senecionine and seneciphylline (Culvenor and Smith, 1954, 1955; Adams and Gianturco, 1956a). Up to 17 recrystallizations failed to separate a mixture containing senecionine, seneciphylline, retrorsine and riddelliine into its components (Adams and Govindachari, 1949c).

2. Via Derivatives

PAs can often be purified by preparing and recrystallizing their picrates or picrolonates. The alkaloids are recovered from these in aqueous solution by treating them with alkali, then extracting the base with chloroform; picrolonates can first be treated with cupric sulphate, which precipitates copper picrolonate (Crowley and Culvenor, 1959). Alternatively, the salts, in methanol solution, can be treated with anion exchange resin (e.g. Dowex 1, in OH form).

3. Column Chromatography

Various chromatographic procedures have been used to obtain pure specimens of PAs from mixtures which could not be resolved by crystallization. Adsorption chromatography on alumina has been of limited value. With it Adams and Govindachari (1949a) separated and purified 'α-longilobine' and 'β-longilobine' (seneciphylline and retrorsine). Development was with chloroform containing a little ethanol, and either alkaloids were eluted with this solvent or the alumina was removed from the column and portions were extracted with aqueous acetic acid. However, mixtures of seneciphylline with senecionine, and retrorsine with riddelliine, run together and cannot be completely separated (Adams and Govindachari, 1949b). Neutral alumina can separate a mixture of the otonecine-based alkaloids otosenine, florosenine, floricaline and floridanine, elution being with combinations of benzene, chloroform and methanol (Cava *et al.*, 1968). A mixture of monocrotaline and crispatine also separates on alumina (Suri and Atal, 1967). By adapting a system used for paper chromatography, Bradbury and Mosbauer (1956) separated ragwort alkaloids on a column of cellulose powder eluted with butanol–acetic acid–water. For similar mixtures, Adams and Gianturco (1956a) used partition chromatography on a Celite 545 column supporting a phosphate buffer; elution was with carbon tetrachloride–chloroform mixtures; Culvenor and Smith (1954, 1955) have used a phosphate buffer, pH 4, supported on Pyrex glass powder. A similar system, at pH 8, can separate the alkaloids of *Heliotropium supinum,* including heliotrine and supinine (Crowley and Culvenor, 1959), and give partial separation of alkaloids from comfrey (Culvenor *et al.*, 1980b). Partition columns of Hyflo supacel moistened with phosphate buffer have been used to separate a mixture from *Heliotropium europaeum* containing lasiocarpine and heliotrine (Culvenor *et al.*, 1954; Culvenor, 1954), and minor alkaloids from *H. indicum* (Mattocks, 1967a).

Otonecine-based PAs, including clivorine and ligularidine, have been separated on a silica gel column eluted with a benzene–ethyl acetate–diethylamine mixture (Asada and Furuya, 1984).

A recent development has been the preparative scale separation of the diastereoisomeric alkaloids intermedine (56) and lycopsamine (70) as their borate complexes (Frahn *et al.*, 1980). The alkaloids contain vicinal glycol groups differing in configuration and thus in the degree to which they complex with borate. The mixture, in chloroform, is passed down a glass column moistened with borax solution. Intermedine is eluted first, followed by lycopsamine. Alternatively, the mixture, dissolved in borax solution is passed through a cation-exchange resin: lycopsamine is eluted first.

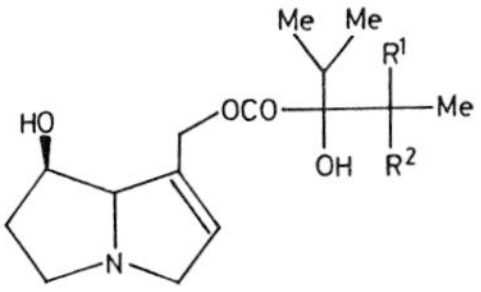

	R^1	R^2
(56)	OH	H
(70)	H	OH

Scheme 2.1.

Indicine *N*-oxide, extracted from plasma, has been purified using alumina (eluted with methanol) and a reverse-phase (C_{18}-bonded) silica gel column (eluted with aqueous methanol) (McComish *et al.*, 1980).

4. Countercurrent Distribution

This technique, little used with PAs, gives partial separation of the mixture of eight alkaloids from *Symphytum* × *uplandicum* (comfrey), the phases being chloroform (stationary) and phosphate buffer, pH 7.5 (Culvenor *et al.*, 1980b); echimidine, acetyl-lycopsamine and uplandicine are obtained pure.

B. Analytical

1. Paper Chromatography

Formerly much used, this technique has been largely superceded by thin-layer chromatography. The most useful system for PAs has been to run ascending or descending chromatograms on Whatman no. 1 paper, using as solvent the upper layer resulting from shaking butan-1-ol with an equal volume of 5% aqueous acetic acid (Adams and Gianturco, 1956a); R_f values for a large number of PAs and bases are given by Chalmers *et al.* (1965). Mattocks (1969b) used paper buffered with sodium acetate (Munier *et al.*, 1952) to obtain better separations of semisynthetic retronecine esters. The above solvent, and also butan-1-ol–aqueous ammonia–water (30 : 1 : 5), have been used to chromatograph some dihydropyrrolizines derived from PAs (Culvenor *et al.*, 1970b,c).

2. Thin-Layer Chromatography (TLC)

For most PA work, silica coated plates have been used. Sharma *et al.* (1965) found a solvent consisting of chloroform–methanol–ammonia solution (85 : 14 : 1) to be the best of many mixtures tried; a similar mixture (86 : 14.5 : 0.5) has been used for dihydropyrrolizine derivatives (pyrroles) from PAs (Culvenor *et al.*, 1970b,c). A mixture of ethyl acetate–acetone–ethanol–ammonia solution (5 : 3 : 1 : 1, by volume) gives excellent results with PAs, synthetic analogues, and pyrroles (Mattocks, 1974, 1981a). Chloroform can be used instead of ethyl acetate (Mattocks, 1967b), but the latter has lower toxicity, is less liable to react with alkaloids, and gives a faster flow on ascending chromatographs. Chalmers *et al.* (1965), using plates prepared from a slurry of silica gel in 0.1 N sodium hydroxide and developed with methanol, examined relationships between pyrrolizidine structures and R_f values. Londareva and Tikhomirova (1971) used alumina plates

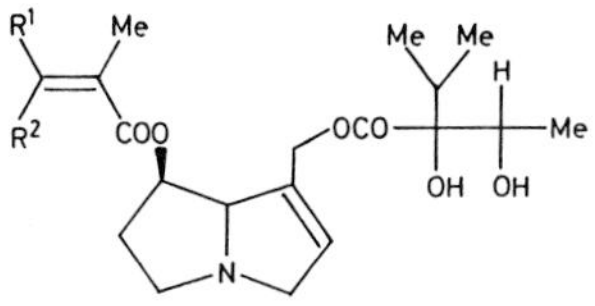

$$\begin{array}{ccc} & R^1 & R^2 \\ (106) & Me & H \\ (107) & H & Me \end{array}$$

Scheme 2.2.

developed with chloroform–methanol (8 : 2) for separating seneciphylline and platyphylline. TLC R_f values of PAs, as with other compounds, may vary slightly with different batches of plates, developing solvent, temperature, degree of chamber saturation and amounts of compounds applied, and it is helpful to include a standard compound (e.g. monocrotaline) on each plate. Chalmers *et al.* (1965) obtained consistent R_f values even with overloaded plates when measurements were made from the bottom edge of the spot. Sometimes different alkaloids have the same R_f value, and with unknown mixtures it is advisable to run chromatograms using more than one solvent system. Closely related isomers may be difficult or impossible to separate by TLC: e.g. symphytine (107) and symlandine (106) (Culvenor *et al.*, 1980b), or intermedine and lycopsamine (Frahn *et al.*, 1980).

Pyrrolizidine *N*-oxides are often much less lipophilic than corresponding bases and have lower R_f values than the latter. Developing solvents suitable for the TLC of *N*-oxides on silica plates include butan-1-ol–acetic acid–water (4 : 1 : 5, volume) (Mattocks, 1967b), and chloroform–methanol–propan-1-ol–water (70 : 50 : 5 : 10) (Wagner *et al.*, 1981). These solvents are also suitable for necine bases.

For chemically reactive pyrrolic and dihydropyrrolizine alcohols and esters on silica plates, either a butanone–acetone mixture (2 : 1) (Mattocks and Driver, 1983) or dioxan can be used; the plates must be thoroughly dried before use.

TLC can be used to purify small amounts of PAs as a preliminary to identification or analysis by other techniques. For example, Culvenor *et al.* (1980b) used preparative silica plates developed with chloroform–methanol–ammonia to separate comfrey (*Symphytum*) alkaloids prior to GC–MS. Huizing and Malingre (1981) described a method of ion-pair adsorption TLC for separating *Symphytum* alkaloids and other PAs. Silica plates are dipped in methanolic solutions of potassium chloride, sodium iodide, or lithium chloride, then dried. Developing solvents are chloroform–methanol mixtures, or chloroform–methanol–ammonia (Sharma *et al.*, 1965). Advantages claimed are the ease with which the impregnant and eluant can be adjusted to give good separations, and that the system is resistant to the effects of atmospheric changes.

Nemorensine and bulgarsenine have been separated on silica plates using chloroform–isobutanol–methanol–ammonia solution (40 : 15 : 15 : 4) (Gencheva, 1978).

For the TLC purification of indicine N-oxide prior to colourimetric determination, D'Silva and Notari (1980) used silica coated aluminum sheets, developed with ether–ethanol–aqueous ammonia–water (5 : 4 : 1 : 1).

3. Visualisation of PAs on TLC and Paper Chromatograms

PAs can be detected on chromatograms by exposure to iodine vapour. This is not specific for alkaloids or even bases—many other compounds appear as brown spots. However, iodine dehydrogenates unsaturated pyrrolizidines (and other pyrrolines) to pyrrolic products; thus a subsequent spray with Ehrlich reagent (see below) detects these compounds as blue or magenta spots (Culvenor et al., 1970c).

The modified iodobismuth reagent of Munier (1953) has been much used for detecting PAs. It is relatively insensitive, and is not specific for these bases. However, it is useful when working with saturated pyrrolizidines, most of which do not respond to Ehrlich-based tests (below). Iodoplatinate reagent (Waldi et al., 1961) is also effective for paper chromatograms, but not TLC (Mattocks, 1964b, 1967b). Dann (1960) found that by treating paper chromatograms with acetic anhydride, then heating them, N-oxides of PAs show up as brown spots, which fluoresce under ultraviolet (UV) light. This procedure is most successful with unsaturated pyrrolizidine N-oxides, which are thereby dehydrated to pyrrolic products capable of giving an intense colour with Ehrlich reagent. This leads to a sensitive procedure for detecting unsaturated PAs and their N-oxides on TLC (Mattocks, 1967b). The plates are treated with hydrogen peroxide (or 3-chloroperbenzoic acid) to convert bases to N-oxides, then with acetic anhydride, then Erhlich reagent, heat being applied at each stage. By omitting the first stage, only N-oxides are detected. Ehrlich reagent alone detects pyrroles (dihydropyrrolizines) derived from PAs. The Ehrlich colours given by some dihydropyrrolizines have been classified by Culvenor et al. (1970c). The procedure for detecting the alkaloid bases can be shortened by dehydrogenating them directly to dihydropyrrolizines. This can be done using iodine (see above) or reactive quinones (Culvenor et al., 1970b; Mattocks, 1969a); thus, spraying the chromatogram with a solution of chloranil in benzene, heating it briefly, then spraying with Ehrlich reagent (Molyneux and Roitman, 1980) is probably the most convenient and sensitive method for visualising unsaturated PAs. Chloranil alone has been used for detecting PAs on TLC (Huizing et al., 1980), but it is less sensitive than the foregoing method, and alkaloids other than pyrrolizidines also give colours.

4. High-Performance Liquid Chromatography (HPLC)

Techniques for separating mixtures of PAs using HPLC have been developed mainly using plant extracts. Thus, PAs from *Senecio vulgaris* have been separated using a Bondapak CN column (300 × 4 mm) eluted with mixtures of tetrahydrofuran (THF) and 0.01 M ammonium carbonate adjusted to pH 7.8 with H_2SO_4 (Qualls and Segall, 1978). An isocratic system (16% THF) is satisfactory; gradient elution (14–26% THF) is better. The eluant is monitored by its absorption at 235 mn; the strong absorption of THF precludes the use of lower wavelengths. Retrorsine, senecionine and seneciphylline can be separated, and can be isolated by evaporating the eluted peaks to dryness and extracting the residues with chloroform for identification by mass spectrometry. Much the same system has been employed for alkaloid mixtures extracted from *Senecio longilobus* (Segall and Molyneux, 1978) and from *S. jacobaea* (Segall, 1978; Segall and Krick, 1979). Segall (1979a) described an alternative reverse phase system using a 10 μBondapak C18 column (300 × 3.9 mm). Solvent mixtures contain methanol and 0.01 M potassium phosphate buffer, pH 6.3. The lower UV absorption of the solvent allows monitoring at 225 nm, and its slight acidity increases the life of the silica based column. To separate *S. vulgaris* alkaloids, isocratic conditions have been used with 50–60% methanol; this has been extended to a semi-preparative scale (Segall, 1979b). From *S. longilobus* extracts, gradient elution (50–85% methanol during 25 min) separates seven alkaloids (with some overlap). For *S. jacobaea* alkaloids this system is less satisfactory, but it is improved by using an isocratic solvent with less methanol (17.5%) and a lower pH (4.79) to increase the retention of seneciphylline and senecionine, and a programmed flow rate (Dimenna *et al.*, 1980).

Ramsdell and Buhler (1981a) have described an improved HPLC system which uses a reverse-phase styrene–divinylbenzene resin column (Hamilton PRP-1; 150 × 4.1 mm). Solvent mixtures comprising acetonitrile and 0.1 M ammonia solution give excellent separations of alkaloids from *S. jacobaea* (gradient, 10–30% acetonitrile, 20 min at 1 ml/min) and *S. vulgaris* (isocratic, 25% acetonitrile). Samples are dissolved in aqueous methanol for injection. Disadvantages of earlier systems, namely peak tailing, short column life and limited sensitivity due to solvent absorption, are overcome; peaks are detected at 220 nm. Because the column is resistant to alkali, the solvent contains no involatile buffer, so that alkaloids can be recovered simply by evaporating it.

A quantitative HPLC analysis of PAs and their *N*-oxides extracted from comfrey (*Symphytum officinale*) roots has been described by Tittel *et al.* (1979). The reduced alkaloids (symphytine and echimidine) are analysed on a 300 × 4 mm MN-Nucleosil-NH$_2$ column, run with dichloromethane–propan-2-ol (300 : 15). The dichloromethane is first equilibrated with 1% ammonium carbonate solution. Detection is at 238 nm. For *N*-oxides, a reverse-phase Nucleosil C18 column is used, with methanol–water (45 : 55); the limits of detection at 220 nm are 40 and 70 ng of echimidine and symphytine N-oxides, respectively. A semi-preparative separation of the latter pair is described by Wagner et al. (1981), using a 300 × 7.8 mm C18 column with the same solvent, and also an improved analytical method, using a Bondapak-NH$_2$ column with isocratic acetonitrile–water (92 : 8) and a graded flow rate (from 2 to 6 ml/min in 18 min). Detection is at 200 nm, which is less sensitive for echimidine and symphytine N-oxides, but enables lycopsamine and acetyl-lycopsamine N-oxides also to be detected. The chromatographic profiles thus produced serve as 'fingerprints' by which Symphytum drug samples of different origins can be recognised. Huizing et al. (1981) described the partial separation of comfrey alkaloids on a preparative scale by ion-pair HPLC based on a method developed for TLC (Huizing and Malingre, 1981). A silica gel column (53 × 2.5 cm) was eluted with 0.075 M lithium chloride in chloroform–methanol (85 : 15). With this an alkaloid mixture (0.75 g) was crudely separated into three groups of isomers: lycopsamine and intermedine; acetyllycopsamine and acetylintermedine; and isomers of symphytine.

Niwa et al. (1983a) have described the reverse-phase HPLC analysis of otonecine-based PAs from Petasites japonicus, using a Cosmosil 5 Ph column, eluted with isocratic methanol plus 0.02 M ammonium carbonate (45 : 55). This system gives a good separation of otosenine, petasitenine, neopetasitenine and senkirkine.

The HPLC of the pyrrolic PA metabolite dehydroretronecine (201) has been described by Ramsdell and Buhler (1981b). A reverse-phase column is used (Brownlee RP-8), and the mobile phase is isocratic methanol (25 or 30%)–0.01 M phosphate buffer, pH 6.85; caffeine is incorporated as an internal standard.

5. Gas Chromatography (GC)

The first appraisal of GC for the characterization of PAs was made by Chalmers et al. (1965), who compared their results with those from paper chromatography and TLC. The main drawback of GC is that the underivatized alkaloids often decompose in contact with packing materials and hot metal components. Simple pyrrolizidine bases can be chromatographed using a copper column, but ester alkaloids decompose. The latter can, however, be run in an all glass system provided that the column, support and components are silanized to eliminate adsorption sites, and the alkaloids are applied in dilute solutions to prevent pyrolysis close to the inlet. The stationary phase used by Chalmers et al. was SE30, on a 1.8 m × 6 mm column. Non-ester alkaloids were run at 140°C, and esters at 205°C: the same system could not be used for both. Retention times were determined for 58 PAs and related bases, and their relationships to molecular structure were discussed. In this study, the sensitivity of detection was less for GC than for TLC. Using a smaller diameter (2 mm) SE30 column, Wiedenfeld *et al.* (1981) chromatographed a mixture of seven PAs. By maintaining 150°C for 45 min, heliotridine-based alkaloids (7-angelyl heliotridine and heliotrine) are eluted first; then a programmed increase to 222°C in 40 min separates esters of retronecine and other bases in a way similar to the isothermal systems of Chalmers *et al.* However, the sensitivity is higher, with detection limits (FID) from 20 ng (senkirkine) to 150 ng (retrorsine).

Closely related isomeric PAs are not easily separated. Huizing *et al.* (1981) chromatographed partially silylated alkaloids from comfrey (*Symphytum* spp.); stereoisomeric pairs were not well resolved on a packed column (OV17). A capillary column (methylsilicone) gives better results: trimethyl silyl (TMS) derivatives of lycopsamine, intermedine and their acetyl derivatives are separated, but isomers of symphytine overlap. Culvenor *et al.* (1980b) separated mixtures of *Symphytum* alkaloids after converting them to methylboronate derivatives. By running these with SE30 or OV17 as the stationary phase and several temperature programmes (maximum 230°C), the isomeric pairs symphytine–symlandine and acetyllycopsamine–acetylintermedine could be resolved. Butylboronate derivatives have been used for the quantitative GC–MS determination of similar alkaloids in honey from *Echium plantagineum,* using an SE30 column (Culvenor *et al.,* 1981).

GC–MS has also been used by Deinzer *et al.* (1977) and by Rothschild *et al.* (1979), to identify PAs extracted from honey and from the pupae of moths, respectively. Another approach to identifying retronecine-based PAs in a plant extract (Deinzer *et al.,* 1978) is to hydrolyse them with barium hydroxide to retronecine (183), which is converted to the moisture-labile bistrifluoroacetate (206). This is identified using GC–MS, with 7% OV210 plus OV17 (4 : 1) as stationary phase, at 100–150°C. Sensitivity is high, but the reproducibility (from the plant) is not good, and the method cannot distinguish individual alkaloids.

Capillary GC gives greatly improved resolution compared with packed columns for underivatized PAs. Luthy *et al.* (1981) used capillary GC–MS for alkaloids extracted from *Senecio alpinus* and from *Senecio*-contaminated hay and silage. Using a 20-m column coated with the silicone SE54, nine alkaloids were separated and identified, including senecionine, seneciphylline (major component), integerrimine, jacobine, jacozine and jaconine. Quantitation was by peak integration.

Pyrrolizidine *N*-oxides, which are often highly polar and labile to thermal

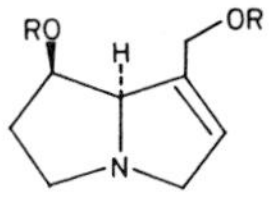

(183) R = H
(206) R = CF₃CO

Scheme 2.3.

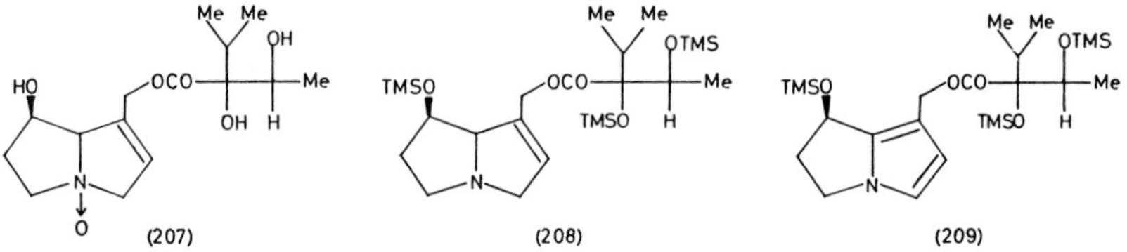

Scheme 2.4.

decomposition, have not been successfully run on GC. Attempts to derivatize them lead (predictably) to de-oxygenation. Ames and Powis (1978) found that indicine *N*-oxide (INO) (207) decomposes on GC to give several peaks, one being indicine base. Attempts at derivatization with fluorinated (anhydride) acylating agents lead to pyrrolic material which polymerizes (cf. Kreher and Pawelczyk, 1964; Mattocks, 1967c). Thus to analyse INO in urine or plasma the samples were reduced to indicine with zinc and acetic acid or ammonium chloride; this was extracted from the basified solutions with chloroform, heliotrine being added as internal standard. Pentafluoropropionyl derivatives were prepared and run on a GC column of OV101 at 185°C, with methane–argon carrier gas and an EC detector. INO was measurable at a level of 100 ng/ml in plasma, with somewhat lower sensitivity in urine. Variable reproducibility in reduction and extraction stages, especially from urine, are drawbacks in methods such as this.

Evans *et al.* (1979, 1980) attempted to make TMS derivatives of INO using the reagents *N*-trimethylsilylimidazole (TSIM) and *N,O*-bis(trimethylsilyl)trifluoroacetamide (BSTFA). GC–MS was used to measure the products. Using TSIM, the *N*-oxide is apparently reduced to base, the mass spectrum being that of the tri-TMS derivative of indicine (208); with BSTFA plus TSIM, the mass spectrum is of the tri-TMS derivative of dehydro-indicine (209). Heliotrine *N*-oxide gives similar results. Using BSTFA plus TSIM for derivatization, INO can be distinguished from indicine base, since these give the TMS derivatives of dehydro-indicine and indicine, respectively. The lower limit of detection is 40 ng of the pyrrolic product injected on the column.

6. Electrophoresis

The paper electrophoresis of 23 PAs and bases, three *N*-oxides and one acid (heliotric) has been described by Frahn (1969). Using electrolytes having pH 4.6–9.4, mobilities of the cations vary according to the degree of protonation and are broadly consistent with assigned ionization constants of the alkaloids (Culvenor and Willette, 1966). At pH 4.6, where the bases should be fully protonated, mobilities are more or less related inversely to molecular weights. Electrolytes capable of forming complexes with glycols were also investigated.

Thus, the cationic mobilities at pH 9.2 of alkaloids having vicinal glycol groups in the acid moiety decrease in the presence of borate, owing to the formation of anionic glycol–borate complexes. The direction of migration of alkaloids having lower base strength, e.g. monocrotaline, is reversed in borate, as the complex anion determines the net ionic charge.

Pairs of alkaloids such as intermedine and lycopsamine, identical except in the stereochemistry of their hydroxyl groups, have mobilities which are the same in non-complexing electrolytes but which differ in borate buffer: the *erythro* configuration in lycopsamine favours complex formation, giving greater anionic mobility (Frahn *et al.*, 1980). Thus, electrophoresis in a complexing medium lends itself to the analytical separation of glycol diastereomers as well as to assignment of sterochemistry. This has been exploited in the identification of some *Symphytum* alkaloids (Culvenor *et al.*, 1980b), and of trachelanthamidine esters from *Heliotropium curassavicum* (Subramanian *et al.*, 1980).

The application of electrophoresis to the analysis of mixtures of PAs which occur in a number of plant species has been discussed (Frahn, 1969). Sodium carbonate electrolytes, pH 8.4 or 9.2, and borate buffer, pH 9.2, are the most useful.

V. DETECTION AND ESTIMATION OF PYRROLIZIDINE ALKALOIDS AND THEIR METABOLITES

A. Methods Employing Ehrlich Reagent

The Ehrlich pyrrole reaction has proved to be of great value for detecting and measuring unsaturated PAs and their *N*-oxides, and in elucidating their metabolism. PAs themselves give no colour with Ehrlich reagent, but those which possess an unsaturated (pyrroline) ring in the base moiety are readily converted to dihydropyrrolizine derivatives ('pyrroles') which give intense Ehrlich colours. The basic alkaloids can be dehydrogenated to pyrroles by reagents such as chloranil (Culvenor *et al.*, 1970b) or converted to *N*-oxides which are readily dehydrated to pyrroles (Mattocks, 1969a).

1. The Ehrlich Reaction

The reagent, an acidic solution of 4-dimethylaminobenzaldehyde, reacts with nucleophilic pyrroles at a position adjacent to the nitrogen, or if such is not available, much less readily at a β position. Electron-withdrawing substituents (such as CO_2Et or CHO) attached to the pyrrole ring inhibit the reaction. The kinetics and mechanism of the reaction of Ehrlich reagent with various substituted pyrroles have been studied by Alexander and Butler (1976). Dihydropyrrolizines derived from PAs are in effect substituted pyrroles and undergo Ehrlich reactions in the same way as the latter.

Pyrroles derived from PAs are usually reactive alkylating agents which can polymerize readily in the presence of acids. Since the Ehrlich reagent is acidic, there is competition between the colour-forming and polymerization reactions. If the latter predominates, much of the pyrrole is precipitated as a polymer, and the Ehrlich colour in the solution is relatively weak. This occurs particularly in aqueous solutions. Polymerization can be prevented by conducting the Ehrlich reaction in the presence of excess of a compound which can react rapidly with the electrophilic centre(s) of the pyrrole. One of the most effective of these is ascorbic acid, and this is exploited in the Ehrlich estimation of PAs and their metabolites in aqueous media (Mattocks and White, 1970). The enhancement by ascorbic acid of the colour intensity from a pyrrole and Ehrlich reagent is evidence that the pyrrole possesses alkylating reactivity (Mattocks and Bird, 1983b).

The chemistry of the above reactions is further discussed in Chapter 5.

2. Spectral Properties of Coloured Ehrlich Derivatives

The pyrrole derivatives from most PAs react with Ehrlich reagent to give a magenta colour with a λ_{max} in the region of 565 nm. This may change if the reactive alcohol or ester group is replaced. The λ_{max} of alkoxy derivatives is the same as for alcohols and esters, but it may be slightly higher when amine substituents are present and 7–10 nm higher for sulphur substituents. This 'sulphur shift' is cancelled by adding mercuric chloride to the coloured solution (Mattocks and White, 1970). An excess of thiol weakens the intensity of Ehrlich colours (Rimington et al., 1956), but the colour is restored by mercuric chloride. The Ehrlich colours from N-substituted pyrroles are considerably weaker than those from O-substituted pyrroles, presumably because of electron-withdrawal by the protonated N.

3. Spectrophotometric Determination of Alkaloids and N-Oxides

This procedure (Mattocks, 1967c, 1968b; Bingley, 1968) is effective for the hepatotoxic PAs, which have an unsaturated necine moiety. The basic alkaloid is oxidized with hydrogen peroxide to its N-oxide; this is converted by acetic anhydride to a pyrrolic derivative which reacts with a modified Ehrlich reagent containing boron trifluoride to give a strong colour. The method is sensitive (limit of detection below 5 μg for many PAs), and it is reproducible provided the original method and later modification (Mattocks, 1968b) are followed carefully.

A quicker but less sensitive method combines the oxidation and acetic anhydride stages.

Most saturated PAs give little or no colour, because their N-oxides are not readily dehydrated to pyrroles. However, rosmarinine consistently gives about 10% of the colour formed from corresponding unsaturated PAs.

To estimate N-oxides only, the oxidation step is omitted. The method was adapted by D'Silva and Notari (1980) to estimate indicine N-oxide in silica gel scraped from TLC plates. The analysis of pyrrolizidine N-oxides has been discussed by Mattocks (1971b).

4. 'Field Test' for Unsaturated PAs

A simplified extraction procedure and qualitative colour test has been devised which is suitable for screening large numbers of plant samples for N-oxides of unsaturated (and thus potentially hepatotoxic) PAs (Mattocks, 1971a). In most plants containing PAs, a proportion of the alkaloids exist in the N-oxide form. However, a modification (unpublished) which has been in use for many years makes it equally easy to detect the alkaloid bases. The plant material (crushed if fresh; powdered if dry) is extracted with methanol. Most of the chlorophyll is removed by partitioning the extract several times with petroleum. The methanol phase is diluted with dimethyleneglycol dimethyl ether and divided into three equal portions. To the first is added about 0.1 ml of acetic anhydride, and to the second, about 0.02 g of chloranil. After being heated in a hot water bath for 1 min, each portion is then heated with Ehrlich reagent containing boron trifluoride. A magenta colour in the first indicates the presence of an unsaturated pyrrolizidine N-oxide; in the second it indicates a PA base; the third fraction is a blank for comparison, since all samples may show a colour if Ehrlich-positive constituents such as indoles or pyrroles are present in the plant.

5. Determination of N-Oxide and Pyrrolic Metabolites

Pyrrolic metabolites formed by dehydrogenation of PAs *in vivo* are readily detected using Ehrlich reagent. Samples of urine or liver tissue of rats given hepatotoxic alkaloids give a mauve or magenta colour when heated with the reagent (Mattocks, 1968a). Several methods have been described for determining these metabolites (Mattocks and White, 1970; Mattocks and Bird, 1983a). Measurements give relative rather than strictly quantitative results, inasmuch as the metabolites are usually mixtures having in common the ability to form colours of unknown absorptivity with Ehrlich reagent.

To obtain colour in soluble form suitable for spectrophotometry from pyrrolic metabolites in tissues such as liver, the sample is homogenized in an alcoholic solution of mercuric chloride. The latter reacts with thiols, which would suppress the Ehrlich colour (Rimington *et al.*, 1956), breaks pyrrole–sulphur bonds, and forms Hg–pyrrole complexes bound to the solid phase. The solids are separated by centrifugation and heated with Ehrlich reagent containing boron trifluoride to give a soluble colour. Absorbance measurements are made at two wavelengths (565 and 625 nm) to compensate for any interference from blood in the tissue, and a corrected result is calculated using the formula

$$abs_{corr} = 1.1\ (abs_{565} - abs_{625})$$

Numerous measurements using this procedure (Mattocks, 1972b, 1981a; Mattocks and White, 1973; White *et al.*, 1973) have proved its value for demonstrating relationships between dosages of hepatotoxic PAs and the amounts of pyrrolic metabolites formed *in vivo*.

Pyrrolic metabolites formed from PAs by microsomal enzyme preparations *in vitro* are measured by diluting the mixture with ethanolic ascorbic acid, then heating it with Ehrlich reagent. In the original procedure (Mattocks and White, 1970, ferric chloride was added before absorbance measurements to oxidize the excess ascorbic acid, which otherwise caused the colour to fade. *N*-Oxides were measured after converting them to pyrroles using an iron complex (nitroprusside; cf. Mattocks, 1968c), after extracting existing pyrroles from the solution. However, reproducibility and sensitivity for *N*-oxides was low, and an improved procedure has been devised (Mattocks and Bird, 1983a), in which the pyrroles are not extracted but the combined pyrroles and *N*-oxides are measured and the latter is determined by difference. Greatly increased sensitivity and linearity are achieved using nitroprusside pretreated with alkali to generate the nitro complex $[Fe(CN)_5NO_2]^{4-}$, an Ehrlich reagent with doubled concentrations of BF_3 (11%) and 4-dimethylaminobenzaldehyde (6%), a lower reaction temperature (50°C) and longer reaction times for these reagents (3 and 5 min, respectively), and iodine in place of ferric chloride for destroying the excess ascorbic acid.

B. Other Methods

1. Colourimetric Analysis Employing Indicator Dyes

The ability of some dyes to form stoichiometric complexes with alkaloid bases is well known. Hayashi (1966) used methyl orange to determine monocrotaline and its metabolites. A sensitive method generally applicable to PA has been described by Birecka *et al.* (1981). It is claimed to detect PA down to 0.5 μg/ml in chloroform solution, but it does not distinguish between saturated and unsaturated alkaloids and would also respond to tertiary bases other than pyrrolizidines.

The alkaloid, dissolved in ethanol-free chloroform, is shaken with aqueous methyl orange and boric or acetic acid. The chloroform phase contains the yellow dye–alkaloid complex; methyl orange is released from this by adding ethanolic sulphuric acid, and is measured spectrophotometrically. The method is unsatisfactory for highly water soluble amino-alcohols such as heliotridine, the dye complexes of which partition into the aqueous phase. Calibration data have been given by Birecka and Catalfamo (1982) for monocrotaline (ϵ_{max} 40,975) and for some necine bases. A similar method was used to determine monocrotaline extracted from TLC (Sha and Tseng, 1980): bromocresol green, transferred from aqueous solution to complex with the alkaloid chloroform, was subsequently released by alkali and measured by spectrometry. Fulvine and crispatine were similarly estimated by Habib and El-Sebakhy (1978) using bromocresol purple, and platyphylline and seneciphylline extracted from TLC have been analysed using the dye tropeolin (Larionov et al., 1980).

2. Polarography

This technique has been little used with pyrrolizidine alkaloids. McComish et al. (1980) employed differential pulse polarography to measure indicine N-oxide recovered from human plasma. Platyphylline and seneciphylline, present together in plant extracts, have been determined polarographically in a solution containing tetraethylammonium iodide (Vdoviko et al., 1979).

3. Nuclear Magnetic Resonance (NMR) Analysis

The NMR signal due to the vinyl H-2 proton in PAs having unsaturated base moieties can be used to determine these alkaloids (Molyneux et al., 1979). Known amounts of p-dinitrobenzene are added to the alkaloids as internal standard. Quantitative measurements of NMR signals due to this standard (δ 8.46 ppm) and to the alkaloid H-2 proton are made by averaging integration scans run in both directions. Macrocyclic pyrrolizidine diesters (δ ca. 6.2 ppm) and other esters (ca. 5.8 ppm) can be distinguished, and estimated by a simple calculation, an 'average' molecular weight being employed when mixed alkaloids are present. Individual alkaloids can sometimes be distinguished and their relative proportions determined from their other characteristic NMR signals. The method cannot be used for saturated PAs. When used to analyse alkaloid mixtures extracted from comfrey (Mattocks, 1980), NMR was quicker than a spectrophotometric method (Mattocks, 1967c, 1968b) but much less sensitive. Sensitivity of course depends on the spectrometer used.

4. Non-specific Methods

General methods for detecting alkaloids and other organic bases are applicable to PAs, and up to 1967 they were the only ones available. Alkaloids in solution give precipitates with Mayer's reagent, silicotungstic acid, and many other reagents; for a review see Farnsworth (1966).

The simplest way to determine total PAs in plant material is to extract the alkaloids exhaustively and weigh the crude basic extracts. Because these can be contaminated with non-alkaloidal material, more accurate results may be obtained by titrating the bases with p-toluene sulphonic acid in chloroform solution (Culvenor and Smith, 1955). Alternatively, titration with a solution of perchloric acid in acetic acid, using crystal violet as indicator (Seaman and Allen, 1951; Pickard and Iddings, 1959) gives good results (Gencheva, 1978; Mattocks and Bird, 1983a). Since alkaloids usually occur as mixtures, an 'average' molecular weight must be assumed when calculating results.

3

Syntheses of Necines, Pyrrolizidine Alkaloids and Analogues and Pyrrolic Derivatives*

*Abbreviations: Bz, benzoyl; DABCO, diazabicyclo[2.2.2]octane; DBU, 1,8-diazabicyclo [5.4.0]undec-7-ene; DCC, N,N'-dicyclohexylcarbodiimide; DDQ, 2,3-dichloro-5,6-dicyano-1,4-benzoquinone; DIBAH, di-isobutylaluminium hydride; DMAP, 4-dimethylaminopyridine; DME, 1,2-dimethoxyethane; Im, imidazole; LAH, lithium aluminium hydride; LDA, lithium di-isopropylamide; MCPBA, *m*-chloroperbenzoic acid; Mes, methanesulphonyl; Py, pyridine; THP, tetrahydropyranyl; Tos, toluenesulphonyl.

I. INTRODUCTION

PAs, like other natural products, have been challenging goals for chemical synthesis. Procedures, often elaborate, have been devised for making necines, necic acids and ester alkaloids. These have been valuable for establishing the precise structures and stereochemistry of many PAs—information of importance to anyone interested in the biological effects of these compounds. However, there are other objectives for synthesis, for example as a source of modified alkaloids or analogues with which to test structure–activity relationships, and for preparing specifically radioactive labelled compounds.

In this chapter particular attention is given to synthetic methods leading to biologically active necines, PAs, analogues and potential metabolites, which may be adapted to the production of such compounds in useful quantities.

Only a few of the natural ester PAs have been synthesized, and plants remain the only practicable source of supply. However, semisynthetic PAs, prepared by esterifying necines of natural origin, as well as wholly synthetic analogues, have proved valuable for studies of metabolism and toxicity.

II. SYNTHESES OF NECINES

Many syntheses of natural necines have been described, but relatively little has been done on the unsaturated necines, which are of greatest toxicological interest.

A. Saturated

For reviews on syntheses of saturated necines (pyrrolizidine alcohols) and other saturated pyrrolizidines, see Kochetkov and Likhosherstov (1965), Warren (1966), and Robins (1979). Syntheses of isoretronecanol, trachelanthamidine, turneforcidine, macronecine, hastanecine, platynecine, rosmarinecine and croalbinecine are listed in Table 3.1. Ohnuma *et al.* (1983) have synthesized a seco-pyrrolizidine nucleus, dihydrodesoxyotonecine (210), previously obtained by hydrogenolysis of otonecine-based PAs.

TABLE 3.1 Syntheses of Necines

Compound	Structure	References
Croalbinecine	(180)	Rueger and Benn (1983b)
Crotanecine	(185)	Yadav *et al.* (1984)
Dihydrodesoxyotonecine	(210)	Ohnuma *et al.* (1983)
Hastanecine	(175)	Hart and Yang (1983)
Heliotridine	(184)	Keck and Nickell (1980); Chamberlin and Chung (1983); Glinski and Zalkow (1985)
Isoretronecanol, trachelanthamidine or both	(170)–(173)	Leonard and Sato (1969); Kochetkov *et al.* (1969); Pizzorno and Albonico (1974); Tufariello and Tette (1975); Borch and Ho (1977); Danishefsky *et al.* (1977); Pinnick and Chang (1978); Nossin and Speckamp (1979); Iwashita *et al.* (1979, 1982); Robins and Sakdarat (1979a, 1981); Flitsch and Wernsmann (1981); Terao *et al.* (1982); Hart and Yang (1982); Robins (1982b); Rueger and Benn (1982); Tatsuta *et al.* (1983); Flitsch and Russkamp (1983); Shono *et al.* (1984)
Macronecine	(177)	Aasen and Culvenor (1969b)
Otonecine	(186)	Niwa *et al.* (1983d)
Platynecine	(176)	Rueger and Benn (1983b)
Retronecine	(183)	Geissman and Waiss (1962); Keck and Nickell (1980); Vedejs and Martinez (1980); Tufariello and Lee (1980); Naraska *et al.* (1982); Rueger and Benn (1983b); Niwa *et al.* (1983c); Vedejs *et al.* (1985)
Rosmarinecine	(181)	Tatsuta *et al.* (1983)
Supinidine	(182)	Tufariello and Tette (1971, 1975); Robins and Sakdarat (1979a,b, 1981); Hart and Yang (1982); Chamberlin and Chung (1982); Terao *et al.* (1982); Rueger and Benn (1982); MacDonald and Narayanan (1983)
Turneforcidine (partial synthesis)	(174)	Ohsawa *et al.* (1983) Aasen and Culvenor (1969a)

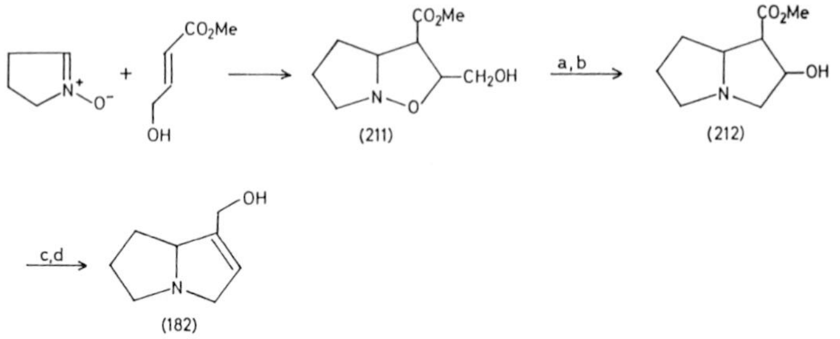

Scheme 3.1. Reagents: a, MesCl; b, H_2, catalytic; c, $POCl_3$, Py; d, LAH, $AlCl_3$.

B. Unsaturated

Syntheses of four unsaturated necines have been described (Table 3.1).

Tufariello and Tette (1975) prepared supinidine according to Scheme 3.1. A 1,3-dipolar addition of pyrroline-1-oxide to methyl 4-hydroxycrotonate gives (211), the mesylate of which is hydrogenated to (212). The latter is converted via dehydration and reduction to (±)-supinidine (182). Extension of this procedure (Tufariello and Lee, 1980), starting with the nitrone (213) (prepared from N-ethylpyrrolidin-3-one), leads to (±)-retronecine (183) via a similar sequence of reactions.

Robins and Sakdarat (1979b) described a useful method for converting saturated pyrrolizidine carboxylates into their 1,2-didehydro derivatives. 1-Ethoxycarbonyl pyrrolizidine (215), prepared according to Pizzorno and Albonico (1974) by cycloaddition of ethyl propiolate to N-formyl-L-proline (214) followed by catalytic hydrogenation of the resulting pyrrolizine, is converted to the phenylselenyl derivative (216), which is reduced to the alcohol (217). Oxidation of (217) with hydrogen peroxide gives (±)-supinidine (182) (Scheme 3.2). Carrying out the LAH reduction before eliminating the phenylseleno group prevents partial reduction of the double bond, which would yield a mixture containing the

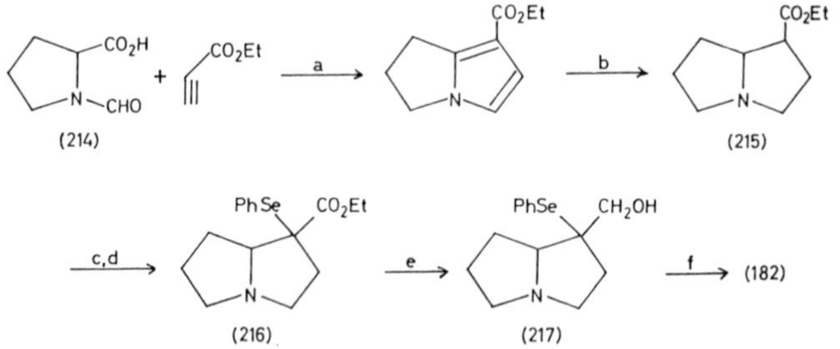

Scheme 3.2. Reagents: a, Ac_2O; b, H_2, catalytic; c, iPr_2NH, BuLi; d, PhSeCl; e, LAH; f, H_2O_2.

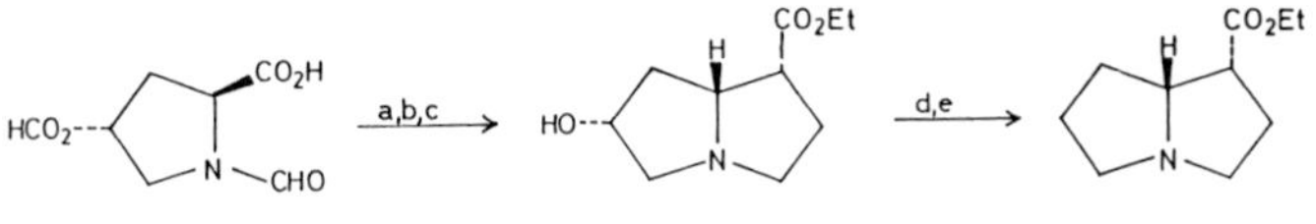

(218)   (219)

Scheme 3.3.   Reagents: a, Ac_2O, ethyl propiolate; b, NH_3, EtOH; c, H_2, catalytic; d, $SOCl_2$; e, H_2, catalytic.

saturated pyrrolizidine (isoretronecanol). This procedure was extended to a chiral synthesis of ($\pm$)- and ($-$)-supinidine, by starting with the N,O-diformyl deriva-tive of ($-$)-4-hydroxyproline (218) (Robins and Sakdarat, 1979a, 1981). The corresponding optically active forms of isoretronecanol and trachelanthamidine were also prepared. The formation of chiral centres at C-1 and C-8 of the pyrrolizidine nucleus in (219) was controlled by the stereochemistry of the pro-line OH group, subsequently removed by chlorination followed by reduction (Scheme 3.3).

Hart and Yang (1982) described another, rather lengthy synthesis of supini-dine. The intermediate (220), prepared in 39% overall yield (seven steps) from 3-methyl-2-butenal, is cyclized in formic acid to three products including (221) (70%), which is converted to supinidine (182) (49% overall) by the reactions shown in Scheme 3.4.

The first synthesis of retronecine was that of Geissman and Waiss (1962). The overall yield of ($\pm$)-retronecine (183) from the intermediate (222) via (223) was less then 2% (Scheme 3.5). A modification, via (224), more than doubles the yield of retronecine (Narasaka *et al.*, 1982).

A recent synthesis by Keck and Nickell (1980) leads to both retronecine and heliotridine (Scheme 3.6). An interesting feature is the introduction of the double

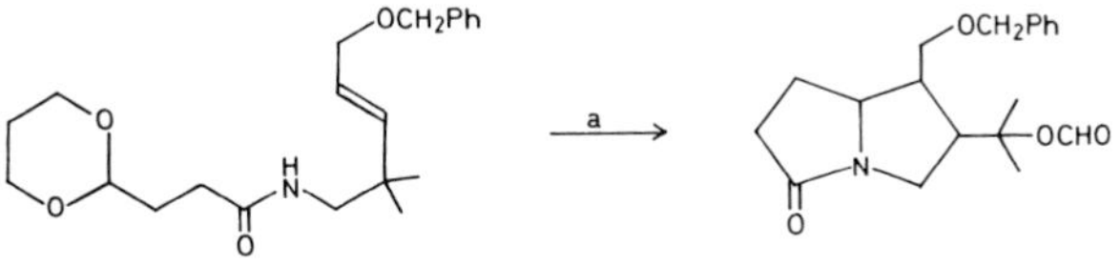

(220)   (221)

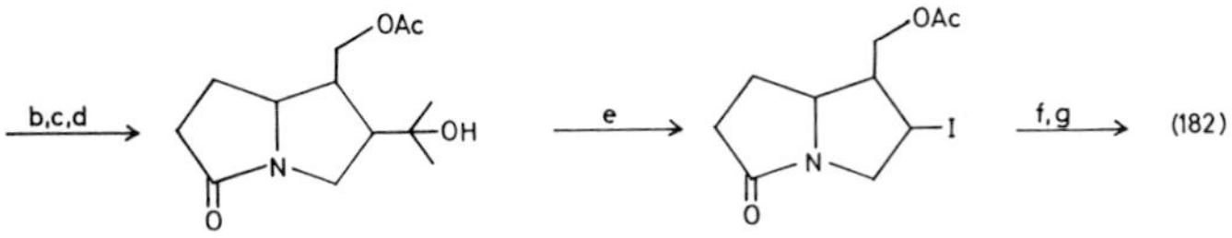

(182)

Scheme 3.4.   Reagents: a, HCO_2H; b, OH^-; c, H_2, catalytic; d, Ac_2O, Py; e, HgO, I_2, CCl_4; f, DBU; g, LAH.

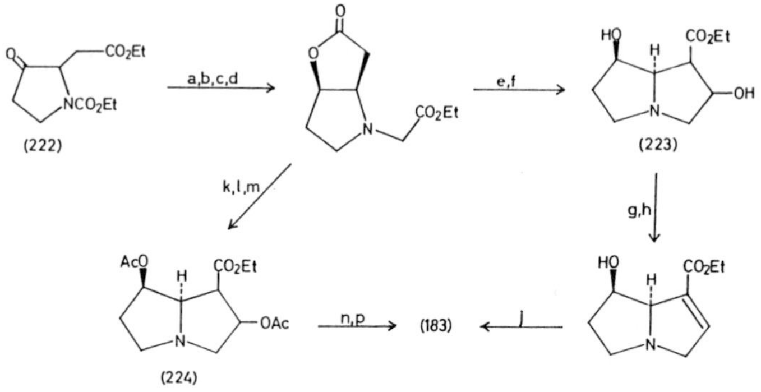

Scheme 3.5. Reagents: a, NaBH$_4$; b, Ba(OH)$_2$; c, HCl; d, ethyl bromoacetate; e, KOEt; f, H$_2$, catalytic; g, Ba(OH)$_2$; h, EtOH, H$^+$; j, LAH; k, NaOEt; l, NaBH$_4$; m, Ac$_2$O, base; n, *t*BuOK; p, DIBAH.

bond and both oxygens of the final products at a very early stage. The carbon, nitrogen and protected oxygen atoms as well as the unsaturation of the final product are contained in the key intermediate (225). To effect cyclization, the mesylate of this is treated with lithium di-isopropylamide, resulting in easily separable lactams (226, 227). The protective groups are removed yielding the corresponding hydroxy lactams, and these are reduced to (±)-heliotridine (184) and (±)-retronecine (183) respectively. The overall yield of these necines appears to be well over 20%, a considerable improvement over the Geissman–Waiss method.

In another recent approach, Vedejs and Martinez (1980) describe a stereospecific synthesis of retronecine from the hydroxy lactam (229) (Scheme 3.7). The latter intermediate can be prepared (yield 60–70%) from the imidate (228). After benzylation and O-methylation of (229) to the methylide (230), 1,3-dipolar cycloaddition of methyl acrylate gives (231) which is hydrogenated, leading to (232). A selenium-based elimination similar to that employed by Robins and

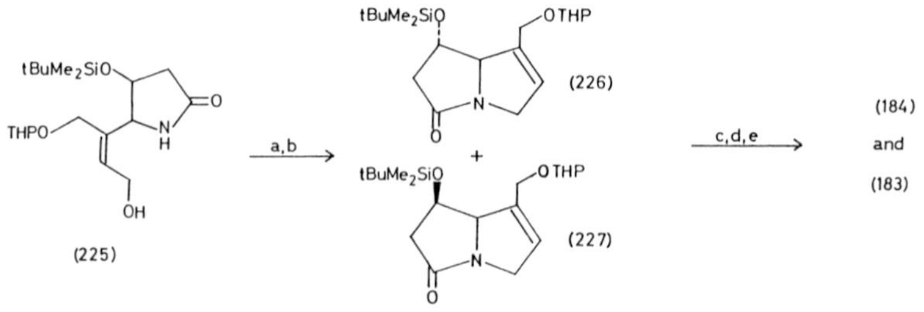

Scheme 3.6. Reagents: a, MesCl, Et$_3$N; b, LDA; c, separation, then MeOH, Py$^+$HTos$^-$; d, Bu$_4$NF; e, LAH.

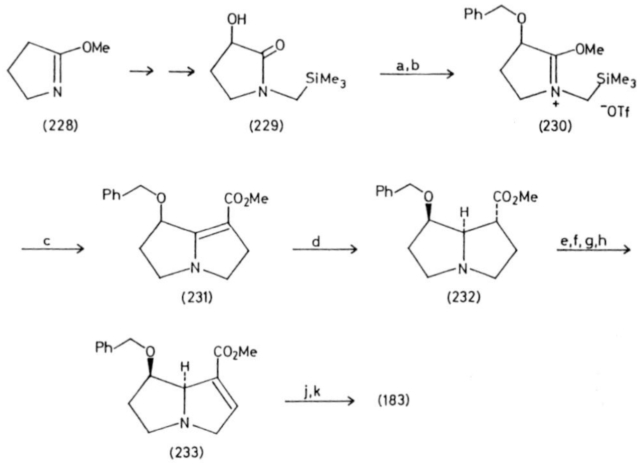

Scheme 3.7.   Reagents: a, PhCH$_2$Br, NaH, DME; b, CF$_3$SO$_3$Me; c, methyl acrylate, CsF; d, H$_2$, catalytic; e, LiNEt$_2$; f, PhSeSePh; g, MCPBA; h, CCl$_4$, reflux; j, DIBAH; k, Li, NH$_3$.

Sakdarat (1979b) leads to (233), which is reduced and debenzylated to ($\pm$)-retronecine. The overall yield from (229) is 20%.

Novel syntheses of ($\pm$)-retronecine have also been described by Ohsawa *et al.* (1983) and by Niwa *et al.* (1983c), the overall yields being poor (4% and 2.4%, respectively). Niwa *et al.* (1983d) also achieved the first total synthesis of otonecine (186), in 3% yield from the intermediate (234) (Scheme 3.8). A convenient method is available for the preparation of heliotridine from the more readily available retronecine (Glinski and Zalkow, 1985).

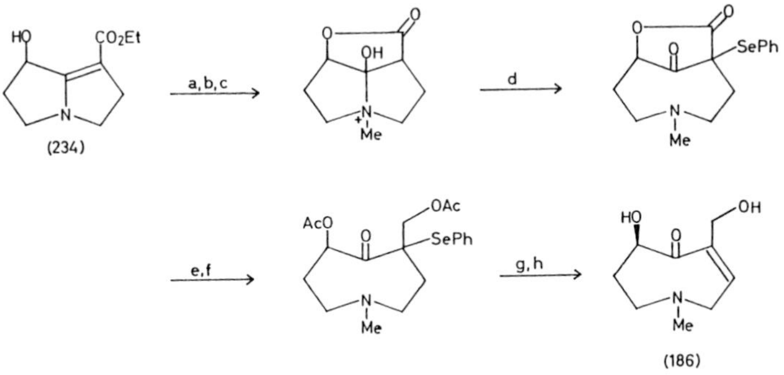

Scheme 3.8.   Reagents: a, LiSPh; b, Hg(OAc)$_2$; c, MeI; d, NaH, PhSeCl; e, DIBAH, Et$_2$AlCl; f, Ac$_2$O, Py; g, H$_w$O$_2$; h, NaOMe.

(235)   (183) or (184)

C. Synthetic Analogues (Synthanecines)

Analogues of necines (called synthanecines) are much more easily synthesized than the natural products, and they make possible more extensive investigations of relationships between structure, toxicity and other biological interactions characteristic of PAs.

Five synthanecines have been prepared (Table 3.2). The resemblance between a synthanecine and a natural necine is evident when their structures are compared. For example, synthanecine A (235) lacks only one carbon atom (C-6) of retronecine (183) or heliotridine (184); synthanecine B (236) is analogous to saturated necines such as platynecine (176); synthanecine E (239) is analogous to supinidine. The homologue synthanecine D (238) has a hydroxyethyl group in place of a hydroxymethyl.

All the synthanecines are prepared by a relatively simple procedure, amenable to large scale operation (Mattocks, 1974, 1978a). A 3-ethoxycarbonyl pyrrolidine, formed by a Dieckmann cyclization, is reduced to the corresponding hydroxy ester, which is dehydrated to a 3-pyrroline ester. Ester groups are then reduced to hydroxymethyl. The choice of reagent for this step is important.

TABLE 3.2 Synthanecines[a]

Synthanecine	Systematic name	Structure	Molecular weight	Picrolonate (m.p., °C)
A	2,3-Bishydroxymethyl-1-methyl-3-pyrroline	(235)	143	176 (dec)
B	2,3-Bishydroxymethyl-1-methylpyrrolidine	(236)	145	192
C	2,3-Bishydroxymethyl-4-methoxy-1-methyl-3-pyrroline	(237)	173	143 (dec)
D	2-(2-Hydroxyethyl)-3-hydroxymethyl-1-methyl-3-pyrroline	(238)	157	155
E	3-Hydroxymethyl-1,2-dimethyl-3-pyrroline	(239)	127	—

[a] Data from Mattocks (1974, 1978a).

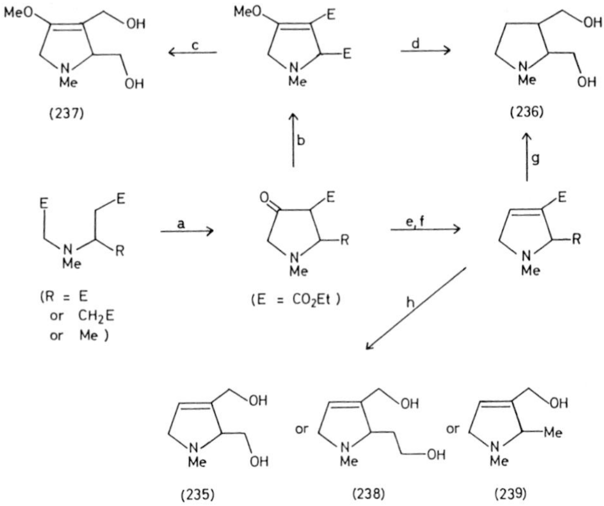

Scheme 3.9. Reagents: a, NaH; b (R = E), CH$_2$N$_2$; c, DIBAH; d, LAH; e, NaBH$_4$; f, TosCl, Py; g (R = CO$_2$Et), LAH; h, DIBAH.

Lithium aluminium hydride is liable to reduce the double bond, giving predominantly a pyrrolidine (e.g. synthanecine B), whereas DIBAH gives good yields of unsaturated alcohols. The synthanecines are oils, usually isolated and purified in the form of their crystalline picrolonate salts. Routes to synthanecines are summarized in Scheme 3.9. The overall yields of synthanecines A–E are about 14%, 24%, 20%, 9% and 14%, respectively.

III. SYNTHESES OF NECIC ACIDS

The most important syntheses and partial syntheses of naturally occurring necic acids are summarized in Table 3.3; they will not be described here in detail. For a review of the earlier literature, see Warren (1966).

IV. SYNTHESES OF ESTER PAs AND OTHER NECINE ESTERS

A few natural PAs or related derivatives (Table 3.4) and many semisynthetic pyrrolizidine esters (Tables 3.5 and 3.6) have been made by esterifying necines. Some synthanecine esters are listed in Table 3.7. 'Open' (non-macrocyclic) esters of necines with simple aliphatic acids are generally easy to prepare; hydroxy esters and 'mixed' diesters are more difficult, and macrocyclic diesters present serious problems.

TABLE 3.3 Synthesis of Some Necic Acids

Acid	References
Crispatic	Matsumoto *et al.* (1973)
Crobarbatic	Huang and Meinwald (1981)
Dicrotalic	Adams and van Duuren (1953a); Klosterman and Smith (1954)
Fulvinic	Vedejs and Larsen (1984)
Heliotrinic (heliotric) (amide only)	Adams *et al.* (1959)
Integerrinecic	Culvenor and Geissman (1961a); Edwards and Matsumoto (1967a); Narasaka and Uchimaru (1982)
Mikanecic	Sydnes *et al.* (1975); Edwards *et al.* (1967a)
Monocrotalic	Adams *et al.* (1952); Matsumoto *et al.* (1979)
Nemorensic	Roeder *et al.* (1980b); Klein (1985)
Retusanecic	Kiyooka and Hase (1973)
Sarracinecic	Edwards *et al.* (1967a)
Senecic	Culvenor and Geissman (1961a); Edwards and Matsumoto (1967a)
Seneciphyllic	Edwards *et al.* (1967b)
Seneciphyllic (isomer)	Gordon-Gray and Whiteley (1977)
Senecivernic	Pastewka *et al.* (1980)
Trachelanthic	Adams and van Duuren (1952); Kochetkov *et al.* (1969)
Trichodesmic	Edwards and Matsumoto (1967b)
Viridofloric	Adams and van Duuren (1952); Adams and Herz (1950); Kochetkov *et al.* (1969)

A. 'Open' Esters

Culvenor *et al.* (1959) re-formed heliotrine (49) (50% yield) by refluxing the 9-chloro derivative of heliotridine with the sodium salt of heliotric acid in aqueous alcohol. Supinine was made in a similar way from supinidine and sodium trachelanthate. The method depends on the high reactivity of allylic halides and is ineffective with saturated necines. Supinidine viridoflorate (Crowley and Culvenor, 1959), intermedine and lycopsamine (Culvenor and Smith, 1966a), 9-angelyl retronecine (Culvenor *et al.* 1970d), 9-angelyl heliotridine and 9-he-

TABLE 3.4 Some Natural PAs or Derivatives
Prepared Synthetically or Semisynthetically

Compound	Structure	Reference
Crispatine; fulvine	(16), (42)	Vedejs and Larsen (1984)
Crobarbatine acetate[a]	(259)	Huang and Meinwald (1981)
Dicrotaline	(29)	Devlin and Robins (1981)
Heliotrine	(49)	Culvenor et al. (1959)
Indicine	(53)	Piper et al. (1981);
		Vedejs et al. (1985);
		Zalkow et al. (1985)
Integerrimine[b]	(55)	Narasaka et al. (1982, 1984)
Intermedine	(56)	Culvenor and Smith (1966a);
		Zalkow et al. (1985)
Lycopsamine	(70)	Culvenor and Smith (1966a);
		Zalkow et al. (1985)
Rinderine	(88)	Culvenor et al. (1976b)
Supinine	(104)	Culvenor et al. (1959)
Trachelanthamine[b]	(158)	Kochetkov et al. (1969)
Viridoflorine[b]	(160)	Kochetkov et al. (1969)

[a] Not converted to crobarbatine.
[b] Total synthesis.

liotryl retronecine (Culvenor et al., 1976b) have also been prepared by this method.

Saturated necines can be transesterified with protected necic acid methyl esters: in this way, total syntheses of trachelanthamine (158), viridoflorine and lindelofine were accomplished by Kochetkov et al. (1969) (Scheme 3.10).

Diesters of retronecine with saturated or unsaturated fatty acids are easily prepared from the necine and excess of the acid chloride (Schoental and Mattocks, 1960; Mattocks, 1969b); thus tigloyl chloride gives ditigloyl retronecine

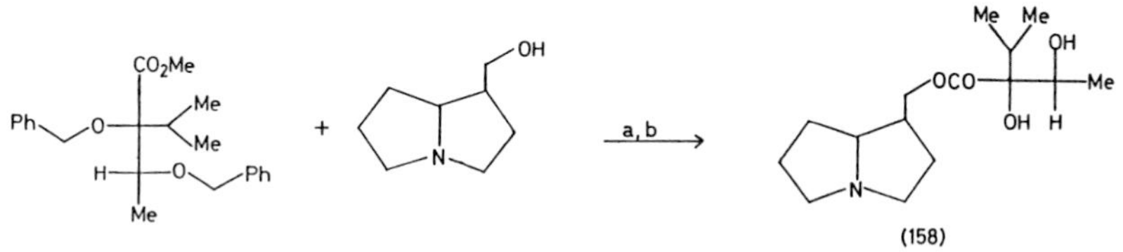

Scheme 3.10.   Reagents: a, NaOMe; b, H_2, catalytic.

TABLE 3.5 Semisynthetic Esters of Retronecine

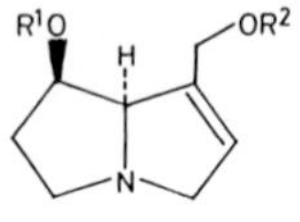

R¹O ... OR² (structure diagram with pyrrolizidine ring, H labels)

R¹	R²	References
Both Ac		Barger et al. (1935)
		Mattocks (1969b)
Both MeCH$_2$CO		Mattocks (1969b)
Both MeCH$_2$CH$_2$CO		Mattocks (1969b)
Both Me$_2$CHCO		Mattocks (1969b)
Both Me(CH$_2$)$_3$CO		Mattocks (1969b)
Both Me$_2$CHCH$_2$CO		Mattocks (1969b)
Both Me$_3$CCO		Mattocks (1969b)
Both Me(CH$_2$)$_4$CO		Mattocks (1969b)
Both MeCH$_2$C(Me$_2$)CO		Mattocks (1969b)
Both Me$_3$CCH$_2$CO		Mattocks (1969b)
Both MeCH$_2$CH$_2$C(Me$_2$)CO		Mattocks (1969b)
Both Me$_2$CH(CH$_2$)$_4$CO		Mattocks (1969b)
Both Me$_2$C=CHCO		Schoental and Mattocks (1960); Mattocks (1969b); Hsu and Allen (1975)
Both MeCH=C(Me)CO (tigloyl)		Mattocks (1969b)
Both CH$_2$(CH$_2$)$_3$CHCO		Mattocks (1969b)
Both Me$_2$CHOCH$_2$CO		Mattocks (1969b)
Both PhCO		Mattocks (1969b); Siddiqi et al. (1978a)
Both p-ClC$_6$H$_4$CO		Kaul and Kak (1974)
Both MeCH(OH)C(Me)(OH)CO		Mattocks (1981a)
Both EtNHCO		Mattocks (1977b)
Both Me$_2$C(OH)CO		Hoskins and Crout (1977)
CO—CH$_2$—CH$_2$—CH$_2$—CO		Devlin et al. (1982)
CO—CH$_2$—C(Me$_2$)—CH$_2$—CO		Devlin et al. (1982)
CO—CH$_2$—C(Ph$_2$)—CH$_2$—CO		Devlin et al. (1982)
CO—CH$_2$—C(CH$_2$)$_4$—CH$_2$—CO		Devlin et al. (1982)
CO—CH$_2$—CH(Me)—CH$_2$—CO		Devlin et al. (1982)
H	MeCH$_2$CO	Hoskins and Crout (1977)
H	Me$_3$CCO	Hoskins and Crout (1977)
Me$_2$CHCO	Me$_3$CCO	Hoskins and Crout (1977)
H	Cl$_3$CCH$_2$OCO(CH$_2$)$_3$CO	Hoskins and Crout (1977)
H	MeCH=C(Me)CO	Hoskins and Crout (1977); Mattocks (1981a)
EtNHCO	MeCH=C(Me)CO	Mattocks (1981a)
H	EtNHCO	Mattocks (1981a)
Ac	EtC(Me)(OAc)CO	Mattocks (1981a)
H	EtC(Ph)(OH)CO	Gelbaum et al. (1982)
MeCH=C(Me)CO (angelyl)	H	Culvenor et al. (1976b)
Heliotryl	H	Culvenor et al. (1976b)

TABLE 3.6 Semisynthetic Esters of Heliotridine

R^1	R^2	References
Both Ac		Culvenor *et al.* (1976b)
		Suri *et al.* (1975a)
Both Me$_3$CCO		Culvenor *et al.* (1976b)
Both Me(CH$_2$)$_3$CO		Culvenor *et al.* (1976b)
Both MeCH=C(Me)CO		Mattocks (1981a)
(tigloyl)		
Both Me$_2$CHCO		K. A. Suri *et al.* (1975a)
Both Me$_2$CHCH$_2$CO		K. A. Suri *et al.* (1975a)
Both Me$_2$C=CHCO		K. A. Suri *et al.* (1975a)
Both PhCO		K. A. Suri *et al.* (1975a)
Both MeOC$_6$H$_4$CO		K. A. Suri *et al.* (1975a)
Both H$_2$NC$_6$H$_4$CO		K. A. Suri *et al.* (1975a)
Both ClC$_6$H$_4$CO		K. A. Suri *et al.* (1975a)
MeCH=C(Me)CO	H	Aasen *et al.* (1969)
(angeloyl)		
H	MeCH=C(Me)CO	Culvenor *et al.* (1976b)
	(angeloyl)	
H	Me$_3$CCO	Culvenor *et al.* (1976b)
Me$_3$CCO	H	Culvenor *et al.* (1976b)

(241) (Scheme 3.11). Under mild conditions, a stoichiometric amount of acid chloride in pyridine gives the 9-monoester (240) (Mattocks, 1981a). Oxidation of the ditiglate (241) provides the dihydroxydiester (242) (Mattocks, 1981a).

When retronecine hydrochloride is heated with 2-methyl-2-acetoxybutyryl chloride (243), the chloro products (244) and (245) are formed (Mattocks, 1964a). With the same reagent in pyridine, retronecine gives the 9-(2-acetoxy-2-methylbutyrate), which can be acetylated to the unsymmetrical diester (246) (Mattocks, 1981a).

(243)   (244)   (245)

TABLE 3.7 Some Diesters of Synthanecine A

Ester moiety	Structure (R)	Reference
Acetyl	Ac	Mattocks (1974)
Tigloyl	MeCH=C(Me)CO	Mattocks (1981a)
N-Ethylcarbamoyl	EtNHCO	Mattocks (1974, 1982a)
N,N-Diethylcarbamoyl	Et$_2$NCO	Mattocks (1974)
Diethylphosphoryl	(EtO)$_2$PO	Mattocks (1974)
2-Methoxy-2-methylbutyryl	EtC(Me)(OMe)CO	Mattocks (1974)
Macrocyclic diesters		
Glutaryl	CO(CH$_2$)$_3$CO	R. H. Barbour and D. J. Robins (1985)
2,4-Dimethylglutaryl	COCHMeCH$_2$CHMeCO	R. H. Barbour and D. J. Robins (1985)
3,3-Dimethylglutaryl	COCH$_2$CMe$_2$CH$_2$CO	R. H. Barbour and D. J. Robins (1985)
3,3-Pentamethylene glutaryl	COCH$_2$C(CH$_2$)$_4$CH$_2$CO	R. H. Barbour and D. J. Robins (1985)
Succinyl	CO(CH$_2$)$_2$CO	R. H. Barbour and D. J. Robins (1985)
2,3-Dimethylsuccinyl	CO(CHMe)$_2$CO	R. H. Barbour and D. J. Robins (1985)
Phthaloyl	CO(C$_6$H$_4$)CO	R. H. Barbour and D. J. Robins (1985)

(246)   (247)   (248)

Hoskins and Crout (1977) investigated alternative methods of esterifying retronecine. By using acyl imidazoles as esterifying reagents, monoesters of retronecine at the more reactive 9-position can be made. This bis-(2-hydroxy-2-methylpropionate) (247) was also made in this way, the coupling reagent being N,N'-carbonyl-di-imidazole. Gelbaum et al. (1982) used this method to prepare 9-(2-hydroxy-2-phenylbutyryl)-retronecine (248), a synthetic analogue of indi-

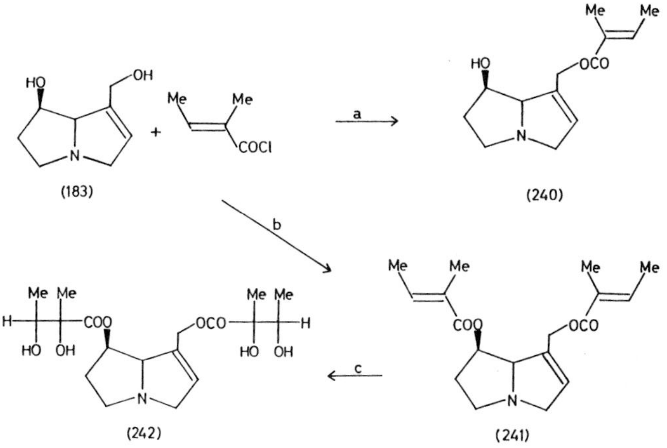

Scheme 3.11.   Reagents: a, Py; b, heat; c, OsO₄.

cine. Hoskins and Crout (1977) used another coupling reagent, *N,N'*-dicyclohex-ylcarbodiimide, to make retronecine esters, and Piper *et al.* (1981) combined trachelanthic acid and retronecine with this reagent to reconstruct the PA indicine.

7-Angelyl heliotridine and 7-angelyl retronecine are prepared by removing the 9-ester group from lasiocarpine and echimidine, respectively, by periodate oxidation (Aasen *et al.*, 1969; Culvenor *et al.*, 1976b).

Glinski and Zalkow (1985) have prepared heliotridine esters starting with retronecine.

B. Carbamate Esters

N-Alkyl carbamates of necines and synthanecines are easily made: thus refluxing retronecine with ethyl isocyanate and a basic catalyst gives the bis-*N*-ethylcarbamate (249) (Mattocks, 1977b). With a short reaction time and no additional catalyst (retronecine base itself being sufficient), the main product is the monocarbamate (250; Scheme 3.12) (Mattocks and Driver, 1983).

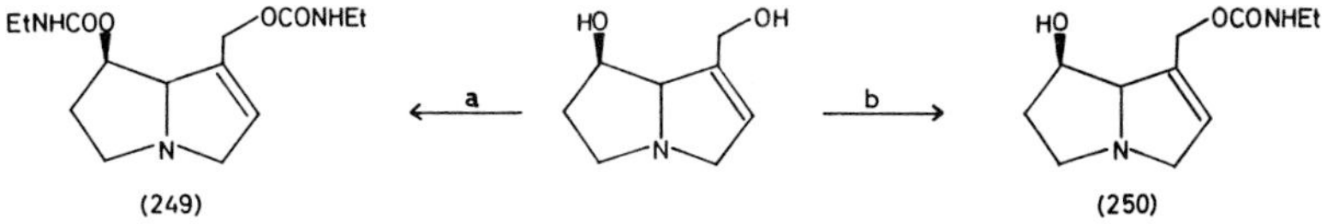

Scheme 3.12.   Reagents: a, EtNCO, DABCO, 6 h; b, EtNCO, 4 min.

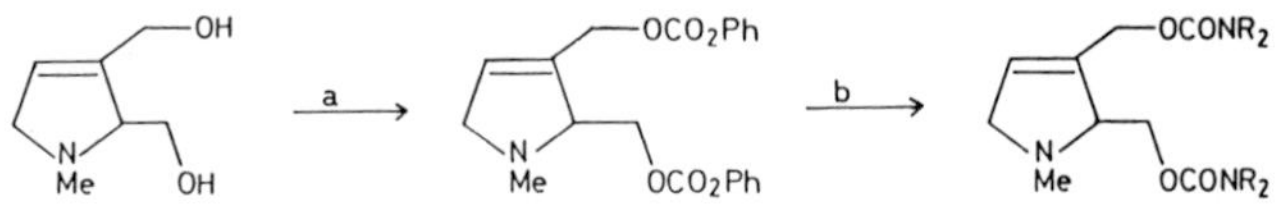

(251)

Scheme 3.13.   Reagents: a, PhOCOCl; b, R_2NH (R = alkyl).

Synthanecine carbamates can also be made by treating the phenylcarbonates, e.g. (251), with amines (Mattocks, 1974) (Scheme 3.13).

C. Macrocyclic Diesters

The preparation of these has provided a considerable challenge to chemists attempting to synthesize PAs. Three syntheses leading to natural macrocyclic PAs or analogous esters have been described. The first was achieved by Robins and Sakdarat (1980), who used a modification of the method of Corey and Nicolaou (1974) to cyclize monoesters of retronecine and 3,3-dimethylglutaric acid (Scheme 3.14). The product, 13,13-dimethyl-1,2-didehydrocrotalanine (252), was the first semisynthetic macrocyclic pyrrolizidine diester to be characterized. The homologues (253)–(256) were later prepared (Devlin *et al.*, 1982), and the first synthesis of a natural macrocyclic PA, dicrotaline (29), was achieved by the same method (Devlin and Robins, 1981; Brown *et al.*, 1983).

Huang and Meinwald (1981) prepared the macrocyclic retronecine diester, crobarbatine acetate (259) (Scheme 3.15). The thioester-acid (257) is O-protected by acetylation, then converted to the imidazolide, and this reacts with retronecine (cf. Hoskins and Crout, 1977) in the presence of alkoxide catalyst

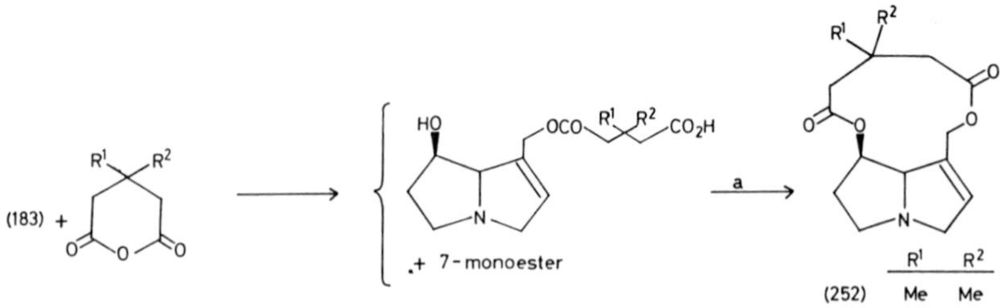

	R^1	R^2
(252)	Me	Me
(253)	H	H
(254)	Ph	Ph
(255)	$-(CH_2)_4-$	
(256)	H	Me

Scheme 3.14.   Reagents: a, Ph_3P, PySSPy.

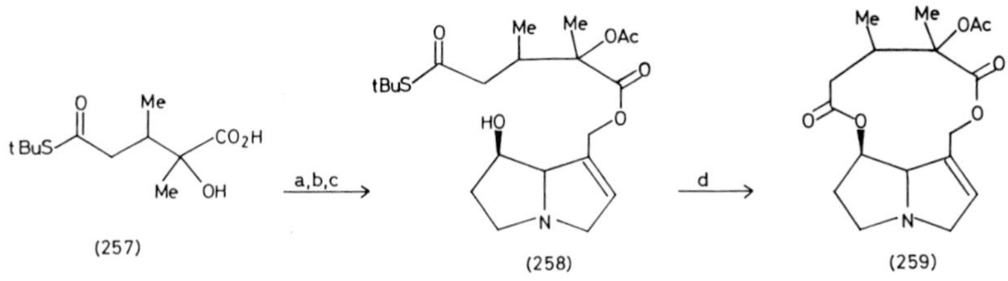

Scheme 3.15.   Reagents: a, Ac_2O, base; b, Im_2CO; c, retronecine, NaH; d, CF_3SO_3Cu.

(generated from retronecine by sodium hydride) to give the allylic half ester (258). Cyclization of the latter is accomplished using copper trifluoromethane sulphonate to give a mixture of two isomers (259), which are separated by alumina chromatography. Deacetylation to crobarbatine itself has not been achieved; hydrolysis leads only to retronecine.

Narasaka *et al.* (1982, 1984) described the first total synthesis of a macrocyclic PA, (±)-integerrimine (55), having synthesized both the acid and necine components (Scheme 3.16). More recently, Vedejs and Larsen (1984) synthesized (±)-crispatine (16) and (±)-fulvine (42) using an alternative cyclization procedure.

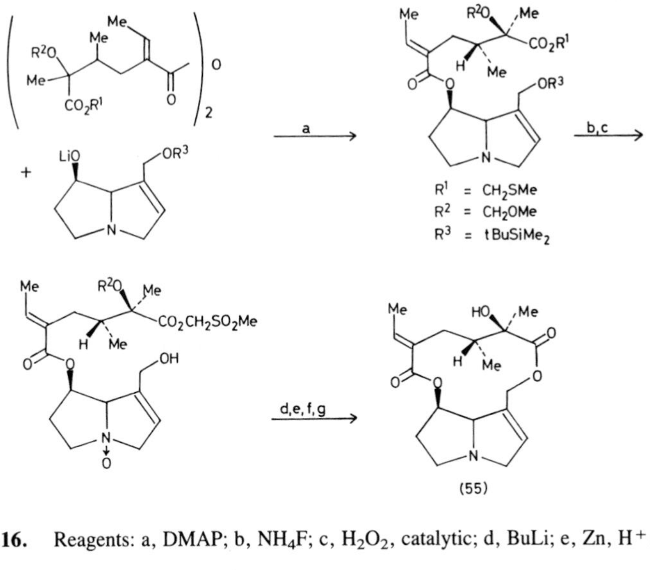

Scheme 3.16.   Reagents: a, DMAP; b, NH_4F; c, H_2O_2, catalytic; d, BuLi; e, Zn, H^+; f, TLC; g, H^+

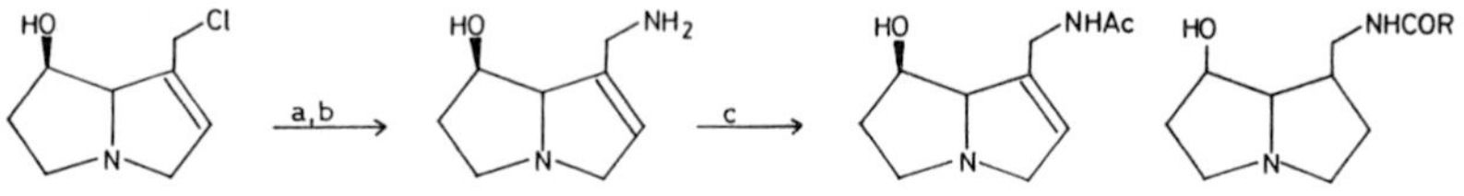

(260)   (261)   (262)

Scheme 3.17.   Reagents: a, NaN$_3$; b, LAH; c, AcOH, DCC.

V. OTHER SEMISYNTHETIC PA DERIVATIVES AND ANALOGUES

A. Amines and Amides Derived from Necines

Mattocks (1969b) prepared retronamine (260), providing a simple route to amide analogues, e.g. (261), of pyrrolizidine esters (Scheme 3.17). Rao *et al.* (1975d) made further amides the same way, for pharmacological studies. Similarly, Jamwal *et al.* (1982) made amide derivatives of the saturated amino alcohol, platynecine (262; R = Ph, Me$_2$CHCH$_2$ or Me$_2$=CH).

B. Quaternary Derivatives

Quaternary salts of PAs have been prepared for their pharmacological interest. For a review, see Atal (1978).

C. Acylated PAs

Free hydroxyl groups on the necine or acid moieties of ester-PAs can be further acylated, giving products with modified toxicity (Table 3.8).

D. Miscellaneous PA Analogues

Willette and Driscoll (1972) prepared esters of 4-amino-2-buten-1-ol (270) as models possessing the amino and allylic ester functions of PAs in an acyclic structure.

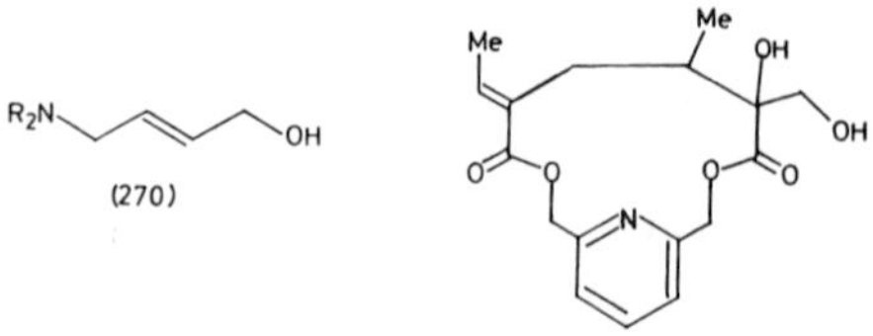

(270)

(271)

TABLE 3.8 Some Acyl Derivatives of Pyrrolizidine Alkaloids

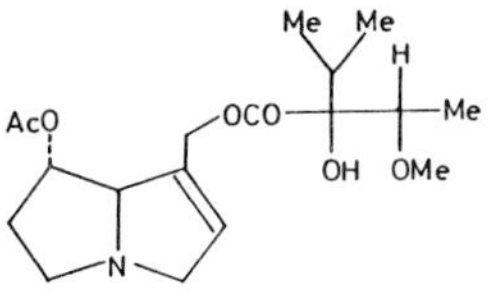

(263)

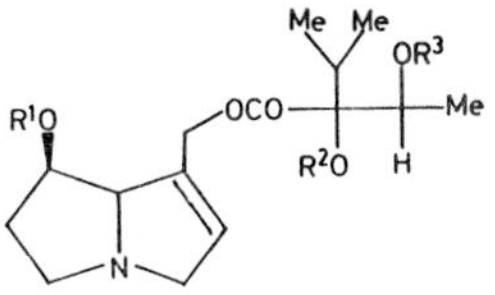

(264) $R^1 = R^3 = Ac$, $R^2 = H$
(265) $R^1 = R^2 = R^3 = Ac$
(266) $R^1 = R^2 = R^3 = Me_2CHCH_2CO$

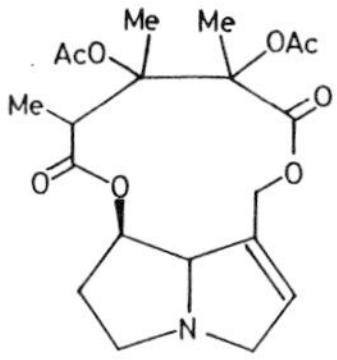

(267)

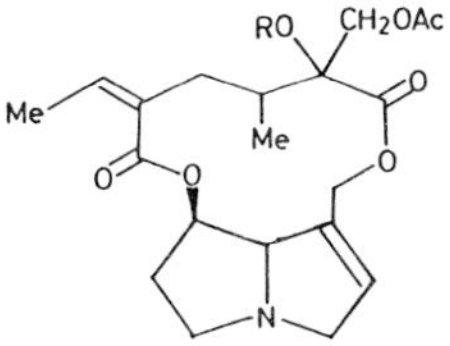

(268) R = H
(269) R = Ac

Compound	Structure	Reference
Diacetyl anacrotine	—	Culvenor and Smith (1972)
7-Acetyl heliotrine	(263)	Culvenor et al. (1976b)
Diacetyl indicine	(264)	Mattocks (1967a)
Triacetyl indicine	(265)	Mattocks (1976a)
Tri-isovaleryl indicine	(266)	Mattocks (1969b)
Acetyl lasiocarpine	(66)	Culvenor et al. (1975b)
Diacetyl lasiocarpine	—	Culvenor et al. (1975b)
3'-Acetyl lycopsamine	—	Roitman (1983a)
3',7-Diacetyl lycopsamine	—	Roitman (1983a)
Diacetyl madurensine	—	Culvenor and Smith (1972)
Diacetyl monocrotaline	(267)	Culvenor and Smith (1957b); Mattocks (1969b)
Acetyl retrorsine	(268)	Mattocks (1969b)
Diacetyl retrorsine	(269)	Mattocks (1969b)

Drewes and Pitchford (1981) made the macrocyclic diester 'pyridine retrone-cate' (271), by combining the dipotassium salt of natural retronecic acid with 2,6-bis(bromoethyl)-pyridine. The yield was poor (9.7%).

VI. DEHYDRO-PYRROLIZIDINE ALKALOIDS AND ANALOGUES

Conversion of the pyrroline ring in unsaturated necines and their esters to a pyrrole ring leads to compounds which are often highly reactive, and thus difficult to isolate and handle. Analogous hydroxymethyl pyrroles and their esters have similar chemical reactivity. Only a few syntheses of dehydronecines have been described; they and their esters are usually prepared by dehydrogenating the parent necines or PAs.

A. Total Syntheses of Dehydro-necines

($\pm$)-Dehydroheliotridine (272) was first synthesized by Viscontini and Gillhof-Schaufelberger (1971) in 19% overall yield (Scheme 3.18).

Bohlmann et al. (1979) described a simpler synthesis starting with (273), which is made from N-formyl proline by the method of Pizzorno and Albonico (1974), and converted via per-acid oxidation to (274), which is then reduced to ($\pm$)-dehydroheliotridine in an overall yield of 21%.

Klose et al. (1980) synthesized the related compound 5,7a-didehydroheliotridin-3-one (275), which is the alcohol nucleus of the senaetnine group of acylpyrrole alkaloids discovered by Bohlmann and his co-workers.

B. Preparation of Pyrrolic (Dehydro) Derivatives from Necines and PAs

Dehydrogenation of 3-pyrrolines, including necines, synthanecines and ester PAs, can be effected either by direct oxidation of these bases or by dehydration

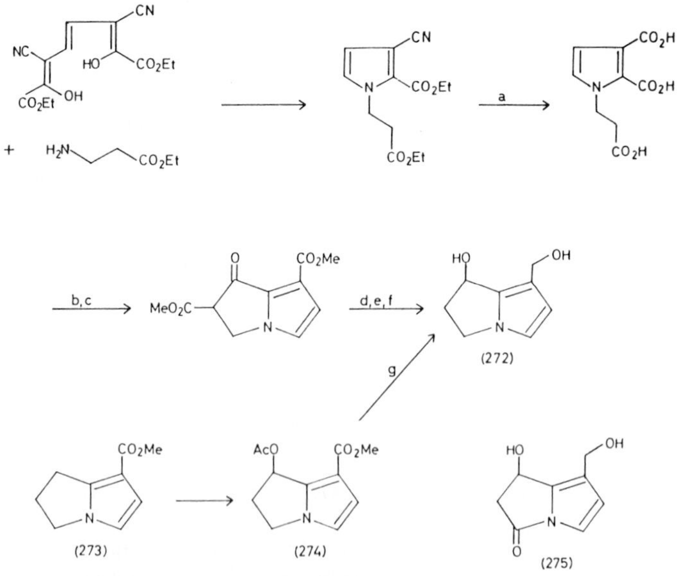

Scheme 3.18.   Reagents: a, NaOH; b, CH$_2$N$_2$; c, NaOEt; d, HCl; e, CH$_2$N$_2$; f, LAH; g, DIBAH.

TABLE 3.9 Preparations of Dehydronecines and Dehydropyrrolizidine Alkaloids

Compound	Starting material	Reagent(s)	Crude yield (%)[a]	References
Dehydroretronecine (retronecine pyrrole)	retronecine N-oxide	FeSO$_4$–MeOH	64	Mattocks (1969a)
Dehydroretronecine	retronecine base	chloranil, NaBH$_4$	92	Culvenor et al. (1970b)
Dehydroretronecine	retronecine (base or -HCl)	Fremy's salt	86	Mattocks (1981c)
Dehydroretronecine	retronecine N-oxide	MnO$_2$, NaBH$_4$	70	Culvenor et al. (1970c)
Dehydroheliotridine	heliotridine base	chloranil, NaBH$_4$	70	Culvenor et al. (1970b)
Dehydrosupinidine	supinidine base	KMnO$_4$–Me$_2$CO	18	Culvenor et al. (1970b)
Dehydrosupinidine	supinidine N-oxide	FeSO$_4$–MeOH	51	A. R. Mattocks (unpublished)
7β-Hydroxy-1-methyl-6,7-dihydropyrrolizine	desoxyretronecine	chloranil, NaBH$_4$	—	Culvenor et al. (1970b)
Dehydroretrorsine	retrorsine N-oxide	FeSO$_4$–MeOH	62	Mattocks (1969a)
Dehydroretrorsine	retrorsine N-oxide	Ac$_2$O	70	Mattocks (1969a)
Dehydromonocrotaline	monocrotaline N-oxide	FeSO$_4$–MeOH	72	Mattocks (1969a)
Dehydromonocrotaline	monocrotaline N-oxide	Ac$_2$O	—	Mattocks (1969a)
Dehydromonocrotaline	monocrotaline N-oxide	Ac$_2$O–CHCl$_3$	63	Culvenor et al. (1970b)
Dehydroheliotrine	heliotrine N-oxide	FeSO$_4$–MeOH	60	Mattocks (1969a)
Dehydroheliotrine	heliotrine N-oxide	Ac$_2$O–CHCl$_3$	77	Culvenor et al. (1970b)
Dehydrolasiocarpine	lasiocarpine N-oxide	FeSO$_4$–MeOH	60	Mattocks (1969a)
Dehydrolasiocarpine	lasiocarpine N-oxide	Ac$_2$O–CHCl$_3$	79	Culvenor et al. (1970b)
Dehydrosenecionine	senecionine N-oxide	Ac$_2$O–CHCl$_3$	100	Culvenor et al. (1970b)

[a] May be much lower after purification.

of their *N*-oxides (see Chapter 5). Methods which are suitable for preparative purposes are described here.

The formation of dehydro derivatives is not difficult to accomplish. The main practical problems are associated with the high chemical reactivity of these compounds, especially the esters, which are liable to polymerize rapidly in the presence of water and especially acids, and which are liable to alkylate nucleophiles such as amines, including the parent alkaloids. They do not keep well, and they are best made in small batches, when they are needed. Table 3.9 lists dehydro-necines and dehydro-PAs which have been prepared. Because of decomposition during recrystallization procedures, the yields of purified products are sometimes considerably lower than the crude yields quoted.

Small quantities of dehydro-PAs are conveniently made by the action of anhydrous methanolic ferrous sulphate on the alkaloid *N*-oxides (Mattocks, 1969a). The reaction is accelerated by a limited amount of fluoride, which removes trivalent iron (Mattocks, 1968c). The dehydration of *N*-oxides is also effected by acetic anhydride, either neat (Mattocks, 1969a) or in chloroform solution (Culvenor *et al.*, 1970b). Scheme 3.19 shows the conversion of monocrotaline (74) to dehydromonocrotaline (monocrotaline pyrrole) (276).

The dehydrogenation of necines is exemplified by preparations of dehydroretronecine (retronecine pyrrole; 6,7-dihydro-7β-hydroxy-1-hydroxymethyl-5*H*-pyrrolizine) (Scheme 3.20). This compound (201) can be made from retronecine *N*-oxide (277) using ferrous sulphate (Mattocks, 1969a) or by oxidizing retronecine (183) with chloranil (Culvenor *et al.*, 1970b) or with potassium nitrosodisulphonate (Fremy's salt) (Mattocks, 1981c). The latter reaction occurs in aqueous solution and retronecine hydrochloride can be used as a starting material. In general, similar methods are used to make other dehydro-necines.

Manganese dioxide oxidations of retronecine, heliotridine and heliotridine *N*-oxide lead to mixtures of products, but MnO_2 oxidation of retronecine *N*-oxide gives mainly the pyrrolic aldehyde (278), which is easily reduced to dehydroretronecine (Culvenor *et al.*, 1970c).

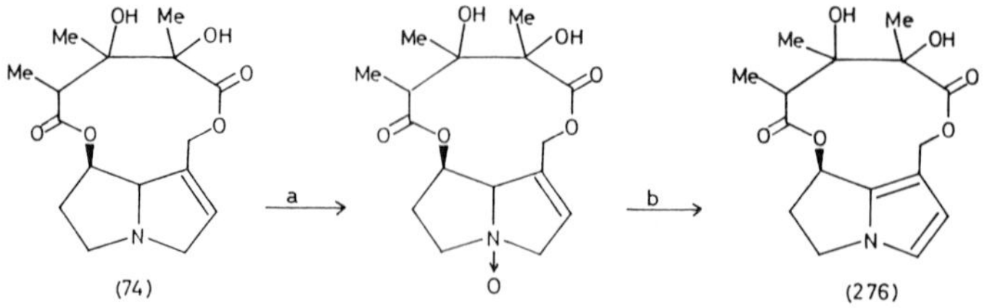

Scheme 3.19. Reagents: a, H_2O_2; b, Fe(II)SO_4, MeOH or Ac_2O.

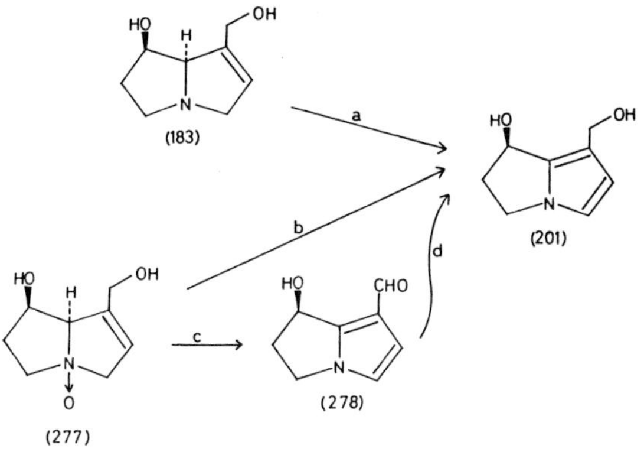

Scheme 3.20   Reagents: a, chloranil or $(KSO_3)_2NO$; b, $Fe(II)SO_4$, MeOH; c, MnO_2; d, $NaBH_4$.

C. Pyrrolic Alcohols Analogous to Dehydro-necines

Synthetic hydroxymethyl pyrroles having chemical properties similar to those of dehydronecines are of great interest for their biological effects. They can be prepared by LAH reduction of the corresponding pyrrole carboxylates, which themselves are readily accessible by dehydrogenation of 3-pyrroline esters available as intermediates in synthanecine syntheses (Mattocks, 1974, 1978a) or by other synthetic routes. The procedure is illustrated by the synthesis of 2,3-bishydroxymethyl-1-methylpyrrole (279) (Scheme 3.21). Synthetic hydroxymethyl pyrroles are listed in Table 3.10.

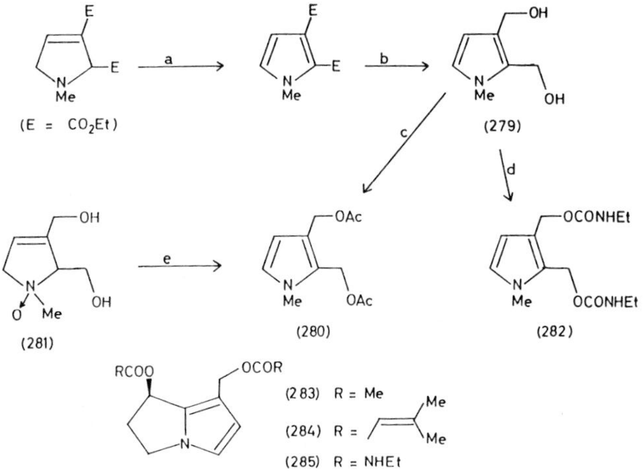

Scheme 3.21.   Reagents: a, DDQ; b, LAH; c, AcCl, Et_2N; d, EtNCO, DABCO; e, Ac_2O, Et_3N.

TABLE 3.10 Syntheses of Hydroxymethyl Pyrroles

Compound					
R^1	R^2	R^3	R^4	R^5	References
Me	CH_2OH	H	H	H	Ryskiewicz and Silverstein (1954)
Me	H	CH_2OH	H	H	Mattocks (1978a)
Me	Me	CH_2OH	H	H	Mattocks (1978a)
Me	CH_2OH	CH_2OH	H	H	Mattocks (1974, 1978a)
Me	CH_2OH	CH_2OH	MeO	H	Mattocks (1974)
Me	CH_2OH	CH_2OH	H	Me	Mattocks (1978a)
Me	CH_2OH	H	CH_2OH	H	Mattocks (1978a)
Me	CH_2OH	H	H	CH_2OH	Mattocks (1978a)
Me	H	CH_2OH	CH_2OH	H	Guengerich and Mitchell (1980)
Ph	CH_2OH	CH_2OH	H	Me	Mattocks (1978a)
Me	CH_2CH_2OH	CH_2OH	H	H	Mattocks (1978a)

D. Semisynthetic Dehydro-necine Esters and Esters of Pyrrolic Alcohols

Table 3.11 lists esters and carbamates which have been prepared from hydroxymethyl pyrroles.

Hydroxymethyl pyrroles can be esterified by acyl chlorides, but the products are highly reactive, and all water and acid must be rigorously excluded during preparation, work-up and storage to prevent rapid polymerization. Thus, the alcohol (279) with acetyl chloride in presence of excess triethylamine gives the diacetate (280) (Mattocks, 1978a). Alternatively, synthanecine A N-oxide (281) can be converted directly to (280) by acetic anhydride which effects both acetylation and dehydration (Mattocks, 1974). In a similar way, diacetyl dehydroretronecine (283) is prepared from retronecine N-oxide (Mattocks and Driver, 1983). Shumaker et al. (1976c) made disenecioyl dehydroretronecine (284) by oxidizing the disenecioyl ester of retronecine with potassium permanganate (cf. Culvenor et al., 1970b).

With ethyl isocyanate and a base catalyst, pyrrole alcohols give good yields of N-ethylcarbamates such as (282). Dehydroretronecine bis-N-ethylcarbamate (285) can be made the same way, whereas the monocarbamate (287) is prepared

TABLE 3.11 Synthetic Hydroxymethylpyrrole Esters and Carbamates

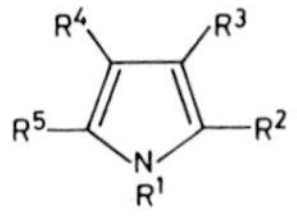

	Structure				
R^1	R^2	R^3	R^4	R^5	References
Me	CH$_2$OCONHEt	H	H	H	Plestina *et al.* (1977)
Me	CH$_2$OAc	H	H	H	Mattocks and Driver (1983)
Me	H	CH$_2$OCONHEt	H	H	Mattocks and Driver (1983)
Me	H	CH$_2$OAc	H	H	Mattocks and Driver (1983)
Me	Me	CH$_2$OCONHEt	H	H	Mattocks (1978a)
Me	CH$_2$OCONHEt	CH$_2$OCONHEt	H	H	Mattocks (1974)
Me	CH$_2$OAc	CH$_2$OAc	H	H	Mattocks (1974, 1978a, 1979)
Me	CH$_2$OCOCMe$_3$	CH$_2$OCOCMe$_3$	H	H	Mattocks (1978a)
Me	CH$_2$OCO(CH$_2$)$_4$Me	CH$_2$OCO(CH$_2$)$_4$Me	H	H	Mattocks (1978a)
Me	CH$_2$CH$_2$OCOCMe$_3$	CH$_2$OCOCMe$_3$	H	H	Mattocks (1978a)
Ph	CH$_2$OAc	CH$_2$OAc	H	Me	Mattocks (1978a)

by treating the *N*-oxide (286) of retronecine-9-*N*-ethylcarbamate (250) with methanolic ferrous sulphate (Mattocks and Driver, 1983) (Scheme 3.22).

The pyrrolic carbamates are somewhat more stable than the acyl esters because their hydrolysis does not give rise to acidic products.

VII. RADIOACTIVE LABELLING

Labelled PAs, analogues, and derived pyrroles are valuable for investigations of distribution, metabolism and tissue binding. For general labelling at low specific activities, suitable source plants can be grown in a $^{14}CO_2$ atmosphere, before

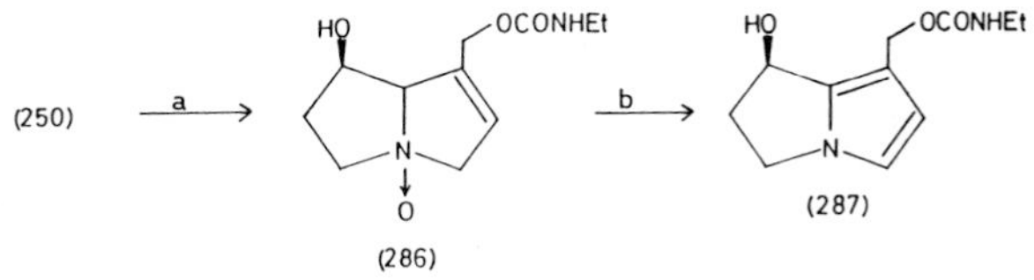

Scheme 3.22.   Reagents: a, H$_2$O$_2$; b, Fe(II)SO$_4$, MeOH.

TABLE 3.12 Specifically Radioactive-labelled Necines, PAs, Analogues and Pyrrolic Derivatives

Group	Compound	Label	Position of label	Reference
Necines	Retronecine	^{3}H	9	Hsu and Allen (1975); Mattocks (1977b); Piper et al. (1981)
	Synthanecine A	^{3}H	3-CH_2	Mattocks and White (1976); Mattocks (1982a)
Ester PAs and analogues	Indicine	^{3}H	9	Piper et al. (1981)
	Disenecioyl retronecine	^{3}H	9	Hsu and Allen (1975)
	Retronecine bis(N-ethylcarbamate)	^{3}H	9	Mattocks (1977b)
	Synthanecine A bis(N-ethylcarbamate)	^{3}H	3-CH_2	Mattocks and White (1976); Mattocks (1982a)
Dehydronecines, dehydro-PAs, and pyrrolic analogues	Senecionine; isosenecionine	^{14}C	18	Mattocks (1983)
	Dehydroretronecine	^{3}H	9	Hsu et al. (1974)
	(±)-Dehydroheliotridine	^{3}H	6	Curtain and Edgar (1976)
	2,3-Bishydroxymethyl-1-methylpyrrole	^{3}H	2-CH_2, 3-CH_2	White and Mattocks (1976)
	2,3-Bisacetoxymethyl-1-methylpyrrole	^{3}H	2-CH_2, 3-CH_2	Mattocks (1979)
	3,4-Bishydroxymethyl-1-methylpyrrole	^{3}H	3-CH_2, 4-CH_2	Guengerich and Mitchell (1980)

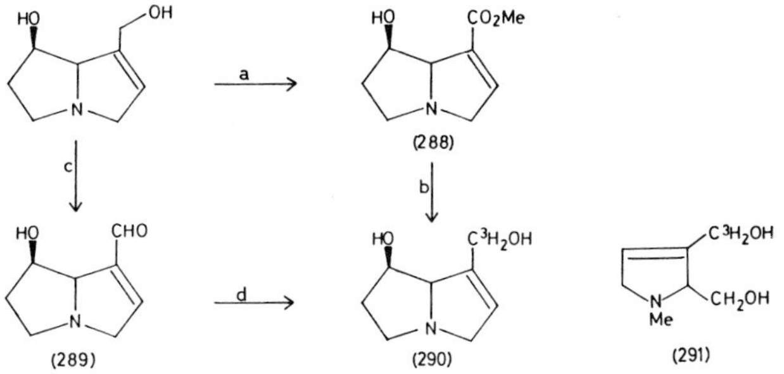

HO̸ ⟋OH   HO̸ CO₂Me

          a →

                (288)

c↓   ↓b

HO̸ CHO   HO̸ C³H₂OH   ⟋C³H₂OH

          d →   ⟍CH₂OH
                                    N
                                    Me

(289)   (290)   (291)

Scheme 3.23.   Reagents: a, MnO₂, HCN, MeOH; b, LiAl³H₄; c, MnO₂; d, NaB³H₄.

isolating the alkaloid [e.g. senecionine, from *Senecio vulgaris* (Brown and Segall, 1982; Segall *et al.*, 1983)], or plants may be fed with labelled amino acids; thus putrescine or isoleucine is incorporated into the necine and necic acid moieties, respectively, of *S. vulgaris* alkaloids (Luthy *et al.*, 1983b). General ³H-labelling of unsaturated PAs by catalytic exchange with tritium gas gives poor results owing to saturation of the double bond.

Alkaloids and related compounds with radioactivity at specific locations in the molecule are more useful; for a summary of those which have been prepared, see Table 3.12. Necines and pyrrolic alcohols can be ³H-labelled by oxidation followed by reduction with a tritiated reagent. Thus, oxidation of retronecine by manganese dioxide in presence of HCN and methanol gives the methyl carboxylate (288), which is reduced to [9-³H]retronecine (290) using tritiated LAH (Hsu and Allen, 1975; Piper *et al.*, 1981) (Scheme 3.23). Alternatively, careful oxidation of retronecine with specially prepared manganese dioxide (Mattocks, 1977a) gives the unstable aldehyde (289), which is reduced by ³H-labelled sodium borohydride to (290) (Mattocks, 1977b). Synthanecine A labelled with tritium in the 3-hydroxymethyl group (291) has been prepared by a parallel procedure (Mattocks and White, 1976; Mattocks, 1982a). Pyrrolic alcohols are tritiated in a similar way. [³H]Dehydroretronecine (292) was prepared by Hsu *et al.* (1974): MnO₂ oxidation of retronecine *N*-oxide (cf. Culvenor *et al.*, 1970c) gives 1-formyl-7-hydroxy-6,7-dihydro-5*H*-pyrrolizine (278), which is reduced by tritiated sodium borohydride to (292). Curtain and Edgar (1976) hydrolysed 1-heliotroxymethyl-7-oxo-6,7-dihydro-5*H*-pyrrolizine (293) (prepared by permanganate oxidation of the PA heliotrine; Culvenor *et al.*, 1970b) with warm water

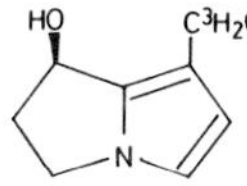

HO̸ C³H₂OH   C³H₂OR

                    ⟍C³H₂OR
                    N
                    Me

(292) 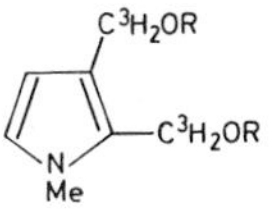   (297) R = H
                   (298) R = Ac

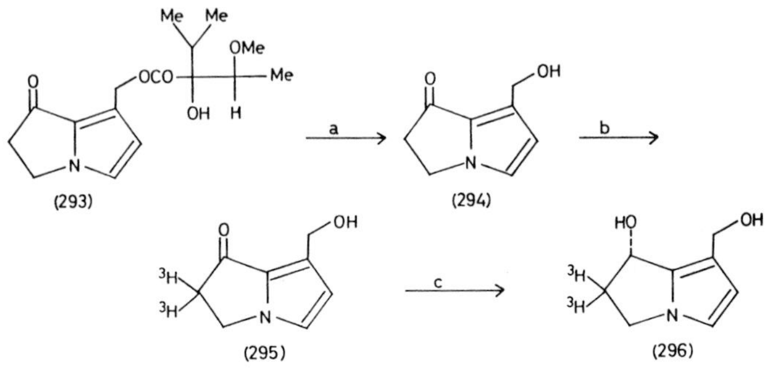

Scheme 3.24.   Reagents: a, H_2O; b, ^3H_2O, OH^-; c, $NaBH_4$.

to the keto-alcohol (294) (Scheme 3.24). The latter undergoes base-catalysed exchange of the 6-protons in tritiated water, giving (295), which is reduced by borohydride to $(\pm)$-[6-^{3}H]dehydroheliotridine (296) (i.e. a racemic mixture of dehydroheliotridine and dehydroretronecine).

Tritiated 2,3-bishydroxymethyl-1-methylpyrrole (297) has been made by oxidizing the unlabelled pyrrole with MnO_2 to a mixture of the corresponding 2- and 3-aldehydes, and reducing this with ^{3}H-labelled sodium borohydride (White and Mattocks, 1976); its diacetate (298) has also been prepared (Mattocks, 1979).

Less attention has been paid to the labelling of PAs in the acid moiety. Mattocks (1983) described the conversion of the natural PA retrorsine into [18-^{14}C]senecionine (Scheme 3.25). Retrorsine (85) is oxidized by periodate to a mixture of two isomeric keto esters (299), which react with [^{14}C]methylmagnesium iodide to give senecionine (92) and its diastereomer isosenecionine (300) (among other products) with ^{14}C in the acid moiety.

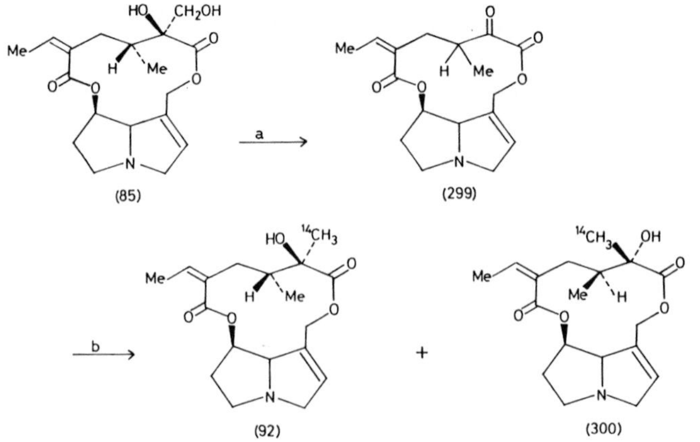

Scheme 3.25.   Reagents: a, IO_4^-; b, $^{14}CH_3MgI$.

4

Physical Properties and Spectroscopy of Pyrrolizidine Alkaloids and Related Compounds

I. INTRODUCTION

This chapter brings together physical data of practical use for identifying known PAs, and for studies on molecular structure and conformation. Physical properties relevant to biological actions of these alkaloids are also discussed. Some of this information has been reviewed previously, particularly by Bull *et al.* (1968); some overlap has been necessary to avoid a disjointed account with too frequent references to earlier reviews which may no longer be readily available.

II. PHYSICAL CHARACTERISTICS OF PAs AND DERIVATIVES

The information in these tables of physical properties is not necessarily from the earliest sources: where possible, up-to-date or more accessible references are cited. In view of the widespread use of mass spectrometry for identification and structural studies, the molecular weights of PAs and necines are given to four decimal places. The groups of compounds tabulated are as follows:

1. Necine bases. See Table 4.1.

2. Unsaturated PAs. Ester PAs having an unsaturated necine moiety are listed in Table 4.2.

3. Saturated PAs. For ester PAs with a saturated necine moiety, see Table 4.3.

4. *N*-Oxides of PAs and necines. Table 4.4 lists those *N*-oxides which have been isolated from plant sources or prepared chemically from the basic alkaloids.

5. Necic acids. The acid moieties most commonly encountered are included in Table 4.5. These are compounds isolated following the hydrolysis or hydrogenolysis of PAs. Some of them are in a form (e.g. lactones) structurally different from that in the intact alkaloid.

III. OTHER PHYSICAL PROPERTIES

A. Ionization

Culvenor and Willette (1966) measured the ionization constants of some necines and PAs. Because many PAs have low water solubility, a mixture of methyl cellosolve and water (8 : 2) was used as solvent. The pK_a values in this mixture were lower (by up to 1.33) than in water owing to a reduction in stabilization of the cation by solvation. Some values are quoted in Table 4.6. Frahn (1969) gives calculated values for some other PAs. In general, substituents on the pyrrolizidine ring have a base-weakening effect (Bull *et al.*, 1968); thus supinidine (182) is a stronger base than heliotridine or retronecine (183), these amino alcohols are stronger bases than their monoesters, and the diesters are still weaker.

B. Partition and Solubility

The distribution of a PA between aqueous and lipid phases will depend on the proportion of unionized base present, i.e. on the pK_a. The ionized form, which predominates in acidic solution, will partition largely into the aqueous phase. The solubility properties of the free base will depend on structural features, such as the presence of OH groups. Thus, some PAs such as echinatine and indicine are highly water soluble; others such as senecionine have poor water solubility, whereas lasiocarpine is exceptionally lipophilic (Bull *et al.*, 1968). The *N*-oxides of PAs are highly soluble in water and many of them cannot be extracted from

TABLE 4.1 Physical Properties of Some Necine Bases

Base	Structure	Formula	Molecular weight	m.p. (°C)	$[\alpha]_D$ (°)[a]	Derivatives with m.p. (°C)	References
Croalbinecine	(180)	$C_8H_{15}NO_3$	173.1052	(gum)		HCl, 165–166; picrate, 155–156	Sawhney and Atal (1973); Sawhney et al. (1974)
Crotanecine	(185)	$C_8H_{13}NO_3$	171.0895	202–203.5	+39(e)	picrate, 131	Atal et al. (1966a); Mattocks (1968d)
Curassanecine	(179)	$C_8H_{15}NO_2$	157.1103	(gum)			Mohanraj et al. (1982)
Hastanecine	(175)	$C_8H_{15}NO_2$	157.1103	113–114	−9.1(m)	HCl, 132–134	Konovalov and Men'shikov (1945); Aasen et al. (1969)
Heliotridine	(184)	$C_8H_{13}NO_2$	155.0946	117–118	+31(m)	HCl, 122–124	Men'shikov (1932); Culvenor et al. (1954)
(+)-Isoretronecanol (lindelofidine)	(170)	$C_8H_{15}NO$	141.1154	40–41	+79.1(e)	Picrate, 194; picrolonate, 182	Warren (1966)
Macronecine	(177)	$C_8H_{15}NO_2$	157.1103	126–128	+49.3(e)	HCl, 152–153	Aasen and Culvenor (1969b)
Otonecine	(186)	$C_9H_{15}NO_3$	185.1052	—	−13(e) (HCl salt)	HCl, 146–148	Briggs et al. (1965)

(*continued*)

TABLE 4.1 (*Continued*)

Base	Structure	Formula	Molecular weight	m.p. (°C)	$[\alpha]_D$ (°)[a]	Derivatives with m.p. (°C)	References
Petasinecine	(178)	$C_8H_{15}NO_2$	157.1103	132–134	−20(e)		Yamada et al. (1978)
Platynecine	(176)	$C_8H_{15}NO_2$	157.1103	151–152	−57(c) −84(e)	picrate, 184	Orekhov et al. (1935)
Retronecine	(183)	$C_8H_{13}NO_2$	155.0946	121–122	+50.2(e)	HCl, 164–165; picrate, 142	Barger et al. (1935)
Rosmarinecine	(181)	$C_8H_{15}NO_3$	173.1052	171–172	−118.5(m)	picrate, 175; MeI, 195	Richardson and Warren (1943); Porter and Geissman (1962)
Supinidine	(182)	$C_8H_{13}NO$	139.0997	(oil)	−10.3(e)	picrate, 142–143	Culvenor (1954)
(−)-Trachelanthamidine	(172)	$C_8H_{15}NO$	141.1154	(oil)	−12.9(e)	HCl, 110–112; picrate, 174	Men'shikov and Borodina (1945); Mattocks (1964b)
(+)-Trachelanthamidine (laburnine)	(173)	$C_8H_{15}NO$	141.1154	(oil)	+15.45(e)		Brandange and Granelli (1973)
Turneforcidine	(174)	$C_8H_{15}NO_2$	157.1103	118.5–120	−12.5(m) −4.3(e)	HCl, 116 (120)	Culvenor and Smith (1957a); Aasen et al. (1969); Aasen and Culvenor (1969a)

[a] Solvents: c, chloroform; e, ethanol, m, methanol.

TABLE 4.2 Physical Properties of Unsaturated Pyrrolizidine Alkaloids

Alkaloid	Structure	Formula	Molecular weight	m.p. (°C)	$[\alpha]_D$ (°)[a]	Derivatives with m.p. (°C)	References
Amabiline	(1)	$C_{15}H_{25}NO_4$	283.1784	(oil)	−7.1(e)	picrate (semi-synthetic), 112–113	Crowley and Culvenor (1959); Culvenor and Smith (1967)
Anacrotine (cis)	(2)	$C_{18}H_{25}NO_6$	351.1682	198	+30(e)	picrate, 224 (dec); MeI, 222	Culvenor and Smith (1972); Mattocks (1968d)
6-Acetyl (trans)	(3)	$C_{20}H_{27}NO_7$	393.1787	106–107	+65(e)		Culvenor and Smith (1972)
6-Angelyl (trans)	(4)	$C_{23}H_{31}NO_7$	433.2100				Culvenor and Smith (1972)
Anadoline (base)	(5)	$C_{20}H_{29}NO_5$	363.2046	(gum)			Culvenor et al. (1975a)
7-Angelyl heliotridine	(6)	$C_{13}H_{19}NO_3$	237.1365	116–117	+10.8(e)	picrate, 165	Crowley and Culvenor (1959)
Trachelanthate	(7)	$C_{20}H_{31}NO_6$	381.2151				Crowley and Culvenor (1959)
Viridoflorate	(8)	$C_{20}H_{31}NO_6$	381.2151				Crowley and Culvenor (1959)
7-Angelyl retronecine	(9)	$C_{13}H_{19}NO_3$	237.1365	76–77	+49(e)		Crowley and Culvenor (1962)
9-Angelyl retronecine	(10)	$C_{13}H_{19}NO_3$	237.1365	(oil)	+2(e)	picrate, 135–137; picrolonate, 169–171	Hikichi et al. (1980)
Asperumine	(11)	$C_{18}H_{25}NO_4$	319.1784				Man'ko and Kotovskii (1970b); Mel'ku-mova et al. (1974)

(continued)

TABLE 4.2 (*Continued*)

Alkaloid	Structure	Formula	Molecular weight	m.p. (°C)	$[\alpha]_D$ (°)[a]	Derivatives with m.p. (°C)	References
Axillaridine	(12)	$C_{18}H_{27}NO_6$	353.1838	148–152	+241(m)	picrate, 235 (dec)	Crout (1969)
Axillarine	(13)	$C_{18}H_{27}NO_7$	369.1787	up to 205 (dec)	+65.1(pyr)	picrate, 214–216 (dec); HCl, 228	Crout (1969)
Bisline	(14)	$C_{18}H_{27}NO_6$	353.1838	169			Coucourakis and Gordon-Gray (1970)
Clivorine	(15)	$C_{21}H_{29}NO_8$	423.1893	147–149	+79(c)	picrate, 173–175	Klasek et al. (1967)
Crispatine	(16)	$C_{16}H_{23}NO_5$	309.1575	137–138	+41(e)		Culvenor and Smith (1963)
Crobarbatine	(17)	$C_{15}H_{21}NO_5$	295.1420	142–143		picrate, 237	Puri et al. (1973)
Cromadurine	(18)	$C_{16}H_{23}NO_5$	309.1576	242–243		picrate, 209–210; HCl, 238	Rao et al. (1975b)
Cronaburmine	(19)	$C_{17}H_{25}NO_5$	323.1733	133–134		picrate, 233 (dec); HCl, 250	Siddiqi et al. (1978b)
Crosemperine	(20)	$C_{19}H_{20}NO_6$	367.1995	117–118	+2.2(e)	HCl, 180 (dec); MeI, 208–209	Atal et al. (1967)
Crotaflorine	(21)	$C_{18}H_{25}NO_7$	367.1631	179–180	+50(e)		Culvenor and Smith (1972)
Crotafoline	(22)	$C_{18}H_{25}NO_5$	335.1733	176–182 (dec)			Crout (1972)
Crotalarine	(23)	$C_{18}H_{27}NO_6$	353.1838	167–168			Rao et al. (1975a)
Crotananine	(24)	$C_{17}H_{25}NO_5$	323.1733	174–175	−80(m)		Siddiqi et al. (1978c)
Crotastriatine	(25)	$C_{19}H_{25}NO_6$	363.1682	133	+9.6(c); +32(e)	perchlorate, 245 (dec)	Gandhi et al. (1968)
Crotaverrine	(26)	$C_{19}H_{27}NO_6$	365.1838	142–144	+32.7(m)		K. A. Suri et al. (1976); O. P. Suri et al. (1976)

Name		Formula					Reference
Acetyl	(27)	$C_{21}H_{29}NO_7$	407.1944	(gum)	+45.5(m)		K. A. Suri et al. (1976); O. P. Suri et al. (1976)
Cynaustine	(28)	$C_{15}H_{25}NO_4$	283.1784	(gum)	+13.2(e)	picrate, 135–136	Culvenor and Smith (1967)
Dicrotaline	(29)	$C_{14}H_{19}NO_5$	281.1263	134–135	+8.1(c)	HCl, 210–211 (dec)	Brown et al. (1983)
Doronenine	(30)	$C_{18}H_{25}NO_5$	335.1733	124–127	+123.4(e)		Roeder et al. (1980a)
Doronine	(31)	$C_{21}H_{30}NO_8Cl$	459.6501				Alieva et al. (1976)
Echimidine	(32)	$C_{20}H_{21}NO_7$	397.2100	(gum)	+13.4(e)	picrate, 142–143	Culvenor (1956)
Echinatine	(33)	$C_{15}H_{25}NO_5$	299.1733	(gum)	+15(e)	picrolonate, 214 (dec)	Crowley and Culvenor (1959)
7-Acetyl	(34)	$C_{17}H_{27}NO_6$	341.1838				O. P. Suri et al. (1975)
Echiumine	(35)	$C_{20}H_{31}NO_6$	381.2151	99–100	+14.4(e)	picrate, 131–132	Culvenor (1956)
Emiline	(36)	$C_{19}H_{27}NO_6$	365.1838	104–105			Kohlmuenzer et al. (1971)
Erucifoline	(37)	$C_{18}H_{23}NO_6$	349.1525	195–197		monoacetyl, 127–129	Sedmera et al. (1972)
Europine (base G)	(38)	$C_{16}H_{27}NO_6$	329.1838	177–178	+10.9(e)		Culvenor (1954)
Floricaline	(39)	$C_{23}H_{33}NO_{10}$	483.2104	195–196	+74.3(c)		Cava et al. (1968)
Floridanine	(40)	$C_{21}H_{31}NO_9$	441.1999	100–103	+66.5(c)		Cava et al. (1968)
Florosenine	(41)	$C_{21}H_{29}NO_8$	423.1893	213–214	+31.9(c)		Cava et al. (1968)
Fulvine	(42)	$C_{16}H_{23}NO_5$	309.1576		-50.8(c)	picrate, 205 (dec); HCl, 285 (dec)	Schoental (1963); Culvenor and Smith (1963)
Globiferine	(120)	$C_{18}H_{27}NO_7$	369.1787	126–129	-8.6(c)		Brown et al. (1984)

(continued)

TABLE 4.2 (*Continued*)

Alkaloid	Structure	Formula	Molecular weight	m.p. (°C)	$[\alpha]_D$ (°)[a]	Derivatives with m.p. (°C)	References
Grahamine	(43)	$C_{21}H_{31}NO_7$	409.2100	163	+100(e)	methiodide, 240–241 (dec)	Atal and Sawhney (1973)
Grantaline	(44)	$C_{18}H_{25}NO_6$	351.1682	219.5–220	+101(e) +33(c)*[b]		Smith and Culvenor (1984); Brown et al. (1984)*
Grantianine	(45)	$C_{18}H_{23}NO_7$	365.1475	209–209.5 (223–224*)	+44(c) +63.8(e)	picrate, 225–228 (dec); HCl, 221–222; methiodide, 242–243	Adams et al. (1942a); Smith and Culvenor (1984); Brown et al. (1984)*
Gynuramine	(124)	$C_{18}H_{25}NO_6$	351.1682	202–204	−16(c)		Wiedenfeld (1982)
Acetyl	(125)	$C_{20}H_{27}NO_7$	393.1787	153–155	−33(c)		Wiedenfeld (1982)
Heleurine (base C)	(46)	$C_{16}H_{27}NO_4$	297.1940	67–68	−12(e)		Culvenor (1954)
Heliosupine	(47)	$C_{20}H_{31}NO_7$	397.2100	(gum)	−4.3(e)	picrate (monohydrate, from H_2O), 103–106	Crowley and Culvenor (1959)
Acetyl	(48)	$C_{22}H_{33}NO_8$	439.2206				Pedersen (1970)
Heliotrine	(49)	$C_{16}H_{27}NO_5$	313.1889	128	+17.6(e); +63.8(c)	methiodide, 108–110	Culvenor et al. (1954)
7-Angelyl	(50)	$C_{21}H_{33}NO_6$	395.2308				O. P. Suri et al. (1975)
Heterophylline	(51)	$C_{23}H_{35}NO_8$	453.2363	190			Edgar et al. (1980)
Incanine	(52)	$C_{18}H_{27}NO_5$	337.1889	97	−38.8(e)	picrate, 246 (dec); HCl, 199	Yunosov and Plekhanova (1953, 1959)

Name	No.	Formula	Mass	mp	[α]	Derivatives	Reference
Indicine	(53)	$C_{15}H_{25}NO_5$	299.1733	97–98	+23(e)	picrate, 90 (from benzene); 66 (hydrate); HCl, 131–132; MeI, 159–160	Mattocks et al. (1961)
Acetyl Integerrimine	(54)	$C_{17}H_{27}NO_6$	341.1838	(gum)	−14.8(e)		Mattocks (1967a)
	(55)	$C_{18}H_{25}NO_5$	335.1733	172	+4.3(m); −19.2(c)	picrate, 213; MeI, 240; nitrate, 204	Rodriguez and Gonzalez (1971); Gonzalez and Calero (1958)
Intermedine	(56)	$C_{15}H_{25}NO_5$	299.1733	140–142	+9.8(e)	picrate, 147	Frahn et al. (1980); Culvenor and Smith (1966a)
7-Acetyl	(57)	$C_{17}H_{27}NO_6$	341.1838	135–136	+43.5(e)	picrate, 204–205	Culvenor et al. (1980b)
Isocromadurine	(58)	$C_{16}H_{23}NO_5$	309.1576	173	−4.8(e)		Rao et al. (1975c)
Isoline	(59)	$C_{20}H_{29}NO_7$	395.1944				Coucourakis and Gordon-Gray (1970)
Jacobine	(60)	$C_{18}H_{25}NO_6$	351.1682	228	−40(c)		Bradbury and Culvenor (1954)
Jacoline	(61)	$C_{18}H_{27}NO_7$	369.1787	221	+48(c)		Bradbury and Culvenor (1954)
Jaconine	(62)	$C_{18}H_{26}NO_6Cl$	387.629	146–147	+52(e)	HCl, 207 (efferv.)	Bradbury and Culvenor (1954)
Jacozine	(63)	$C_{18}H_{23}NO_6$	349.1525	228	−140(c)		Bradbury and Culvenor (1954)
Junceine	(64)	$C_{18}H_{27}NO_7$	369.1787	191–192	−3(c)		Adams and Gianturco (1956b)

(continued)

97

TABLE 4.2 (*Continued*)

Alkaloid	Structure	Formula	Molecular weight	m.p. (°C)	$[\alpha]_D$ (°)^a	Derivatives with m.p. (°C)	References
Lasiocarpine	(65)	$C_{21}H_{33}NO_7$	411.2257	96.5–97	−3(e); +0.9(c)		Culvenor et al. (1954)
Acetyl Latifoline	(66) (67)	$C_{23}H_{35}NO_8$ $C_{20}H_{27}NO_7$	453.2363 393.1787	(gum) 102–103	−0.9(e) +57(e)		Culvenor et al. (1975b) Crowley and Culvenor (1962)
Ligularidine Ligularine Ligularizine	(68) (69) (126)	$C_{21}H_{29}NO_7$ $C_{23}H_{31}NO_9$ $C_{21}H_{29}NO_8$	407.1944 465.1999 423.1893	196 (gum)	−49.8(e) −34(c)	picrate, 210–211; $[\alpha]_d$ −24.5°(c)	Hikichi et al. (1979) Klasek et al. (1971) Asada and Furuya (1984)
Lycopsamine	(70)	$C_{15}H_{25}NO_5$	299.1733	132–134	+5.7(e)		Culvenor and Smith (1966a); Frahn et al. (1980)
7-Acetyl Madurensine (*trans*)	(71) (72)	$C_{17}H_{27}NO_6$ $C_{18}H_{25}NO_6$	341.1838 351.1682	175–176	+55.7(e)		Culvenor et al. (1980b) Culvenor and Smith (1972)
7-Acetyl (*trans*)	(73a)	$C_{20}H_{27}NO_7$	393.1787	179–180	+68(e)		Culvenor and Smith (1972)
7-Acetyl (*cis*)	(73b)	$C_{20}H_{27}NO_7$	393.1787	156–157	+102.5(e)		Culvenor and Smith (1972)
Monocrotaline	(74)	$C_{16}H_{23}NO_6$	325.1525	198 (202–203)	−54.7(c); −15(e)	picrate, 237–238; HCl, 184 (dec); MeI, 205–206 (dec)	Culvenor and Smith (1957a); Adams and Rogers (1939)
Monocrotalinine	(75)	$C_{18}H_{25}NO_6$	351.1682	160–161 (dec)	+125(e)		Rajagopalan and Batra (1977a)

Name	No.	Formula	Mass	mp (°C)	[α]	Derivative mp (°C)	Reference
Neoligularidine	(127)	$C_{21}H_{29}NO_7$	407.1944	117–119	−58(c)		Asada and Furuya (1984)
Neosenkirkine	(76)	$C_{19}H_{27}NO_6$	365.1838	216	+26.8(c)		Asada and Furuya (1982)
Nilgirine	(77)	$C_{17}H_{23}NO_5$	321.1576	131	+31.5(e)	picrolonate, 208 (dec); HCl, 232–233	Sawhney and Atal (1972); Atal et al. (1968)
Onetine	(78)	$C_{19}H_{29}NO_8$	399.1893	192–193	+73(c)		Bull et al. (1968)
Otosenine	(79)	$C_{19}H_{27}NO_7$	381.1787	221	+20.8(c)		Cava et al. (1968)
Parsonsine	(80)	$C_{22}H_{33}NO_8$	439.2206	158 (from benzene); 198 (from toluene)	+19.8(m)		Edgar et al. (1980)
Petasitenine	(81)	$C_{19}H_{27}NO_7$	381.1787	129–130	+44(e)	picrate, 223–225; HCl, 212 (dec)	Yamada et al. (1976b)
Acetyl	(82)	$C_{21}H_{29}NO_8$	423.1893	(amorph.)	+49(e)		Yamada et al. (1976b)
Retroisosenine	(83)	$C_{18}H_{25}NO_5$	335.1733	127	+118(c)		Nghia et al. (1976)
Retronecine dibenzoate	(84)	$C_{22}H_{21}NO_4$	363.1470	(oil)	−8.4(e)	picrate, 136; HCl, 162	Mattocks (1969b)
Retrorsine	(85)	$C_{18}H_{25}NO_6$	351.1682	214–215 (dec)	−17.6(e)	picrate, 195–197; MeI, 256 (dec)	Manske (1931); Barger et al. (1935)
Retusamine	(86)	$C_{19}H_{25}NO_7$	379.1631	174	+12(e)		Culvenor and Smith (1957a); Culvenor et al. (1967b)
Riddelliine	(87)	$C_{18}H_{23}NO_6$	349.1525	195–196	−109(c)	HCl, 225–226 (dec); MeI, 260–262 (dec)	Adams et al. (1942b); Adams and van Duuren (1953c)
Rinderine	(88)	$C_{15}H_{25}NO_5$	299.1733	100–101	+24.6(e)		Locock et al. (1966)

(continued)

TABLE 4.2 (*Continued*)

Alkaloid	Structure	Formula	Molecular weight	m.p. (°C)	$[\alpha]_D$ (°)[a]	Derivatives with m.p. (°C)	References
Sceleratine	(89)	$C_{18}H_{27}NO_7$	369.1787	178	+54(e)		de Waal and Pretorius (1941)
Sceleratinyl chloride	(90)	$C_{18}H_{26}NO_6Cl$	387.629	196	+32.4	picrate, 213–215	Gordon-Gray (1967)
Scorpioidine	(121)	$C_{20}H_{31}NO_6$	381.2151	(oil)	–5(m)		Resch et al. (1982)
7-Acetyl	(122)	$C_{22}H_{33}NO_7$	423.2257	(oil)	–8(m)		Resch et al. (1982)
Sencalenine	(128)	$C_{18}H_{25}NO_5$	335.1733	(oil)	–8(e)		Roeder et al. (1984a)
Senecicannabine	(91)	$C_{18}H_{23}NO_7$	365.1474	198	–8.9(c)		Asada et al. (1982a)
Senecionine	(92)	$C_{18}H_{25}NO_5$	335.1733	232 (up to 245); (dec)	–56(c)	picrate, 191; MeI, 249	Barger and Blackie (1936)
7-Senecioyl retronecine	(114)	$C_{13}H_{19}NO_3$	237.1365	(oil)	–2(e)		Roeder et al. (1984)
7-Senecioyl-9-sarracinyl retronecine	(115)	$C_{18}H_{25}NO_5$	335.1733	(oil)	+4(e)		Roeder et al. (1984)
Seneciphylline (jacodine)	(93)	$C_{18}H_{23}NO_5$	333.1576	217	–139(c)		Bradbury and Culvenor (1954)
Senecivermine	(94)	$C_{18}H_{25}NO_5$	335.1733	105–107 (dec)	–34.9(e)		Roeder et al. (1979)
Senkirkine	(95)	$C_{19}H_{27}NO_6$	365.1838	196.5–197.5	–16(m); –2(c)	picrate, 225; picrolonate, 138; MeI, 192–194 (dec)	Briggs et al. (1965)

Acetyl	(96)	$C_{21}H_{29}NO_7$	407.1944	195–196	−34(m)	picrate, 208–209; picrolonate, 222	Briggs et al. (1965); Atal and Sawhney (1973)
Hydroxy	(97)	$C_{19}H_{27}NO_7$	381.1787	124–125	+5.3(e)	picrate, 242 (dec); picrolonate, 201 (dec); HCl, 229–230 (dec)	Crout (1972)
Sincamidine	(98)	$C_{16}H_{27}NO_5$	313.1889	(gum)			Culvenor and Smith (1966a)
Spartioidine	(99)	$C_{18}H_{23}NO_5$	333.1576	178	−83.5(e)		Bull et al. (1968)
Spectabiline	(100)	$C_{18}H_{25}NO_7$	367.1631	186	+121(c); +143(e)		Culvenor and Smith (1957b, 1958)
Spiracine	(101)	$C_{23}H_{35}NO_{10}$	485.2261				Edgar et al. (1980)
Spiraline	(102)	$C_{22}H_{33}N_9$	455.2155				Edgar et al. (1980)
Spiranine	(103)	$C_{23}H_{35}NO_9$	469.2312				Edgar et al. (1980)
Supinine (base F)	(104)	$C_{15}H_{25}NO_4$	283.1783	148–149	−13(e)	picrate, 144	Culvenor (1954); O'Kelly and Sargeant (1961)
Swazine	(105)	$C_{18}H_{23}NO_6$	349.1525	165	−104(e)	methiodide, 202–205 (dec)	Gordon-Gray and Wells (1974)
Symlandine	(106)	$C_{20}H_{31}NO_6$	381.2151				Culvenor et al. (1980b)

(continued)

TABLE 4.2 (*Continued*)

Alkaloid	Structure	Formula	Molecular weight	m.p. (°C)	$[\alpha]_D$ (°)[a]	Derivatives with m.p. (°C)	References
Symphytine	(107)	$C_{20}H_{31}NO_6$	381.2151				Furuya and Hikichi (1971); Culvenor et al. (1980b)
Syneilesine	(108)	$C_{19}H_{29}NO_7$	383.1944	195			Hikichi and Furuya (1976)
Acetyl	(109)	$C_{21}H_{31}NO_8$	425.2050	(oil)			Hikichi and Furuya (1976)
Trichodesmine	(110)	$C_{18}H_{27}NO_6$	353.1838	160–161	+38(e)	picrate, 228; HCl, 205; MeI, 202	Adams and Gianturco (1956b); Yunusov and Plekhanova (1959)
Uplandicine	(111)	$C_{17}H_{27}NO_7$	357.1787	(gum)	+0.1(e)		Culvenor et al. (1980b)
Usaramine	(112)	$C_{18}H_{25}NO_6$	351.1682	183	+7.1(e)	picrate, 231–232 (dec); picrolonate, 148–150 (dec); HCl, 246 (dec)	Culvenor and Smith (1966b); Mattocks (1968e); Sawhney et al. (1967)
Yamataimine	(113)	$C_{18}H_{27}NO_5$	337.1889	181–182	+63.6(e)		Hikichi et al. (1978)

[a] Solvents: c, chloroform; e, ethanol; m, methanol; pyr, pyridine.
[b] Data marked with asterisk are from reference similarly marked.

TABLE 4.3 Physical Properties of Saturated Pyrrolizidine Alkaloids

Alkaloid	Structure	Formula	Molecular weight	m.p. (°C)	$[\alpha]_D$ (°)[a]	Derivatives with m.p. (°C)	References
Angularine	(130)	$C_{18}H_{25}NO_6$	351.1682	200–201	−98(e)		Porter and Geissman (1962)
Bulgarsenine	(131)	$C_{18}H_{27}NO_5$	337.1889	115	−54(c)		Nghia et al. (1976)
Coromandalin	(133)	$C_{15}H_{27}NO_4$	285.1940	(gum)	−6.9(e)		Subramanian et al. (1980)
Croalbidine	(134)	$C_{18}H_{29}NO_7$	371.1944	208–209		picrate, 224–225; HCl, 154–155	Sawhney and Atal (1973)
Crocandine	(135)	$C_{16}H_{25}NO_5$	311.1733	244–246	+130(m)		Siddiqi et al. (1979b)
Cropodine	(136)	$C_{16}H_{25}NO_6$	327.1682	226–228	+70(m)		Haksar et al. (1982)
Curassavine	(137)	$C_{16}H_{29}NO_4$	299.2096	(gum)	+0.9(e)		Subramanian et al. (1980)
Cynaustraline	(138)	$C_{15}H_{27}NO_4$	285.1940	(gum)	+48(e)	picrolonate, 149–150	Culvenor and Smith (1967)
Farfugine	(139)	$C_{13}H_{21}NO_3$	239.1521	(gum)	+23(e)	picrate, 157	Niwa et al. (1983b)
Fuchsisenecionine	(140)	$C_{13}H_{21}NO_3$	239.1521		−120	aurichloride, 156	Roeder and Wiedenfeld (1977)

(continued)

TABLE 4.3 (*Continued*)

Alkaloid	Structure	Formula	Molecular weight	m.p. (°C)	$[\alpha]_D$ (°)[a]	Derivatives with m.p. (°C)	References
Hastacine	(141)	$C_{18}H_{27}NO_5$	337.1889	171	−72.3		Konovalov and Men'shikov (1945); Culvenor et al. (1968)
Helifoline	(142)	$C_{13}H_{21}NO_4$	255.1490	131–132	+25.4(e)		Mohanraj et al. (1981)
Heliovicine	(143)	$C_{15}H_{27}NO_4$	285.1940	(gum)	−2.7(e)		Subramanian et al. (1980)
Hygrophylline	(144)	$C_{18}H_{27}NO_6$	353.1838	173–174	−67.3(e)		Schlosser and Warren (1965)
Isocrocandine	(145)	$C_{16}H_{25}NO_5$	311.1733	172–174	+36(m)		Siddiqi et al. (1979b)
Ligularinine	(145a)	$C_{18}H_{27}NO_5$	337.1889	103–104	−88(c)		Asada and Furuya (1984)
Lindelofine	(147)	$C_{15}H_{27}NO_4$	285.1940	107	+50(e)		Warren (1955)
Macrophylline	(148)	$C_{13}H_{21}NO_3$	239.1521	42–44	+34.5(e)		Warren (1966)
Macrotomine	(149)	$C_{15}H_{27}NO_5$	301.1889	95–97	−6.9(e)		Warren (1955)
Nemorensine	(150)	$C_{18}H_{27}NO_5$	337.1889	132–133	−59(c)		Klasek et al. (1973a)
Neoplatyphylline	(151)	$C_{18}H_{27}NO_5$	337.1889	131–133 (129)	+2(c); −4(e)		Culvenor et al. (1968); Roeder et al. (1982b)

Name	No.	Formula	Mass	mp	[α]	Derivative, mp	References
Petasinine	(152)	$C_{13}H_{21}NO_3$	239.1521	amorph.	+16(e)	phenylurethane, 168–171	Yamada et al. (1978)
Platyphylline	(153)	$C_{18}H_{27}NO_5$	337.1889	129	−56(c); −59(e)	percholate, 226	Richardson and Warren (1943); Warren (1955)
Procerine	(154)	$C_{13}H_{21}NO_5$	271.1419	238–239 (dec)			Jovceva et al. (1978)
Retusine		$C_{16}H_{25}NO_5$	311.1733	174–175	+16.2(e)		Culvenor and Smith (1957a)
Rosmarinine	(155)	$C_{18}H_{27}NO_6$	353.1838	209	−91.5(m)	picrolonate, 176	Richardson and Warren (1943)
Sarracine	(156)	$C_{18}H_{27}NO_5$	337.1899	45–46	−121(e)		Culvenor and Geissman (1961b)
Strigosine	(157)	$C_{14}H_{25}NO_4$	271.1783	(gum)	−19.3(e)	picrate, 141; MeI, 135–136	Mattocks (1964b)
Trachelanthamine	(158)	$C_{15}H_{27}NO_4$	285.1940	92–93	−18.1(H_2O)		Men'shikov (1946); Warren (1955)
Turneforcine	(159)	$C_{13}H_{21}NO_3$	239.1521	oil			Warren (1955)
Viridoflorine	(160)	$C_{15}H_{27}NO_4$	285.1940	102.5–103.5	−11.7(e)		Men'shikov (1948); Warren (1955)

[a] Solvents: c, chloroform; e, ethanol; m, methanol.

TABLE 4.4 Physical Properties of Pyrrolizidine *N*-Oxides

Compound	Formula[a]	m.p. (°C)[b,c]	$[\alpha]_D$ (°)[c]	Notes[d]	References
N-Oxides of necines					
Heliotridine *N*-oxide	$C_8H_{13}NO_3$	201 (dec) (ac + m)		P	Culvenor *et al.* (1954)
Platynecine *N*-oxide	$C_8H_{15}NO_3$	217–218 (dec) (ac + e)		P; contains H_2O or EtOH; picrate, m.p. 160–162°C	Koekemoer and Warren (1951)
Retronecine *N*-oxide	$C_8H_{13}NO_3$	up to 215 (dec) (ac + m)	+25(H_2O); +42(e)	E or P; picrate, m.p. 145°C	Leisegang and Warren (1949); Culvenor and Smith (1957a)
Alkaloid *N*-oxides					
Anadoline (*N*-oxide of retronecine tigloyl-trachelanthate)	$C_{20}H_{29}NO_6$	186 (dec)		E	Culvenor *et al.* (1975a)
Curassavine *N*-oxide	$C_{16}H_{29}NO_5$	186–188 (c + ac)	−6.6(e)	E; contaminated with coromandalin and heliovicine *N*-oxides.	Subramanian *et al.* (1980)
Echimidine *N*-oxide	$C_{20}H_{31}NO_8$	165 (dec) (ac + m)		P	Culvenor (1956)
Europine *N*-oxide	$C_{16}H_{27}NO_7$	170 (dec) (ac)	+27	E	Culvenor (1954); Zalkow *et al.* (1978)
Heliotrine *N*-oxide	$C_{16}H_{27}NO_6$	171–172 (ac + m)	+26.6(e)	E or P	Culvenor (1954); Culvenor *et al.* (1954)
Indicine *N*-oxide	$C_{15}H_{25}NO_6$	119–120 (ipa + EtOAc) 130–131 (ac + m)	+34.8(e)	E; solvent-free P; containing EtOH (not MeOH as in ref.)	Kugelman *et al.* (1976) Mattocks *et al.* (1961)
Lasiocarpine *N*-oxide	$C_2H_{33}NO_8$	134–135 (dec) (EtOAc or ac + m)	+13.1(e)	E or P; may contain solvent (hard to remove without dec) resulting in lower m.p.; soluble in chloroform	Culvenor (1954); Culvenor *et al.* (1954)

Compound	Formula[a]	m.p. (°C)[b]	[α][c]	Notes[d]	Reference
Monocrotaline N-oxide	$C_{16}H_{23}NO_7$	192–196 (dec) (ac + m)		P; analysis fits methanol solvate better than $5H_2O$ given in ref; picrate, m.p. 165–175°C (dec)	Culvenor and Smith (1957a)
Nemorensine N-oxide (oxynemorensine)	$C_{18}H_{27}NO_6$	160–163 (EtOAc)	−35(c)	E	Klasek et al. (1980)
Platyphylline N-oxide	$C_{18}H_{27}NO_6$	180–184 (dec) (ac + e)	−59(H_2O)	P; very hygroscopic	Koekemoer and Warren (1951)
Retrorsine N-oxide (isatidine)	$C_{18}H_{25}NO_7$	141(e)	−8(H_2O)	E or P	Christie et al. (1949)
Rosmarinine N-oxide	$C_{18}H_{27}NO_7$	amorph. (dec 169)		P; hygroscopic	Koekemoer and Warren (1951)
Sarracine N-oxide	$C_{18}H_{27}NO_6$	125–126 (ac)	−94(e)	E; crystals with $1H_2O$; dried sample has m.p. 140–141°C but reabsorbs H_2O from air in 10 min; soluble in chloroform	Culvenor and Geissman (1961b)
Senecionine N-oxide	$C_{18}H_{25}NO_6$	141–142 (dec) (c + pet)	−22(c)	E; solvate with ⅔$CHCl_3$; hygroscopic	Kupchan and Suffness (1967)
		125 (ac + m)		P; solvate with $2H_2O$	Culvenor et al. (1970b)
Supinine N-oxide	$C_{15}H_{25}NO_5$	167–168 (ac)		P; hygroscopic	Crowley and Culvenor (1959)

[a] Formula is that of the solvent-free N-oxide. Crystalline N-oxides frequently contain solvent, which may be in nonstoichiometric amounts and may not be the solvent last used for recrystallization. Attempts to remove this solvent sometimes result in decomposition.

[b] Recrystallization solvents in parentheses.

[c] Solvents: ac, acetone; c, chloroform; e, ethanol; ipa, isopropanol; m, methanol; pet, light petroleum.

[d] E, Extracted from plant material; P, prepared chemically from base.

TABLE 4.5 Physical Properties of Necic Acids

Acid	Structure	Formula	Molecular weight	m.p. (°C)[a]	$[\alpha]_D$ (°)[b]	References
Angelic (cis-2,3-dimethylacrylic)		$C_5H_8O_2$	100.0524	45	−208(e)	Klasek et al. (1969, 1970)
Clivonecic (lactone)	(187)	$C_{10}H_{12}O_4$	196.0736	142–144 (bnz)		
Crispatic	(193)	$C_8H_{14}O_5$	190.0841	133–134 (bnz)	0	Culvenor and Smith (1963)
Crobarbatic (lactone)		$C_7H_{10}O_4$	158.0579	177–178		Puri et al. (1973)
Cromaduric		$C_8H_{14}O_5$	190.0841	138–139 (eth + pet)	−14.5(m)	Rao et al. (1975b)
Dicrotalic		$C_6H_{10}O_5$	162.0528	108–109		Brown et al. (1983)
Echimidinic		$C_7H_{14}O_5$	178.0841	(glass)	+17.5(e)	Culvenor (1956)
Fulvinic	(192)	$C_8H_{14}O_5$	190.0841	113–114 (bnz)	0	Culvenor and Smith (1963)
Heliotric (heliotrinic)	(189)	$C_8H_{16}O_4$	176.1049	94–95	−12(H_2O)	Culvenor et al. (1954)
Hygrophyllinecic (monolactone)		$C_{10}H_{14}O_5$	214.0841	181 (bnz)	−187.9(e)	Schlosser and Warren (1965)
Hygrophyllinecic (dilactone)		$C_{10}H_{12}O_4$	196.0736	103–105 (subl.)	−97.6(e)	Schlosser and Warren (1965)
Incanic		$C_{10}H_{18}O_4$	202.1205	161–163		Warren (1966)
Isatinecic (cis)	(196)	$C_{10}H_{16}O_6$	232.0947	148 (EtOAc)	0	Christie et al. (1949)
Integerrinecic (trans-senecic)	(197)	$C_{10}H_{16}O_5$	216.0998	150 (H_2O)	+18(e)	Kropman and Warren (1950)

Name	(No.)	Formula	Mass	mp (solvent)	$[\alpha]$	Reference
Integerrinecic (lactone)		$C_{10}H_{14}O_4$	198.0892	153 (bnz)	+39(e)	Culvenor and Geissman (1961a)
Isocromaduric		$C_8H_{14}O_5$	190.0841	129–130 (eth + pet)	+14.9(m)	Rao et al. (1975c)
Isolinecic		$C_{10}H_{18}O_6$	234.1103	140–142 (ac + pet)	+65.9(e)	Coucourakis and Gordon-Gray (1970)
Isolinecic (dilactone)		$C_{10}H_{14}O_4$	198.0892	73 (subl.)		Coucourakis and Gordon-Gray (1970)
Jaconecic (lactone)		$C_{10}H_{16}O_6$	232.0947	182	+31.7(e)	Barger and Blackie (1937)
Junceic (lactone)		$C_{10}H_{16}O_6$	232.0947	180–182		Adams and Gianturco (1956d)
Lasiocarpic	(191)	$C_8H_{16}O_5$	192.0998	96–97	+8.4(e)	Culvenor et al. (1954)
Monocrotalic (lactone)	(194)	$C_8H_{12}O_5$	188.0685	182	−5.3(H_2O)	Adams and Rogers (1939)
Nemorensic (cyclic form)		$C_{10}H_{16}O_5$	216.0998	176–180 (ac + pet)	−59.6(e)	Roeder et al. (1980a)
Nilgiric		$C_9H_{14}O_5$	202.0841	126–127		Atal and Sawhney (1973)
Retronecic (trans-isatinecic)	(198)	$C_{10}H_{16}O_6$	232.0947	181	−11.4(H_2O)	Christie et al. (1949)
Retronecic (lactone)		$C_{10}H_{14}O_5$	214.0841	186 (EtOAc)		Christie et al. (1949)
Retusaminic (lactone)		$C_{10}H_{14}O_6$	230.0790	164	−45(e)	Culvenor et al. (1967b)
Retusanecic (α)		$C_8H_{12}O_4$	172.0736	130–131 (EtOAc + pet)	+3.3(e)	Culvenor and Smith (1957a)
Retusanecic (β)		$C_8H_{12}O_4$	172.0736	118 (eth + pet)	−60(e)	Culvenor and Smith (1957a)

(continued)

TABLE 4.5 (*Continued*)

Acid	Structure	Formula	Molecular weight	m.p. (°C)[a]	$[\alpha]_D$ (°)[b]	References
Riddelliic	(200)	$C_{10}H_{14}O_6$	230.0790	102–103 (hydrated: 62)	−2.65(e)	Adams *et al.* (1942b)
Sarracinic (sarracinecic)		$C_5H_8O_3$	116.0473	57–58 (bnz)		Culvenor and Geissman (1961b)
Sceleranecic (dilactone)		$C_{10}H_{14}O_5$	214.0841	156	−9.3(H_2O)	de Waal and Pretorius (1941)
Sceleratinic (dilactone)		$C_{10}H_{14}O_4Cl$	233.5422	207		Gordon-Gray (1967)
Senecic (*cis*)	(195)	$C_{10}H_{16}O_5$	216.0998	151	+11.8(e)	Richardson and Warren (1943)
Senecic (lactone) (*cis*)[c]		$C_{10}H_{14}O_4$	198.0892	129–130 (bnz–pet)		Culvenor and Geissman (1961a)
Senecioic (3,3-dimethylacrylic)		$C_5H_8O_2$	100.0524	69–70		Culvenor (1954)
Tiglic (*trans*-2,3-dimethylacrylic)		$C_5H_8O_2$	100.0524	64		
(+)-Trachelanthic	(188)	$C_7H_{14}O_4$	162.0892	93–94	+2.3(e)	Culvenor (1954)
Trichodesmic		$C_{10}H_{16}O_5$	216.0998	219–220 (ac + pet) 209–211 (eth + pet)		Devlin and Robins (1984); Adams and Gianturco (1956c)
(−)-Viridofloric	(190)	$C_7H_{14}O_4$	162.0892	142 (bnz)[d]	−1.3(H_2O)	Crowley and Culvenor (1959)

[a] Crystallization solvents (in parentheses): ac, acetone; bnz, benzene; eth, diethyl ether; pet, light petroleum.
[b] Solvents: e, ethanol; m, methanol.
[c] Treatment of senecic acid with mineral acid gives the *trans*-lactone (integerrinecic lactone).
[d] Can be considerably lowered by minute amounts of impurities.

TABLE 4.6 Ionization Constants of Some Necine Bases
and PAs[a]

Compound	Structure	Type of ester[b]	$pK_a{}^c$
Supinidine	(182)	N	8.86
Heliotridine	(184)	N	8.45
Retronecine	(183)	N	8.38
Supinine	(104)	M	8.44
Heliotrine	(49)	M	7.82
Monocrotaline	(74)	CD	6.93
Senecionine	(92)	CD	6.73
Lasiocarpine	(65)	D	6.55

[a] Data from Culvenor and Willette (1966).

[b] N, Non-esterified amino alcohol; M, monoester; CD, macrocyclic diester; D, diester.

[c] In 80% methyl cellosolve.

aqueous solution with organic solvents such as chloroform. Necine bases, especially those with more than one OH group, are also very water soluble.

The partition coefficient is particularly sensitive to pH changes near to the pK_a of the alkaloid; for many PAs the latter is well below 9, hence to be of relevance to biological effects the partition coefficient should be measured at physiological pH. Partition coefficients of some natural and semisynthetic PAs in the system octan-1-ol–aqueous buffer, pH 7.4, have been measured by Mattocks and Bird (1983a) (see Chapter 6, Table 6.4). A practical application of partition has been the separation of eight PAs by counter-current distribution (Culvenor et al., 1980b).

C. Electrophoretic Mobility

Frahn (1969) made a detailed study of the paper electrophoresis of PAs and related compounds in seven different electrolytes. In non-complexing electrolytes, PAs migrate as cations: at pH 4.6, when they are fully protonated, their mobilities are generally inversely related to molecular weight. However, N-oxides (of supinine, heliotrine and lasiocarpine), being incompletely ionized at this pH, migrate much more slowly than their parent bases. Mobilities are also related to pH: thus to an extent dependent on their pK_a, PAs migrate more slowly as the PH of the electrolyte is raised and ionization is reduced.

In buffer solutions containing borate or arsenite, some PAs containing vicinal glycol groups form complex anions, to an extent determined by their stereochemistry. These alkaloids show lowered mobility or even a reversal of the direction of migration in such electrolytes.

Electrophoresis thus provides a useful means of characterizing and separating PAs.

D. Crystallography and Conformation

X-Ray crystallography has been used to determine the sterochemistry of a number of PAs (Table 4.7). Spectroscopy, including infrared (IR), nuclear magnetic resonance (NMR) and circular dichroism, has been useful in conformation studies: for reviews see Bull et al. (1968) and Robins (1979, 1982a).

The five-membered rings in retronecine and heliotridine-based PAs are inclined together at angles of 115–130°, and the saturated (pyrrolidine) ring is puckered; for instance endo-puckering is seen in heliotridine-based PAs and exo-puckering in retronecine-based macrocyclic ester PAs such as fulvine (Culvenor and Woods, 1965; Wodak, 1975).

The macrocyclic diesters of retronecine are relatively inflexible structures; in all the 12-membered macrocycles the ester carbonyl groups are roughly anti-parallel, whereas in the 11-membered macrocyclic diesters studied, with the exception of trichodesmine, the carbonyls are syn-parallel.

Culvenor (1966) defined the conformation of senecionine on the basis of NMR studies; recently X-ray analyses have been made of its reactive pyrrolic metabolite, dehydrosenecionine (Mackay et al., 1983) and that of monocrotaline, dehydromonocrotaline (Mackay et al., 1984). X-Ray studies of otonecine (186) esters show that the transannular interaction between N and C-8 in the 12-membered macrocycles clivorine and senkirkine is much weaker, and the 8-carbonyl groups is more ketonic in character than in the 11-membered macrocyclic diester retusamine (Birnbaum, 1974; Perez-Salazar et al., 1977).

(186)   (49)

The monoester heliotrine (49) may have a conformation resembling that of a macrocyclic diester in some solvents (e.g. chloroform), owing to H-bonding between the 7-hydroxyl and the methoxy group in the acid moety (Bull et al., 1968); however, X-ray analysis shows that this is not so in crystalline heliotrine (Zalkow et al., 1979).

TABLE 4.7 Crystallographic Studies of Pyrrolizidine Alkaloids
and Related Compounds

Compound	Reference
Axillarine (HBr–EtOH solvate)	Stoeckli-Evans and Crout (1976)
Clivorine	Birnbaum (1972)
Dehydromonocrotaline	Mackay et al. (1984)
Dehydrosenecionine	Mackay et al. (1983)
1,2-Didehydrocrotalanine (picrate)	Stoeckli-Evans and Robins (1983)
Doronine (+ benzene)	Wong and Roitman (1984)
Fulvine	Sussman and Wodak (1973)
Grantaline	Mackay and Culvenor (1983)
Heliotrine	Wodak (1975);
	Zalkow et al. (1979)
Incanine	Tashkhodzhaev et al. (1979a)
Jacobine (bromohydrin)	Fridrichsons et al. (1963);
	Perez-Salazar (1978);
	Perez-Salazar et al. (1978)
Junceine	Stoeckli-Evans (1982)
Monocrotaline	Stoeckli-Evans (1979a);
	Wang and Hu (1981)
Monocrotaline sulphite (HCl)	Wang (1980a)
Otosenine	Perez-Salazar et al. (1977)
Parsonsine	Eggers and Gainsford (1979)
Retrorsine	Stoeckli-Evans (1979b);
	Coleman et al. (1980)
Retusamine	Wunderlich (1962, 1967)
Retusine	Wang (1980b)
Seneciphylline	Wiedenfeld et al. (1984)
Senkirkine	Birnbaum (1974)
Swazine (methiodide)	Laing and Sommerville (1972)
Trichodesmine	Tashkhodzhaev et al. (1979b)
Yamataimine	Hikichi et al. (1978)

IV. SPECTROSCOPY

A. Ultraviolet

PAs show relatively weak UV absorption, with λ_{max} below 220 nm, and
ultraviolet (UV) spectroscopy has been little used for structural studies. In early
studies UV spectra were used to distinguish between the conjugated cis- and

trans-ethylidene groups in pairs of geometric isomers such as senecionine and integerrimine, the trans isomer (e.g. integerrimine) having a slightly lower λ_{max} than the cis (Adams and van Duuren, 1953b; Kropman and Warren, 1950; Schlosser and Warren, 1965; Mattocks 1968e). Simanek *et al.* (1969) have recorded UV data for 21 PAs (including several isomeric pairs), six necic acids, and two necines (retronecine and otonecine). Gupta *et al.* (1975a) have published UV spectra of retronecine, platynecine, and data for nine semisynthetic esters of these necines.

The dihydropyrrolizine derivatives of PAs have stronger maxima, at slightly higher wavelengths, than their parent alkaloids (Mattocks, 1969a).

B. Circular Dichroism

Culvenor *et al.* (1971b) and Hrbek *et al.* (1972) have measured circular dichroism (CD) spectra for a large number of PAs and their necines and necic acid components, and have provided detailed discussions of the data. Saturated 1-substituted pyrrolizidines such as isoretronecanol show a negative Cotton effect; saturated 1,7-disubstituted derivatives such as platynecine and 1-substituted unsaturated necines such as supinidine show a positive Cotton effect. A larger positive Cotton effect is given by the unsaturated 1,7-disubstituted necines retronecine and heliotridine.

C. Infrared

Apart from the characterization of individual alkaloids by 'fingerprinting', the most useful information to be gained from the IR spectra of PAs is concerned with unsaturation, hydroxyl and ester groups.

Culvenor and Dal Bon (1964) measured carbonyl- and hydroxyl-stretching frequencies of some PAs in Nujol mulls and in CCl_4 and $CHCl_3$ solutions. We have recorded IR spectra of PAs in KBr discs or as liquid films, and in $CHCl_3$ (Table 4.8). PA esters with an α,β-unsaturated acid moiety typically show a conjugated carbonyl-stretching peak at $1710–1720$ cm^{-1}. Saturated ester carbonyls normally absorb at around $1730–1740$ cm^{-1}. Thus, in KBr (or Nujol) spectra, PAs such as retrorsine or lasiocarpine with both types of ester show two carbonyl peaks. However, in chloroform solution usually only one carbonyl peak is seen, in the lower frequency range: this is attributed to a lowering of the frequency of the saturated ester-carbonyl, owing to its hydrogen bonding with an adjacent hydroxyl group (Culvenor and Dal Bon, 1964). The two ester bands in the KBr spectrum of fulvine, which has a saturated acid moiety, appear to be due to an H-bonded and a non-H-bonded carbonyl, respectively. Most PAs with an unsaturated acid moiety show a $C\!=\!\!C$ stretching band between 1640 and 1660 cm^{-1}. Most unsaturated necines and their esters show a weak band between 3025 and 3090 cm^{-1} in KBr spectra, due to C—H stretching at C-2 (Table 4.8).

TABLE 4.8 Infrared Absorption Bands from Some PAs

Compound	Vehicle	IR stretching frequencies (cm^{-1})			
		Ester CO	OH	C=C (acid moiety)	C=C—H[a] (necine moiety)
PAs and derivatives					
Anacrotine	KBr	1715;1730	3540s		
Anacrotine	CHCl$_3$	1725	3525		
Clivorine	CHCl$_3$	1740	3450br		3090
Fulvine	KBr	1720;1740	3400br		3075
Fulvine	CHCl$_3$	1730	3500br		
Heliotrine	KBr	1730	3370		3090
	CHCl$_3$	1720	3460		
Integerrimine	KBr	1710;1730	3400w	1655m	3050
Isosenecionine	KBr	1710;1730	3120w.br	1640	3060
Lasiocarpine	KBr	1710;1730	3520; 3170br	1660w	—
Monocrotaline	KBr	1730	3500s		3060
Monocrotaline	CHCl$_3$	1730	3540		
Monocrotaline N-oxide	KBr	1735	3540; 3300br		
Oxo-retrorsine	KBr	1715br		1640	3070
Platyphylline	KBr	1705;1720	3400w	1640w	
Retrorsine	KBr	1710;1740	3580s	1660	3060
Retrorsine	CHCl$_3$	1720	3520	1660	
Retrorsine N-oxide	KBr	1720;1740	3450; 3150br	1655	
Rosmarinine	KBr	1715;1740	3400br	1650	
Senecionine	KBr	1715;1740	3580w	1660	3060
Senkirkine	KBr	1710;1730	3400br	1655	3025
Supinine	KBr	1750	3350		3060
Semi-synthetic esters of retronecine					
Bis-N-ethylcarbamoyl	film[b]	1710br	3330(NH)		3060
Di-isovaleroyl	film	1740			3070
Disenecioyl (HCl salt)	KBr	1720vs		1640s	3070
7-Tigloyl	film	1710		1650m	3070
Ditigloyl	film	1710vs		1650s	3060
9-Pivaloyl	KBr	1730			3060

[a] Weak band due to C—H stretching at C-2.
[b] Liquid film on NaCl plate.

TABLE 4.9 Infrared Absorption Bands from Some Dehydropyrrolizidine Alkaloids (Pyrrolic Derivatives of PAs) and Dehydronecines

	IR stretching frequencies (cm^{-1})			
Compound	Vehicle	Ester CO	OH	Pyrrolic CH[a]
Dehydroretronecine	$CHCl_3$		3400br	
Dehydroretronecine dimethyl ether	film[b]			3090; 3110
Diacetyl dehydroretronecine	film	1735		3100; 3115
Dehydroheliotrine	film	1725br	3500br	3095; 3125
Dehydrolasiocarpine	film	1710br	3500br	3095; 3125
Dehydromonocrotaline	KBr	1715	3460s; 3520	3100; 3130
Dehydroretrorsine	Kbr	1715;1725	3290br; 3540	3090; 3110
Dehydrosupinidine	film		3350br	

[a] Weak bands due to C—H stretching at C-2 and C-3 in the dihydropyrrolizine moiety.
[b] Liquid film on NaCl plate.

Hydroxyl-stretching absorptions vary from sharp bands above 3500 cm^{-1}, associated with intramolecular hydrogen bonding, to broad bands at lower frequencies due to intermolecular H bonding.

Gupta *et al.* (1975b) have given detailed interpretations of IR spectra of platynecine, retronecine, and some semi-synthetic esters of these necines. Bands at around 610, 750 and 965 cm^{-1} were found to be characteristic of the pyrrolidine ring. An inverse relationship was found between the C=O stretching frequencies in the esters. Birnbaum (1974) examined the ketone carbonyl absorption in otonecine esters, and showed that increasing transannular N · · · CO interaction causes a lowering in the frequency of the carbonyl peak.

Synthanecines (hydroxymethyl-*N*-alkyl-pyrrolidines and -pyrrolines) show a characteristic band at 2790–2800 cm^{-1} not given by corresponding pyrroles or by pyrrolizidines: this is attributed to the 5-methylene group (Mattocks, 1974).

Dehydro-PAs (Table 4.9) show ester carbonyl absorptions similar to their parent alkaloids; they also show a pair of weak bonds in the range 3090–3125 cm^{-1} representing CH-stretching at the pyrrolic C-2 and C-3 positions, and an aromatic band near to 1580 cm^{-1} due to the pyrrolic structure (Mattocks, 1969a).

D. Nuclear Magnetic Resonance

NMR spectrometry has proved valuable in PA structure determination. Most publications on new PAs now include NMR data. The information is now extensive, although widely scattered in the literature, and only a brief summary is possible here.

1. Proton NMR

a. PAs and Necines. For published spectra of some necines and PAs, see Tables 4.10 and 4.11. Culvenor *et al.* (1965) made a detailed study of the NMR spectra of retronecine and heliotridine. Culvenor and Woods (1965) have provided useful tables of chemical shifts and coupling constants for the pyrrolizidine nucleus in a large number of PAs, and Bull *et al.* (1968) have listed the ranges of chemical shifts of methyl and olefinic protons in the acid moiety of many PAs. Some of these data are summarized in Table 4.12. Segall and Dallas (1983) have analyzed the proton NMR spectra of retrorsine, seneciphylline, senecionine and their *N*-oxides. Mohanraj and Herz (1982) have given a detailed discussion of the NMR spectra of some saturated pyrrolizidine monoesters.

The necine H-3, H-5, H-6 and H-7 signals are often difficult to distinguish, especially in 60-MHz spectra, because of multiplicity and overlapping. The H-2 signal in unsaturated PAs is a poorly resolved multiplet usually appearing as a singlet near to δ 6 ppm, well distinguished from other signals; this has been used for the quantitative analysis of unsaturated PAs (Molyneux *et al.*, 1979), and its position can be used to distinguish between 'open' and 11- and 12-membered macrocyclic diester structures.

TABLE 4.10 Published ^{1}H-NMR Spectra
of Necines

Compound	Reference
Croalbinecine	Sawhney *et al.* (1974)
Crotanecine	Atal *et al.* (1966a)
Curassenecine	Mohanraj and Herz (1982)
Hastanecine	Aasen *et al.* (1969)
Heliotridine	Culvenor *et al.* (1965)
Isoretronecanol	Mohanraj and Herz (1982)
Macronecine	Aasen and Culvenor (1969b); Aasen *et al.* (1969)
Otonecine (HCl)	Atal *et al.* (1967)
Platynecine	Aasen *et al.* (1969)
Retronecine	Culvenor *et al.* (1965)
Supinidine	Mohanraj and Herz (1982)
Trachelanthamidine	Mohanraj and Herz (1982)
Turneforcidine	Aasen *et al.* (1969)

TABLE 4.11 Published ^{1}H-NMR Spectra
of Some PAs

Compound	Reference
Anacrotine	Atal *et al.* (1966a)
Clivorine	Klasek *et al.* (1970)
Crispatine	Culvenor and Smith (1963)
Crosemperine	Atal *et al.* (1967)
Curassavine	Mohanraj and Herz (1982)
Fulvine	Culvenor and Smith (1963)
Intermedine	Culvenor and Smith (1966a)
Ligudentine	Klasek *et al.* (1971)
Ligularine	Klasek *et al.* (1971)
Lycopsamine	Culvenor and Smith (1966a)
Madurensine	Atal *et al.* (1966a)
Monocrotaline	Culvenor and Smith (1963)
Nilgirine	Atal *et al.* (1968)
Sarracine *N*-oxide	Culvenor and Geissman (1961b)
Senecionine	Atal *et al.* (1966a)
Senkirkine	Briggs *et al.* (1965)
Swazine	Gordon-Gray and Wells (1974)
Symphytine	Furuya and Hikichi (1971)

The H-7 signal, also a multiplet, usually appears as a broad triplet; its width in 7-esterified PAs can be used as an indication of conformation (Culvenor and Woods, 1965). The shielding of H-7 in relation to conformation of 3-oxopyrrolizidines is discussed by Aasen *et al.* (1971).

The signals due to the two H-9 protons are of particular interest for conformation studies; they are often magnetically non-equivalent when the 9-oxygen is substituted. The mean value for H-9 in unsaturated PAs lies between 4.2 and 5 ppm; for many esters it is a singlet, but for complex esters including most macrocyclic diesters it is split into two sets of signals differing by amounts varying from 0.14 (echinatine) and 0.16 ppm (monocrotaline) to 1.47 (senecionine) and 1.53 ppm (jacobine) (Culvenor and Woods, 1965). The effect is discussed in relation to conformation by Culvenor (1966).

b. Synthanecines. The NMR spectra of synthetic necine analogues having a 3-pyrroline ring are simpler than spectra of unsaturated necines. The range of chemical shifts for free and esterified synthanecine A is given in Table 4.13. It is interesting that in a few synthanecine A esters the 2-CH$_2$ protons become non-equivalent, although to a smaller degree than the 9-protons in some PAs.

TABLE 4.12 [1]H Chemical Shifts for Unsaturated Ester-PAs[a]

Proton	Chemical shift (δ, ppm, in $CDCl_3$)	Multiplicity
In necine moiety		
H-2	5.7–5.9[b]; 5.9–6.07[c]; 6.15–6.27[d]	s(m)
H-3	3.2–3.6 and 3.8–4.0	m
H-5	2.5–2.7 and 3.1–3.3	m
H-6	1.8–2.3	m
H-7	ca. 1.8[e]; ca. 4.1[f]; ca. 4.9–5.45[g]	t(m)
H-8	ca. 4	
H-9	4.2–4.4[h]; 4.7–5.0[j,k]	s or 2d
Me in acid moiety		
MeCH	0.85–1.16	d
MeCH—CO_2R	1.14–1.29	d
MeC—OH	1.18–1.44	s
MeC(OH)—CO_2R	1.3–1.55	s
MeC—OCOR	1.57–1.7	s
MeC(CO_2R)—OCO	1.66	s
MeCH—C (with O epoxide)	1.23–1.27	d
MeCH=C—CO_2R	1.74–1.78[l]; 1.83–1.9[m]; 1.92–1.96[n]	d
C=C(CO_2R)Me	1.82–1.94[n]	s

[a] Data from Culvenor and Woods (1965) and Bull et al. (1968).
[b] In 9-monoesters and 'open' diesters.
[c] In 11-membered macrocyclic diesters.
[d] In 12-membered macrocyclic diesters.
[e] No 7-OH present.
[f] In 7-OH compounds.
[g] When 7-ester present.
[h] In 9-OH compounds.
[j] In 9-esters.
[k] Often a pair of doublets, separated by up to 1.53 ppm; mean value given.
[l] Trans to carboxyl (as in integerrimine).
[m] Cis to carboxyl (as in senecionine).
[n] Cis (in angelate esters).

c. **Dihydropyrrolizines and Pyrroles.** Culvenor et al. (1970b) have listed chemical shifts for dehydropyrrolizidine alkaloids, dehydronecines and other dihydropyrrolizine derivatives. The pyrrolic protons, H-2 and H-3, show distinctive signals, usually doublets in the range 6.1–6.6 ppm. The H-9 protons in macrocyclic diesters show non-equivalence; the degree of this is not necessarily

TABLE 4.13 [1]H Chemical Shifts
for the Synthanecine A Nucleus[a]

| Proton | Chemical shift (δ, ppm, in $CDCl_3$)[b] | |
	X = OH	X = OCOR or OCONHR
N-Me	2.46s	2.48–2.52s
H2	ca. 3.1	ca. 3.2–3.8m
2-CH_2	3.66d	4.1–4.23d[c]
3-CH_2	4.20s	4.61–4.75s
H4	5.75	5.75–5.9
H5	ca. 3.4	ca. 3.4–3.8m

[a] Data from Mattocks (1974) and unpublished results.

[b] X = Substituents at positions 2-CH_2 and 3-CH_2.

[c] Occasionally split to a pair of doublets: e.g. when X = OCONHEt, δ = 4.01, 4.27; and when X = OCOC(Me)(OMe)Et, δ = 4.12, 4.39.

related to that in corresponding PAs, owing to the different conformation associated with the heterocyclic nucleus.

Chemical shifts for the pyrrolic nucleus in dehydroretronecine, its esters and some of its alkylation products are shown in Table 4.14. For the NMR spectrum of dehydroheliotridine, see Viscontini and Gillhof-Schaufelberger (1971). Table

TABLE 4.14 [1]H Chemical Shifts for the Dihydropyrrolizine Nucleus
in Dehydroretronecine Derivatives[a]

| Proton | Chemical shift (δ, ppm in $CDCl_3$)[b] | | | |
	X = OH	X = OR	X = OCOR	X = NR_2
H-2	6.18d	6.18	6.14–6.31	6.14–6.24
H-3	6.56d	6.58	6.48–6.7	6.5–6.7
H-5	ca. 4	ca. 4	ca. 4	ca. 4
H-7	5.17q	4.72	ca. 5–6.2	ca. 4.1–4.4
H-9	4.56s	4.35	5.03–5.17 (mean[c])	3.39–3.7

[a] Data from results of Culvenor et al. (1970b) and our unpublished results.

[b] X = Substituents at positions C-7 and C-9.

[c] Split to two singlets in macrocyclic diesters.

TABLE 4.15 [1]H Chemical Shifts
for Some 2,3-Disubstituted 1-Methylpyrrole Derivatives[a]

Proton	Chemical shift (δ, ppm, in $CDCl_3$)[b]				
	X = OH	X = OCOR	X = OR	X = NMe$_2$	X = SR
N-Me	3.62s	3.61–3.63	3.62–3.63	3.62	3.62–3.65
2-CH$_2$	4.59s	5.14–5.19	4.47–4.5	3.7	3.74–3.81
3-CH$_2$	4.51s	5.04–5.09	4.38–4.4	3.62	3.60–3.65
H-4	6.10d	6.13–6.18	6.09–6.12	6.03	6.05–6.07
H-5	6.57d	6.60–6.64	6.57–6.6	6.5	6.53–6.57

[a] Data from Mattocks (1974, 1978a) and unpublished results.
[b] X = Substituents at positions 2-CH$_2$ and 3-CH$_2$.

4.15 shows chemical shifts of pyrrolic derivatives formed by dehydrogenation of synthanecine A, and their O-, N- and S-alkylation products.

2. Deuterium NMR

Rana and Robins (1983, 1984) have used ^{2}H-NMR in a study of the biosynthesis of the retronecine moiety of retrorsine derived from [2]H-labelled putrescines. Samples of retrorsine with deuterium in either the 2, 6 and 7 or the 3 and 9 positions were studied. Spectra in pyridine at 90° gave sharper signals than spectra in chloroform at 60°C. Signals (δ, ppm in $CHCl_3$) for deuterium at various locations (parentheses) in the necine moiety were as follows: 2.15 (6α); 2.40 (6β); 3.35 (3β); 3.90 (3α); 5.0 (7α); 5.49 (9); and 6.20 (2). Grue-Sørensen and Spenser (1983) made a similar ^{2}H-NMR study of the biosynthesis of retronecine in the PAs of *Senecio vulgaris* plants administered (*R*)- or (*S*)-[1-^{2}H]putrescine.

3. ^{13}C-NMR

Carbon NMR spectrometry is becoming much used for structural studies on PAs. It will not be dealt with here, except to refer to the principal sources of information.

A useful ^{13}C-NMR study has been made by Jones *et al.* (1982), with data on 11- and 12-membered macrocyclic diester PAs and a summary of previous results on other alkaloids. Drewes *et al.* (1981) have analyzed the spectra of retrorsine, swazine, isoline and hygrophylline. Roeder *et al.* (1982a) gave ^{13}C-NMR data for the alkaloids of comfrey (*Symphytum officinale*). Molyneux *et al.* (1982) gave ^{13}C-NMR data for retronecine and for nine macrocyclic diester PAs

TABLE 4.16 Sources of ^{13}C-NMR Data for Some Necines and PAs

Compound	References
Necines	
Curassanecine	Mohanraj and Herz (1982)
Isoretronecanol	Mohanraj and Herz (1982)
Platynecine	Mody *et al.* (1979); Klasek *et al.* (1980)
Retronecine	Mody *et al.* (1979); Molyneux *et al.* (1982); Zalkow *et al.* (1979)
Supinidine	Mohanraj and Herz (1982)
Trachelanthamidine	Mohanraj and Herz (1982)
Alkaloids	
Anacrotine	Jones *et al.* (1982)
7-Angeloyl heliotridine	Roeder *et al.* (1980c)
7-Angeloyl retronecine	Rueger and Benn (1983a)
Bulgarsenine	Roeder *et al.* (1980a)
Crispatine	Jones *et al.* (1982); Mody *et al.* (1979)
Curassavine	Mohanraj and Herz (1982)
Doronenine	Roeder *et al.* (1980a)
Europine	Zalkow *et al.* (1979)
Floridanine	Roeder *et al.* (1983)
Florosenine	Roeder *et al.* (1983)
Fulvine	Jones *et al.* (1982)
Grantaline	Jones *et al.* (1982)
Gynuaramine	Wiedenfeld (1982)
Helifoline	Mohanraj *et al.* (1981)
Heliosupine	Zalkow *et al.* (1979)
Heliotrine	Mody *et al.* (1979)
Heliovicine	Mohanraj and Herz (1982)
Hygrophylline	Jones *et al.* (1982); Molyneux *et al.* (1982); Drewes *et al.* (1981)
Integerrimine	Jones *et al.* (1982)
Intermedine	Roeder *et al.* (1982a); Roitman (1983a)
Isoline	Drewes *et al.* (1981)
Jacobine	Jones *et al.* (1982)
Jacozine	Jones *et al.* (1982)
Lasiocarpine	Mody *et al.* (1979); Zalkow *et al.* (1979)
Lycopsamine	Roeder *et al.* (1982a); Roitman (1983a)
Madurensine	Mody *et al.* (1979)
Monocrotaline	Jones *et al.* (1982); Mody *et al.* (1979); Molyneux *et al.* (1982); Zalkow *et al.* (1979)
Diacetyl	Jones *et al.* (1982)
Nemorensine	Klasek *et al.* (1980)
Neoplatyphylline	Jones *et al.* (1982); Roeder *et al.* (1982b)

TABLE 4.16 (*Continued*)

Compound	References
Otosenine	Jones *et al.* (1982); Roeder *et al.* (1983); Yamada *et al.* (1976a)
Petasitenine	Yamada *et al.* (1976a)
Platyphylline	Jones *et al.* (1982); Molyneux *et al.* (1982); Roeder *et al.* (1982b)
Retrorsine	Jones *et al.* (1982); Molyneux *et al.* (1982); Drewes *et al.* (1981)
Riddelliine	Jones *et al.* (1982); Molyneux *et al.* (1982)
Sencalenine	Roeder *et al.* (1984a)
Senecionine	Jones *et al.* (1982); Molyneux *et al.* (1982)
7-Senecioyl retronecine	Roeder *et al.* (1984a)
7-Senecioyl-9-sarracinyl retronecine	Roeder *et al.* (1984a)
Seneciphylline	Jones *et al.* (1982); Molyneux *et al.* (1982)
Senecivernine	Roeder *et al.* (1979)
Senkirkine	Jones *et al.* (1982); Molyneux *et al.* (1982)
Spectabiline	Jones *et al.* (1982); Molyneux *et al.* (1982)
Supinine	Mody *et al.* (1979)
Swazine	Drewes *et al.* (1981)
Symlandine	Roeder *et al.* (1982a)
Symphytine	Roeder *et al.* (1982a)
Syneilesine	Jones *et al.* (1982)
Triangularine	Roitman (1983b)

and their *N*-oxides; Mohanraj and Herz (1982) have studied a series of necine bases and saturated PA esters, and Mody *et al.* (1979) gave chemical shifts for retronecine, platynecine and a number of ester PAs. Sources of ^{13}C-NMR data for PAs and necines are summarized in Table 4.16.

E. Mass Spectroscopy

Sources of data on the mass spectra of individual PAs and necines are listed in Table 4.17. Fragmentation patterns are discussed briefly here.

1. Electron-impact (EI) Spectra

Saturated necines such as retronecanol (169), laburnine (173) and platynecine (176) typically give fragments in the ranges *m/z* 95–97, 122–123 and 138–140;

TABLE 4.17 Sources of Mass Spectra and MS Data on Some Necines and PAs

Compound	References
Necines	
Curassanecine	Mohanraj *et al.* (1982)
Hastanecine	Culvenor *et al.* (1968)
Heliotridine	Pedersen and Larsen (1970)
Laburnine	Neuner-Jehle *et al.* (1965)
Platynecine	Neuner-Jehle *et al.* (1965); Aasen and Culvenor (1969a)
Retronecanol	Neuner-Jehle *et al.* (1965)
Retronecine	Neuner-Jehle *et al.* (1965); Pedersen and Larsen (1970); Deinzer *et al.* (1978)[a]
Alkaloids	
Anacrotine	Atal *et al.* (1966a); Dreifuss *et al.* (1983)[b]
7-Angeloyl heliotridine	Pedersen and Larsen (1970)
Axillarine	Crout (1969)
Bulgarsenine	Nghia *et al.* (1976)
Crocandine	Dreifuss *et al.* (1983)[b]
Cromadurine	Rao *et al.* (1975b)
Echimidine	Pedersen and Larsen (1970)
Echinatine	Pedersen and Larsen (1970)
Ehretinine	Dreifuss *et al.* (1983)[b]
Floridanine	Cava *et al.* (1968)
Fulvine	Dreifuss *et al.* (1983)[b]
Hastacine	Culvenor *et al.* (1968)
Heliosupine	Pedersen and Larsen (1970)
Acetyl	Pedersen (1970)
Heliotrine	Neuner-Jehle *et al.* (1965); Pedersen and Larsen (1970); Evans *et al.* (1980)[a]; Dreifuss *et al.* (1983)
Indicine	Evans *et al.* (1979, 1980)[a]
Integerrimine	Luthy *et al.* (1981)
Intermedine	Dimenna *et al.* (1980)
Jacobine	Segall (1978); Rothschild *et al.* (1979); Luthy *et al.* (1981)
Jacoline	Segall (1978); Dreifuss *et al.* (1983)[b]
Jaconine	Segall (1978); Segall and Krick (1979); Luthy *et al.* (1981); Dreifuss *et al.* (1983)[b]
Jacozine	Klasek *et al.* (1968b); Segall (1978); Segall and Krick (1979); Luthy *et al.* (1981)
Junceine	Atal *et al.* (1966b)
Lasiocarpine	Pedersen and Larsen (1970)
Ligularidine	Hikichi *et al.* (1979)

TABLE 4.17 (*Continued*)

Compound	References
Lycopsamine	Dreifuss *et al.* (1983)[b]
Monocrotaline	Neuner-Jehle *et al.* (1965); Rothschild *et al.* (1979)
Neoplatyphylline	Culvenor *et al.* (1968)
Platyphylline	Abdullaev *et al.* (1974a); Culvenor *et al.* (1968)
Retrorsine	Segall and Molyneux (1978)
Riddelliine	Segall and Molyneux (1978)
Rosmarinine	Dreifuss *et al.* (1983)[b]
Seneciphylline	Abdullaev *et al.* (1974a); Deinzer *et al.* (1977); Qualls and Segall (1978); Luthy *et al.* (1981)
Senecionine	Atal *et al.* (1966a); Abdullaev *et al.* (1974a); Qualls and Segall (1978); Luthy *et al.* (1981); Dreifuss *et al.* (1983)[b]
19-Hydroxy-	Eastman and Segall (1982)
Symphytine	Furuya and Araki (1968)
Trichodesmine	Atal *et al.* (1966b); Rothschild *et al.* (1979)
Yamataimine	Hikichi *et al.* (1978)

[a] Derivatized.
[b] Negative-ion mass spectrum.

there is a characteristic base peak at m/z 82 (Neuner-Jehle *et al.*, 1965; Luthy *et al.*, 1981):

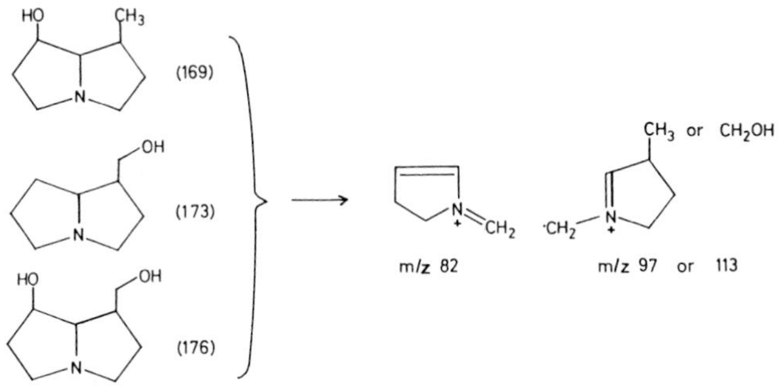

The necine moiety in saturated PAs such as hastacine and platyphylline gives a similar pattern (Luthy *et al.*, 1981; Culvenor *et al.*, 1968). Corresponding fragments from unsaturated necines are 2 mass units lower; thus retronecine (183) and heliotridine (184) give identical mass spectra with major fragments at m/z 80 (base), 94, 111 and 155 (M^+) (Pedersen and Larsen, 1970; Neuner-Jehle *et al.*, 1965):

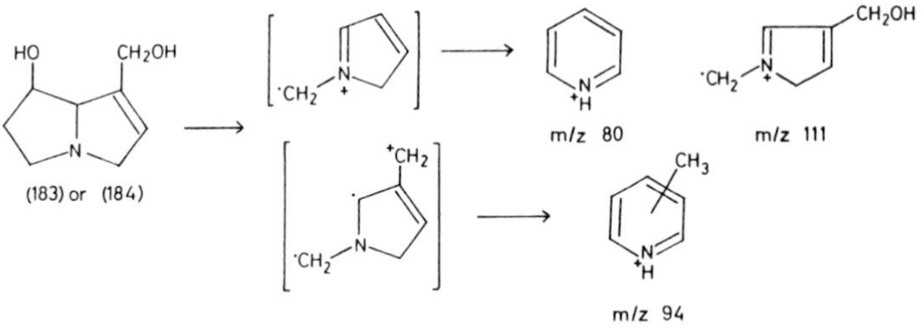

HO   CH$_2$OH

(183) or (184)

$\cdot$CH$_2$ — N$^+$

$^+$CH$_2$

$\cdot$CH$_2$ — N$^+$

N$^+$H   *m/z* 80

CH$_2$OH   *m/z* 111

$\cdot$CH$_2$ — N

CH$_3$ N$^+$H   *m/z* 94

9-Monoesters of unsaturated necines give hydroxylated fragments (especially *m/z* 138) as a result of C-9–O fission:

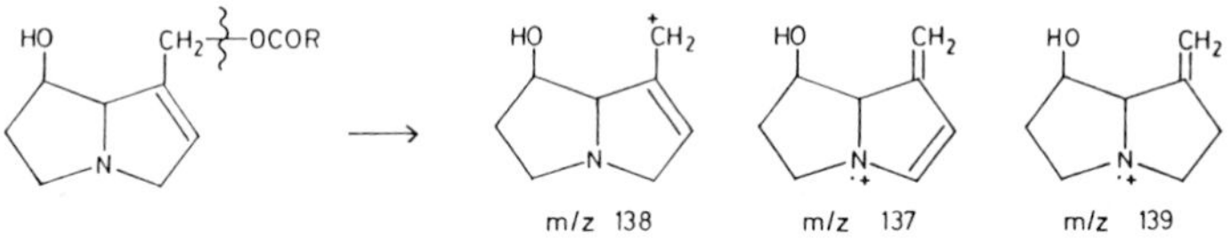

HO   CH$_2$—OCOR

HO   CH$_2$ *m/z* 138

HO   CH$_2$ *m/z* 137

HO   CH$_2$ *m/z* 139

Esterification at C-7, as in 7-angeloyl heliotridine (6), results in intense ions at *m/z* 137 and 106 corresponding to loss of the ester (Pedersen and Larsen, 1970):

RCOO   OH

OH   *m/z* 137

m/z 106

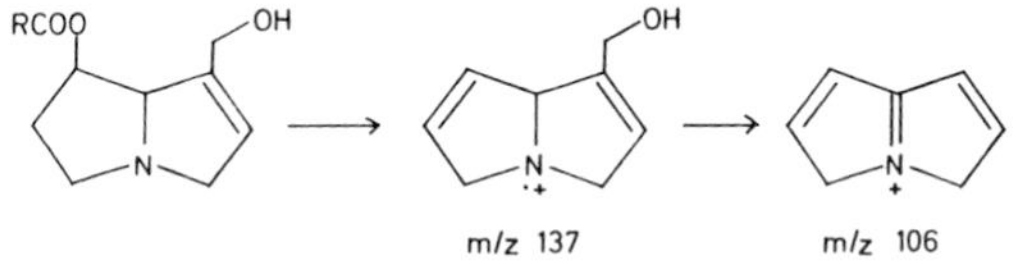

The necine moiety in unsaturated pyrrolizidine diesters gives groups of fragments at *m/z* 93–95, 119–121, and 136–139 (Pedersen and Larsen, 1970; Luthy *et al.*, 1981). According to Crout (1969), prominent fragments at *m/z* 80, 93, 119, 120 and 136 are characteristic of unsaturated diester PAs. Atal *et al.* (1966a) have suggested splitting patterns for senecionine and anacrotine, which are macrocyclic esters of retronecine (1983) and crotanecine (185), respectively, and shown that anacrotine gives a series of fragments 16 mass units higher than those from senecionine due to the extra necine hydroxyl (e.g. 152, 154). PAs based on otonecine (186) show an M − 15 peak due to loss of *N*-methyl, and give characteristic fragments, *m/z* 94, 96, 110, 122–123 and 149–151, from the necine moiety (Cava *et al.*, 1968).

Ester PAs generally show a progressive loss of the acid moiety, starting at the primary or allylic linkage (C-9–O). Groups commonly lost from the acid moiety are OH (17 mass units) or H_2O (18), CH_2OH (31), acetyl (59) if present, Cl if present, CH_3 (15) and CO_2 (44). The molecular ion is usually seen, although it is often very weak.

Picrate salts show similar mass spectra to those of the parent PAs superimposed on that of picric acid; however, the molecular ion of heliosupine is not observed under these conditions (Crout, 1969).

2. Chemical Ionization (CI)

CI mass spectra are much simpler than EI spectra, and give useful complementary information. Thus the CI spectrum of jacozine (63) with methane as reactant gas shows an $(M + 1)^+$ fragment (m/z 350) and only two others, at m/z 138 and 120 (base) (Luthy et al., 1981). Karchesy et al. (1984b) have compared both positive and negative ion CI mass spectra of mono-, di- and macrocyclic diesters of retronecine.

3. Negative-ion Chemical Ionization

The negative-ion CI mass spectrometry of PAs has been reported by Dreifuss et al. (1983), using OH^- as the reactant ion. Thus senecionine (92) gives a strong $(M - H)^-$ ion (m/z 334), a weak $(M + OH)^-$ ion, and fragments representing both the practically intact necic acid anion and the necine moiety. Other macrocyclic diester PAs behave similarly. The necic acid ion is particularly useful for structural studies, because in the normal EI spectra of PAs the acid moiety is greatly fragmented. The appearance of an ion, m/z 154, is characteristic of the unsaturated necine moiety, retronecine (MW 155) and is thus useful for screening plant extracts for potentially hepatotoxic PAs. Negative-ion methane CI mass spectra of semisynthetic and natural PAs based on retronecine have been studied by Karchesy et al. (1984b).

4. Mass Spectroscopy of N-Oxides

The N-oxides of PAs often give unsatisfactory mass spectra owing to their low volatility and ease of thermal decomposition. Abdullaev et al. (1974b) found that major fragmentations of PA N-oxides were associated with N—O cleavage, dehydrations and dehydrogenations, and ring rearrangements. The mass spectrum of anadoline in its N-oxide form (301) shows a weak molecular ion (m/z 397) and fragments at $M - 16$ and $M - 18$ representing the tert-base (5) and its pyrrolic derivative (302) (Culvenor et al., 1975a):

Indicine *N*-oxide (207) does not give a satisfactory mass spectrum (Evans *et al.*, 1979): derivatization by trimethylsilylating reagents apparently removes the *N*-oxide function and converts the compound to either the trimethylsilyl (TMS) derivative of indicine (208) or that of dehydro-indicine (209) (see Chapter 2). Both of these, and corresponding heliotrine derivatives, give satisfactory mass spectra (Evans *et al.*, 1980).

Karchesy *et al.* (1984a) studied fast atom bombardment mass spectra of *N*-oxides of macrocyclic diesters of retronecine. These compounds give a characteristic $(M + H)^+$ pseudomolecular ion base peak, a $(M + H - 16)^+$ ion, and a prominent series of fragments with *m/z* 136, 120, 118, 106, 94 and 80.

5. *Pyrrolic Derivatives of PAs*

The EI mass spectrum of dehydroretronecine (201) (DHR) shows a quite intense molecular ion (*m/z* 153). Loss of OH or 2(OH) gives fragments at *m/z* 136 and 119, respectively, with associated fragments at 134–135, 117 (base) and 118 (Mattocks, 1969a):

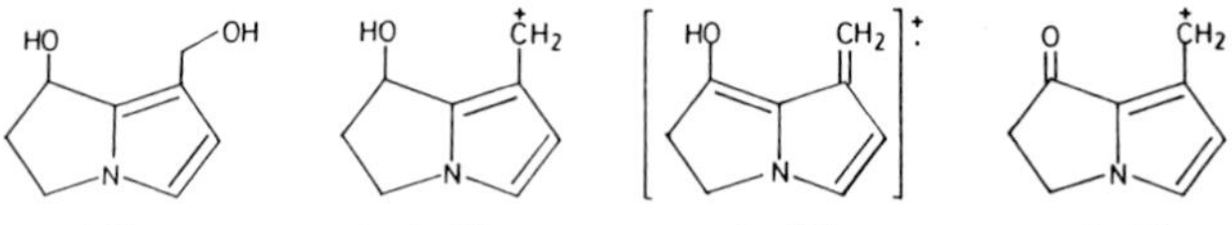

(201) m/z 136 m/z 135 m/z 134

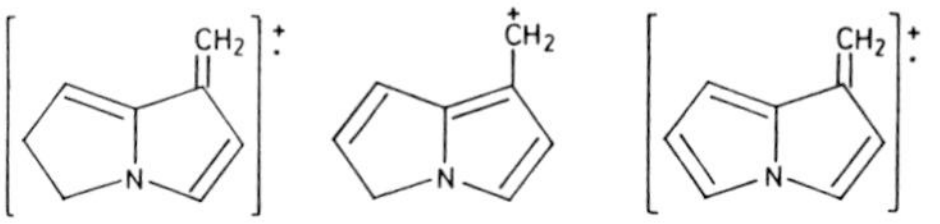

m/z 119 m/z 118 m/z 117

The dimethyl ether (303) of DHR gives peaks at *m/z* 181 (M^+), 150 (base; $M - CH_3O$), and a group at 117–120, the latter (which is the most abundant) being attributed to a rearrangement:

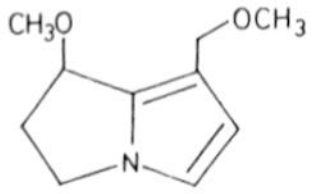

(303)

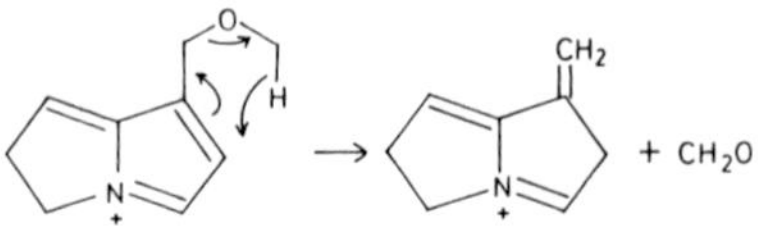

m/z 120

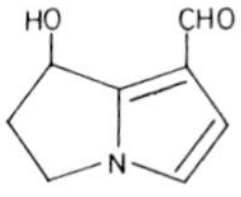

(278)

For the mass spectra of DHR 7-methyl ether and the aldehyde (278), see Segall *et al.* (1984).

Mass spectroscopy of dehydro-PAs (pyrrolic esters) is tricky because of their extreme instability. Dehydromonocrotaline (276) gives a molecular ion, a fragment at m/z 254 representing partial cleavage of the acid moiety, and a group of fragments, m/z 116–120, derived from the dihydropyrrolizine nucleus (Mattocks, 1969a). Disenecioyl DHR (284) fails to give a molecular ion. The heaviest fragment, m/z 249, is due to the loss of C_4H_4O from one senecioyl chain accompanied by a rearrangement; peaks at m/z 83, 117–120 and 134–136 (base, 135) are again characteristic of the dihydropyrrolizine moiety (Shumaker et al., 1976c). The TMS derivatives of dehydro-indicine (209) and dehydroheliotrine both show a weak molecular ion and somewhat simplified mass spectra compared with the corresponding PAs. Both give a base peak, m/z 208, representing the ion (304).

(276)   (284)   (304)

Mass spectral data on some other dihydropyrrolizine derivatives and dehydro-PAs are given by Culvénor et al. (1970b).

5

Chemical Properties of Pyrrolizidine Alkaloids and Their Dehydro-derivatives

I. INTRODUCTION

The purpose of this chapter is to survey those chemical properties and reactions of PAs and related compounds, including necines and pyrrolic derivatives, which may be of interest in relation to their isolation, analysis and biological actions. Structure determinations of PAs, necines and necic acids have been reviewed elsewhere (Leonard, 1950, 1960; Warren, 1955, 1966; Bull *et al.*, 1968; Robins, 1982a), and will not be covered here. Syntheses of PAs are dealt with in Chapter 3. Pyrrolizidine chemistry, including syntheses, stereochemistry and reactions, has also been reviewed by Kochetkov and Likhosherstov (1965) and by Robins (1979).

II. CHEMISTRY OF PAs

A. Stability

This has been discussed in Chapter 2 (Section II). It is sufficient here to say that many crystalline PAs remain stable for years when kept in a cool, dark place, but those which are non-crystalline often darken progressively at room temperature. Much of the decomposition is probably associated with oxidation and the breakdown of ester linkages. Dark coloured products, especially from the breakdown of PAs with an unsaturated necine moiety, may in part consist of pyrrolic polymers. The N-oxides of PAs, even when crystalline, are less stable and may steadily decompose, as shown by a darkening of colour and the presence of pyrrolic products.

B. Hydrolysis

Ester PAs can undergo acid- or base-catalysed hydrolysis. In diesters, the primary ester linkage (at C-9) is normally the most susceptible to hydrolysis, especially the allylic ester in unsaturated PAs. Thus, lasiocarpine (65) can be partially hydrolysed by alkali to 7-angelyl heliotridine (6) (Crowley and Culvenor, 1959). The rates of base-catalysed hydrolysis are greatly influenced by steric hindrance around the ester groups. This is demonstrated with a series of retronecine diesters in which chain branching in the acid moiety is a major factor inhibiting hydrolysis (Mattocks, 1982b). Esters of unsaturated acids are also more resistant: thus the senecioate of retronecine (305) is hydrolysed at one-sixteenth the rate of the isovalerate (306). Bull *et al.* (1968, p. 61) found that the half-lives of various PAs in 0.5 N NaOH vary enormously. Alkaloids with a highly branched acid moiety, such as heliotrine (49), are very resistant to hydrolysis. However, hydrolysis of some β-hydroxy esters [e.g. lasiocarpine (65)] is accelerated owing to hydrogen-bonding of the OH with the carbonyl group. Monocrotaline (74) is relatively open to hydrolysis because the macrocyclic ring restricts steric hindrance of the ester groups.

Hydrolysis of a PA will not necessarily yield the intact acid component(s): the action of base may break down the acid moiety, either before or after ester

(65)   (307)

hydrolysis. Thus, lasiocarpine (65) can lose acetone in a retro-aldol reaction; this is followed by rapid hydrolysis of the resulting 2-hydroxy-3-methoxybutyrate (307), thus contributing to the relative lability of lasiocarpine to alkali (Bull *et al.*, 1968). Some macrocyclic PAs can also undergo retro-aldol reactions: these include monocrotaline, which gives butan-2-one, and trichodesmine and junceine, which both give 4-methylpentan-2-one (Adams and Gianturco, 1956d). On the other hand monocrotalic acid (308) is decomposed by alkali after the hydrolysis of monocrotaline: loss of CO_2 and then H_2O leads to monocrotic acid (309) (Adams *et al.*, 1939).

(308)   (309)

Another feature of some macrocyclic diesters is that following either hydrolysis or hydrogenolysis of the more sensitive primary ester, cleavage of the remaining ester linkage may be accelerated by intramolecular transesterification. This can occur when there is a hydroxyl group in a 4- or 5-position in the acid moiety relative to the secondary ester. Thus, acid hydrolysis of senecionine (92) leads to senecic acid lactone (which may be further hydrolysed to senecic acid) with no trace of the intermediate amino acid (310) (Bull *et al.*, 1968).

(92)   (310)

C. Hydrogenolysis and Reduction

Hydrogenation (Pt, H_2, 1 atm) of unsaturated necine esters can split the allylic ester and saturate the necine double bond (and that in the acid, if present). The secondary ester is unaffected, but in some macrocyclic PAs it may be split by intramolecular transesterification. Thus the initial hydrogenolysis product (311) from monocrotaline gives monocrotalic lactone (194) and retronecanol (169); limited hydrogenation gives desoxyretronecine (312), showing that the allylic oxygen is more susceptible than the double bond. However, hydrogenation of monocrotaline using Raney nickel catalyst gives small yields (up to 4%) of dihydromonocrotaline (313) along with retronecanol (Constantine *et al.*, 1967). Hydrogenation using Raney nickel also converts retronecine (183) to platynecine (176) (Adams and Rogers, 1941).

Otonecine esters (314) are hydrogenolysed (H$_2$, Pt) in a similar way to retronecine esters, the necine moiety being converted to dihydrodesoxyotonecine (210) (Hikichi and Furuya, 1976). Using Raney nickel as catalyst, only the acid moiety of clivorine (15) is hydrogenated (Hikichi *et al.*, 1979). Reduction of PAs with lithium aluminium hydride cleaves both 7- and 9-ester groups giving the reduced acid moiety: thus monocrotaline gives the alcohol (315). Such reactions are useful in structure elucidation (Robins and Crout, 1969).

D. Alkylating Action

Those PAs which are allylic esters can act as alkylating agents through alkyl-oxygen fission of the ester linkage. Thus heliotrine (49) reacts with the strong nucleophile, benzyl mercaptan, to give the sulphide (316) (Culvenor *et al.*, 1962).

E. Reactions of Hydroxyl Groups

Hydroxyls in either the acid or necine moieties of PAs can be acylated. Acetyl derivatives of PAs often occur in nature. Acylations of some PAs are listed in Chapter 3, Table 3.8. When more than one OH group is present, acylation can be selective: primary hydroxyls react most easily, tertiary hydroxyls are more re-sistant. Thus, mild acetylation of indicine (53) gives the diacetyl derivative (264) (see Chapter 3); triacetyl indicine (265) is formed under more severe conditions (Mattocks, 1967a). Anacrotine (2) with acetic anhydride gives the 6-acetyl deriv-ative (3); the *tert*-OH in the acid moiety is also acetylated if acetyl chloride is used (Culvenor and Smith, 1972). Similarly, tosylation of rosmarinine (155) under mild conditions gives mainly the 2-tosylate (Koekemoer and Warren, 1955). The treatment of otonecine-based PAs with acetic anhydride can lead to pyrrolic products (see below).

PAs having a glycol with cis configuration in the acid moiety (e.g., mono-crotaline) form cyclic sulphite and phenylboronate esters; this can assist in as-signments of conformation (Robins and Crout, 1969). Cyclic methylboronate derivatives have facilitated the gas chromatographic separation of closely related PAs (lycopsamine and intermedine) having stereoisomeric glycol groups (Edgar and Culvenor, 1975).

Necine hydroxyls can be esterified by conventional methods (see Chapter 3, Tables 3.5 and 3.6). Thionyl chloride replaces the allylic primary hydroxyl in unsaturated necines by Cl; thus heliotridine, retronecine and supinidine give the chlorides (317), (318) and (319), respectively (Adams and van Duuren, 1954; Culvenor *et al.*, 1959; Culvenor and Smith, 1961); more prolonged action gives the dichloro derivatives of heliotridine (320) or retronecine (321) (Culvenor and Smith, 1961; Mattocks, 1969b). These halides are stable only as salts (e.g. hydrochlorides): the free bases form quaternary polymers. Platynecine with cold thionyl chloride gives the cyclic sulphite (322), whereas in hot $SOCl_2$ the di-chloride (323) and anhydroplatynecine (324) are formed (Adams and van Du-uren, 1954).

Sodium methoxide reacts with chloro heliotridine (317) to form only the methyl ether (325) (Culvenor *et al.*, 1967a), whereas chloro retronecine (318), with its favourably placed 7β-OH, gives some anhydroretronecine (326) (Culvenor and Smith, 1962); the latter is better prepared from chloro retronecine and potassium *tert*-butoxide (our unpublished results). The methyl ether (327) of retronecine can be made from the latter using potassium *tert*-butoxide and methyl iodide (Culvenor and Smith, 1962).

Sodium carboxylate salts react with allylic necine chlorides to form esters: thus heliotrine (49) can be prepared from (317) and sodium heliotrate (Culvenor *et al.*, 1959). The allylic chloride of retronecine can be replaced by azide, reduction of which forms retronamine (260) (Mattocks, 1969b).

F. Reactions of the Basic Nitrogen

1. Salt Formation

As tertiary bases, most PAs form stable salts with acids. These often crystallize readily and are thus useful for the isolation, purification or characterization of the alkaloids. Salts commonly prepared include hydrochlorides, picrates and picrolonates; perchlorates, nitrates and reinecke salts (Brandange and Granelli, 1973) have also been made.

2. Quaternization

PAs react with alkyl halides to form quaternary salts: methiodide salts have been used to characterize some alkaloids (see Chapter 4, Tables 4.2 and 4.3). The otonecine base moiety, e.g. in senkirkine, also forms a methiodide (Briggs *et al.*, 1965). Other quaternary PAs and necines are reviewed by Atal (1978).

3. N-Oxidation

Necine and PA bases are readily converted to *N*-oxides by aqueous or alcoholic hydrogen peroxide (Leisegang and Warren, 1949; Christie *et al.*, 1949; Koekemoer and Warren, 1951). N-Oxidation is also brought about by 3-chloroperbenzoic acid (Piper *et al.*, 1981) at room temperature; the reaction is more rapid than epoxidation of the allylic double bond by this reagent. N-Oxidation is easily reversed: PA *N*-oxides are reduced to *tert*-bases by zinc and mineral acid (Christie *et al.*, 1949; Koekemoer and Warren, 1951), by zinc and acetic acid or NH_4Cl (Ames and Powis, 1978), or by a redox polymer (Serdoxit) (Huizing and Malingre, 1979a).

N-Oxides of PAs often crystallize in association with water or an organic solvent. Attempts to remove this solvent (which may be present in non-stoichiometric amounts), using heat or reduced pressure, from PAs having an unsaturated necine moiety can lead to pyrrolic decomposition products.

G. Conversion to Dehydro-alkaloids

Dehydro derivatives (dihydropyrrolizine alcohols and esters) are readily formed by dehydrogenation of the pyrroline ring present in unsaturated necines and PAs containing them. These compounds, also known as 'pyrroles' or 'pyrrolic' derivatives, are also formed by dehydration of the corresponding *N*-oxides (Scheme 5.1). Various procedures and reagents convert PAs into pyrrolic deriva-

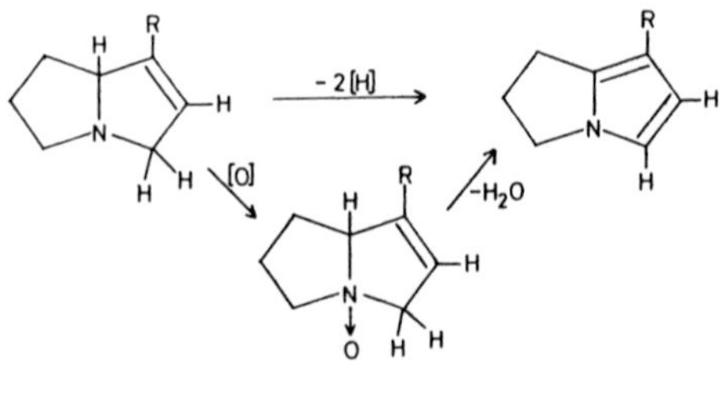

Scheme 5.1.

tives. Because of their instability in presence of water, the dehydro-alkaloids must be prepared under anhydrous conditions; the dehydro-necines can survive for a limited time in basic aqueous solution.

1. Dehydrogenation of PAs and Necines

a. With Manganese Dioxide. PAs in chloroform solution stirred with 'active' manganese dioxide are oxidized, but side reactions result in poor yields of dehydro-alkaloids; oxidation also occurs in aqueous solution, although of course dehydro-pyrrolizidine esters would be hydrolysed under these conditions (Mattocks, 1969a). Culvenor et al. (1970c) studied the reaction in detail and confirmed that MnO_2 dehydrogenation is not generally satisfactory for preparing the dehydro-alkaloids; however, pyrrolizidine amino alcohols (necines), and some PAs carrying free hydroxyl groups on the necine moiety are readily oxidized to dihydropyrrolizines. Thus, with MnO_2 in chloroform, heliotridine (184) gives the 1-aldehyde (329) as main product, along with dehydroheliotridine (272) and the aldehydes (331) and (332). The 7-oxo-derivative (330) is a further oxidation product from (329). The aldehydes (331) and (332) are evidently formed by ring cleavage before pyrrole formation, since they are not formed from (329) or (272). Reaction mechanisms proposed by Culvenor et al. (1969b, 1970c) involve oxidation at the nitrogen to form (272), and oxidation via manganate ester intermediates, at the secondary hydroxyl to give (331), and at the allylic hydroxyl to give initially the unstable pyrroline aldehyde (328). The latter has not been isolated but its existence is deduced from the formation of the ester (333) when HCN and methanol are present during the MnO_2 oxidation of (184) (Aasen and Culvenor, 1969a). The aldehyde (289), isomeric with (328), has since been prepared by reaction of retronecine with a specially prepared acidic manganese dioxide (Mattocks, 1977a): it appears that the allylic OH is oxidized while the N is immobilized at an acidic site on the oxide surface, thus preventing dehydrogenation.

Necine *N*-oxides are also oxidized to pyrroles by MnO_2: retronecine *N*-oxide give mainly the aldehyde (278), whereas heliotridine *N*-oxide (334) gives a mixture of (329) and (331) (Culvenor *et al.*, 1970c). These aldehydes are easily reduced by borohydride: thus (329) affords dehydroheliotridine (272).

Diester PAs, such as lasiocarpine, are relatively resistant to MnO_2, but heliotrine (49), with a free 7-hydroxyl on the necine moiety, is dehydrogenated, mainly to the oxo-derivative (309), and 7-angelylheliotridine (6) gives the aldehyde (335) (Culvenor *et al.*, 1970c).

b. With Potassium Permanganate. Treatment of PAs with $KMnO_4$ in acetone gives dehydro-alkaloids in variable yields, along with carbonyl derivatives from unesterified hydroxyls when these are present on the necine moiety (Culvenor *et al.*, 1970b).

c. Catalytic Dehydrogenation. Treatment of PAs with platinum oxide or Raney nickel gives mixtures of products similar to those from MnO_2 oxidation (Culvenor *et al.*, 1970b).

d. With Quinones. 2,3-Dichloro-5,6-dicyanobenzoquinone (DDQ) efficiently dehydrogenates PAs to pyrrolic derivatives (Mattocks, 1969a): hydroxylic solvents interfere with the reaction, and necine bases give only traces of pyrroles. The less reactive quinone, chloranil, is also effective and has been used in chloroform solution for the preparative dehydrogenation of necines (Culvenor *et al.*, 1980a), and in the detection of PAs on chromatograms (Molyneux and Roitman, 1980).

e. With 1,1-Diphenyl-2-picrylhydrazyl. This stable free radical can dehydrogenate PAs in chloroform solution. Pyrrolic products from several PAs have been detected using the Ehrlich colour test, but not identified (Mattocks, 1969a).

f. Using Potassium Nitrosodisulphonate (Fremy's Salt). This compound $[(KSO_3)_2NO\cdot]$ is also a radical reagent (Zimmer *et al.*, 1971). It is stable in alkaline aqueous solution, in which it converts retronecine to dehydroretronecine (Mattocks, 1981c).

g. Using Iodine. After PAs in solution or on chromatograms have been exposed to iodine, pyrrolic products are detectable with the Ehrlich reaction (cf. Culvenor *et al.*, 1970c). The reaction has not been studied in detail.

h. With Oxygen. Culvenor *et al.* (1970b) have pointed out that PAs after long storage in air often contain Ehrlich-positive impurities, attributed to oxidation of the alkaloids to dihydropyrrolizines.

i. With Aryl Thiols. An excess of thiophenol in aqueous acetone converts retronecine to dehydroretronecine (Juneja *et al.*, 1984). PAs such as monocrotaline are similarly dehydrogenated, the dehydro alkaloid then reacting with excess thiol to give the bisthiophenyl adduct of dehydroretronecine. Other aryl thiols, including 6-mercaptopurine and benzyl mercaptan, but not alkyl thiols such as cysteine, bring about a similar reaction.

2. Dehydration of Pyrrolizidine N-Oxides

Pyrrolic derivatives are formed from the *N*-oxides of unsaturated necines and PAs by processes which amount to the removal of the elements of water. This can be accomplished in several ways.

a. By Heat. PA *N*-oxides decompose more readily than the parent alkaloids when they are heated or even when stored in the dry state. They become discoloured and give a positive Ehrlich reaction; this may account in part for the presence of pyrrolic compounds in PA-containing plant materials after storage. This reaction is accelerated when the *N*-oxides are heated in non-hydroxylic solvents: thus when retrorsine *N*-oxide is heated at 100°C in diethylene glycol dimethyl ether, it forms pyrrolic products which give Ehrlich and alkylation reactions characteristic of dehydroretrorsine (336) (Mattocks, 1969a).

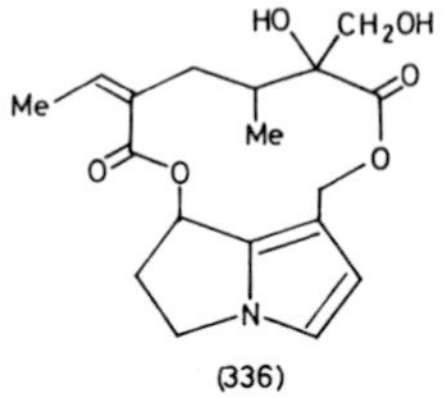

(336) 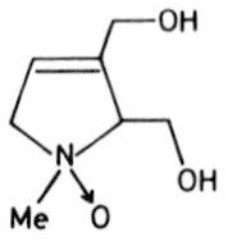   (337)

b. With Acetic Anhydride and Other Acylating Agents. Extending the work of Dann (1960), Mattocks (1967b,c) showed that unsaturated PA N-oxides when treated with acetic anhydride form pyrrolic products which give a colour reaction with Ehrlich reagent. The reaction can also be used to make dihydro-pyrrolizines on a preparative scale (Mattocks, 1968a, 1969a; Culvenor et al., 1970b). The process is an application of the Polonovsky reaction to N-oxides of N-alkyl-3-pyrrolines (Kreher and Pawelczyk, 1964), and is effective also with synthanecine N-oxide (337) (Mattocks, 1974). The reactions are thought to pro-ceed via unstable N-acetoxy intermediates, e.g. (338), which rapidly lose a molecule of acetic acid and a proton (Mattocks, 1969a) (Scheme 5.2).

N-Oxides of PAs are similarly dehydrated by other acyl anhydrides. The N-oxides of indicine and heliotrine are also converted to pyrrolic derivatives by a combination of the trimethylsilylating reagents bis(trimethylsilyl)trifluoro-acetamide (BSTFA) and N-trimethylsilylimidazole (TSIM), presumably via a similar mechanism (Evans et al., 1979, 1980) (see Chapter 2).

c. By Iron(II) Complexes. The N-oxides of necines and PAs are converted to pyrroles by complexes of ferrous iron (Mattocks, 1968c, 1969a). Little of the ferrous sulphate is consumed; its action is catalytic and the reaction probably proceeds via a radical mechanism (cf. Ferris et al., 1967) (Scheme 5.3). The complex form of the iron is important, and it is probably closely associated with the alkaloid throughout the reaction. Thus, radicals in the reaction mixture can-not be detected by chemical means (Mattocks, 1969a), and Fe(III) involved in

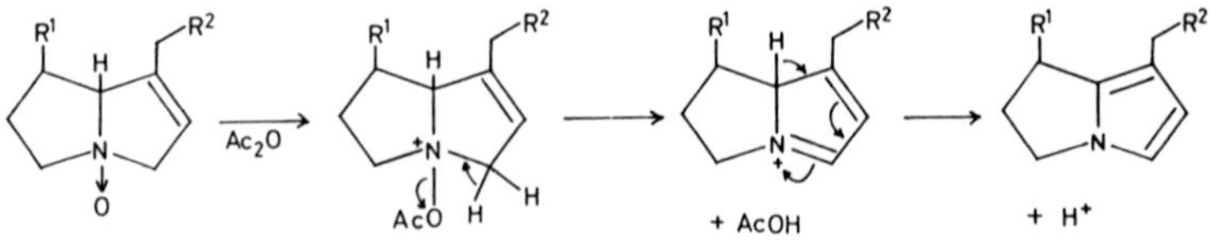

(338)

Scheme 5.2.

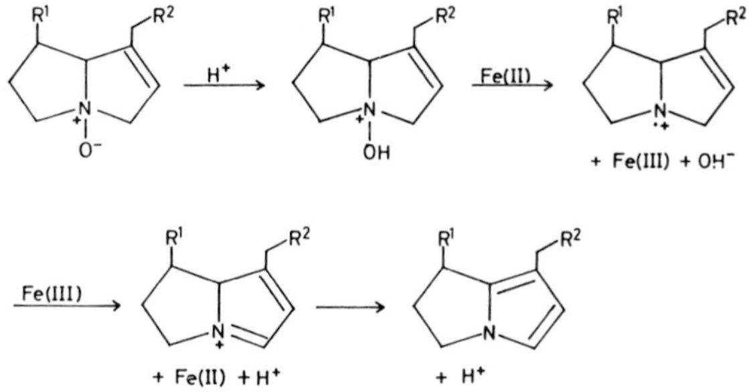

Scheme 5.3.

the catalytic cycle is not inhibited by fluoride (unless in large amounts), even though the latter can complex with any free ferric iron in solution (Mattocks, 1968c).

The iron-catalysed reaction also occurs in aqueous solution (pyrrolic ester products would be hydrolysed under these conditions). It is brought about by a ferrous–EDTA complex in presence of ascorbic acid: some reduction of N-oxide to the basic alkaloid also occurs (Mattocks, 1968c). Other iron complexes are also effective, including nitroprusside; here the effective catalyst is probably the nitro complex, $[Fe(CN)_5NO_2]^{4-}$, formed by oxidation of nitroprusside, rather than the nitroprusside ion itself (Mattocks and Bird, 1983a).

3. Pyrroles from Otonecine-based PAs

Klasek et al. (1975a) heated clivorine (15) with acetic anhydride in pyridine and obtained a product formulated as the quaternary pyrrole, dehydroclivorinium acetate (339), which could also be converted to its chloride. Attempts to convert the otonecine-based PAs senkirkine, otosenine and florosenine to pyrrolic analogues in a similar way failed; however, Hikichi et al. (1979) obtained a quaternary pyrrole from ligularidine. Scheme 5.4 shows a possible mechanism for the de-oxygenation of the otonecine moiety.

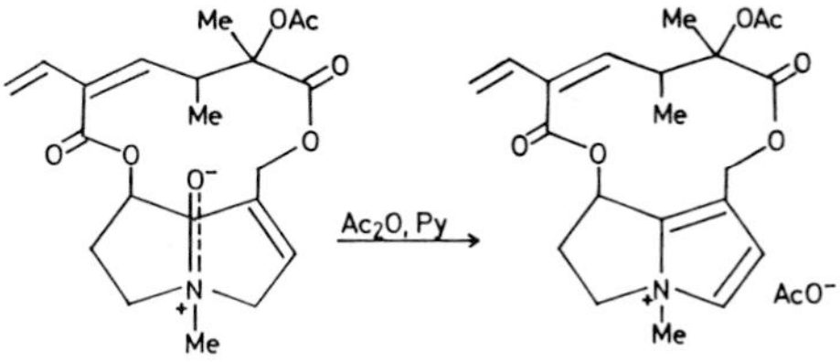

(15)   (339)

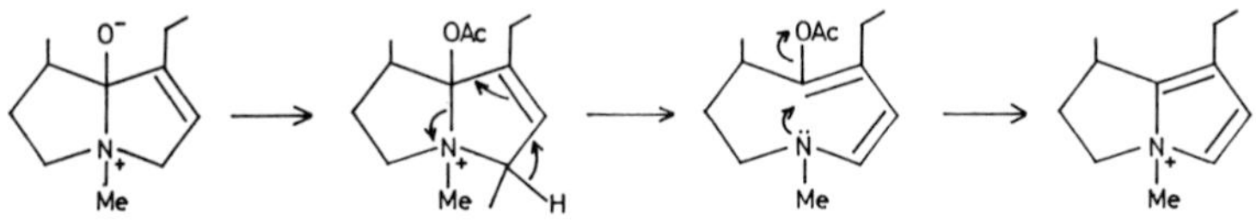

Scheme 5.4.

4. Pyrroles from PAs with a Saturated Necine Moiety

The N-oxides of rosmarinine (155) and hygrophylline (144) give small but significant amounts of pyrrolic products (about one-tenth those from unsaturated PAs) when treated with acetic anhydride (Mattocks, 1967c). Methanolic ferrous sulphate also converts both these *N*-oxides to pyrroles (our unpublished observations). The reaction has not been studied in detail, but the pyrrolic products resemble those from unsaturated PAs, having similar reactivity as alkylating agents and towards Ehrlich reagent.

H. Preparation of Epoxides

The double bond in unsaturated necines can be expoxidized. Two isomers are possible. Supinidine (182), with ice cold perbenzoic acid (2 weeks) and reduction of *N*-oxides, gives mainly the α-epoxide (340), with a smaller amount of β-epoxide (341) (Culvenor *et al.*, 1967a). The more hindered necine double bond of monocrotaline is resistant to perbenzoic acid, although the *N*-oxide is readily formed. However, this alkaloid with hydrogen peroxide in trifluoroacetic anhydride gives the *O*-trifluoroacetyl derivative (342) of monocrotaline α-epoxide, along with a smaller amount of the corresponding β-epoxide, both as tri-fluoroacetate salts; surprisingly, no *N*-oxide is formed (Culvenor *et al.*, 1970a). The trifluoroacetyl ester is hydrolysed during work-up under aqueous conditions, leading to the free monocrotaline epoxides.

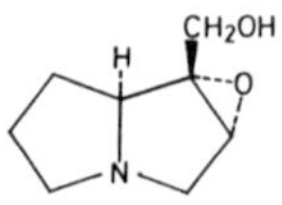

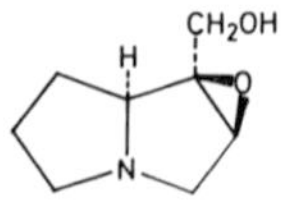

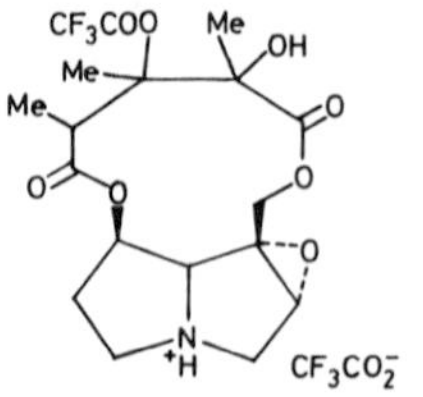

The β forms of both supinidine and monocrotaline epoxides are the more reactive, being very rapidly ring-opened to chlorohydrins by HCl, whereas the α-epoxides are resistant to acid.

The ethylidene group in the acid moiety of macrocyclic otonecine and retro-necine esters can be converted to the β-epoxide using performic acid (Asada and Furuya, 1984a,b).

I. Interconversions of PAs

A few natural PAs have been chemically converted to other known alkaloids. Apart from acetylations of hydroxyl groups, these interconversions include the following. The epoxide in jacobine (60) is converted by HCl to the chlorohydrin (jaconine) (62); the reverse is brought about by aqueous sodium hydroxide or silver oxide (Bradbury, 1954). Dehydration of the necine moiety of rosmarinine (155), accomplished by heating its *p*-toluenesulphonate with pyridine, gives senecionine (92) (Koekemoer and Warren, 1955). The *trans*-ethylidene group in the acid moiety of usaramine (112) is isomerized by UV light to the cis form, retrorsine (85) (Culvenor and Smith, 1966b); bromination of the latter and reduction of the product gives usaramine (Mattocks, 1968e). Periodate oxidation of the glycol group in the acid moiety converts retrorsine to an unstable keto-ester (299), which with methylmagnesium iodide forms isomeric products including senecionine (92) (see Chapter 3).

Partial hydrogenation of clivorine (15) gives ligularidine (68) (Hikichi *et al.*, 1979). De-oxygenation of the epoxide group in the acid moiety of petasitenine (81) gives senkirkine (95) as the major product (Yamada *et al.*, 1976b). The ethylidene group in another otonecine-based PA, neoligularidine (127), has been epoxidised giving ligularizine (126) (Asada and Furuya, 1984a).

III. CHEMISTRY OF DEHYDRO-NECINES, DEHYDRO-PAs, AND ANALOGOUS PYRROLES

The chemical properties of dehydro-necines (dihydropyrrolizine alcohols) resemble those of hydroxymethyl pyrroles. In contrast with their parent amino alcohols, the nitrogen has extremely weak basicity. Like other pyrroles they show nucleophilic carbon reactivity, especially adjacent to the N, exemplified by their reaction with Ehrlich reagent. In addition, they have electrophilic (alkylating) reactivity, and they are polymerized by acids. The dehydro-PAs (dihydropyrrolizine esters) are still more reactive, and form red polymers in presence of acids or moisture. Dehydro-necines and their esters are also decomposed by short wavelength UV radiation.

A. Nucleophilic Reactivity

Dihydropyrrolizine alcohols and esters give strongly coloured derivatives with Ehrlich reagent (4-dimethylaminobenzylaldehyde) in acid solution: Lewis acids

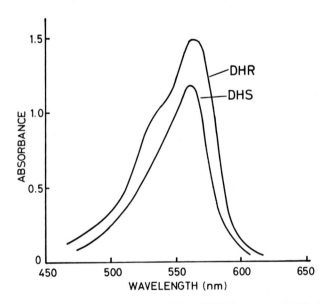

(343)

such as boron trifluoroide are the most effective (Mattocks, 1967c). The reaction is characteristic of pyrroles with an unsubstituted α carbon (Alexander and Butler, 1976), and the coloured product is presumably of the form (343) (Mattocks and Bird, 1983b). The spectra of the Ehrlich colours from the pyrrolic dehydrogenation products of many PAs are very similar, with λ_{max} typically 563–564 nm with an inflection near 530 nm, which is accentuated in hydroxylic solvents (Fig. 5.1) but is absent in the colour from dehydrosupinidine (345).

The compounds formed by alkylation of nucleophiles by dehydro-PAs also react with Ehrlich reagent. The colours from alkylation of amines are weaker than O-alkylation products; S-alkylation products give a strong colour: e.g. the molar absorptivities of colour from the compounds of dehydroretronecine with ascorbic acid, nicotinamide, azide and thiosulphate are approximately in the ratio $1 : 0.28 : 0.5 : 1.25$. The colour from S-alkylation has a λ_{max} ca. 6–10 nm higher than that from O-derivatives; this shift is cancelled if the S-derivative is treated with $HgCl_2$, either before or after the reaction with Ehrlich reagent.

The relative nucleophilic reactivity of some dihydropyrrolizine-alcohols and

Fig. 5.1. Spectra of coloured products from reaction of Ehrlich reagent with dehydroretronecine (DHR) and dehydrosupinidine (DHS).

pyrrolic alcohols is indicated by a comparison of their rates of reaction with Ehrlich reagent (Table 5.1). The results demonstrate the similarities between dihydropyrrolizines and analogous pyrroles, and the enhancement of reactivity by electron donating substituents (Me, MeO). Under certain conditions, pyrrolic alcohols and esters give a much weaker Ehrlich colour than expected. This is especially noticeable using pyrroles with high alkylating reactivity in predominantly aqueous solution. The colour may then be a small fraction of that formed in ethanol solution. This is because in acidic solution a proportion of the pyrrole polymerizes before it can react with Ehrlich reagent: the polymer often appears as a precipitate. In ethanol, the alcohol can undergo rapid acid-catalysed reaction with the pyrrole to give an ethoxy-derivative which is resistant to polymerization but is still able to undergo the Ehrlich reaction. A similar situation occurs in aqueous solution if a large excess of a nucleophile (Nu) such as ascorbic acid, able to react with the pyrrole, is dissolved in it. The ascorbic acid effectively 'protects' the pyrrole against acid catalysed polymerization and thus allows the formation of maximum Ehrlich colour. In a typical experiment, the Ehrlich

TABLE 5.1 Pseudo First-order Reaction Rates of Some Pyrrolic Alcohols with Excess of Ehrlich Reagent[a]

HO (344)

R OH (345) R = H (201) R = OH

N Me (346)

Compound		Ehrlich colour: λ_{max}	K (min^{-1})	$t_{1/2}$ (min)
(344)		562	0.069	10.0
(345) Dehydrosupinidine		562	1.16	0.6
(201) Dehydroretronecine		565	0.041	16.9
(346) Position of CH$_2$OH:	Other substituents			
2		566	0.071	9.7
3		566	0.04	17.3
3	2-Me	565	0.99	0.7
2,3		568	0.029	23.9
2,3	4-MeO	546	0.90	0.77

[a] Reagent was chosen for slow reaction times. Mixtures consisted of 2.5 μmol pyrrole and 124 μmol 4-dimethylaminobenzaldehyde in ethanol containing $0.2\,M$ HCl and $1.8\,M$ acetic acid, total volume 3.5 ml, at 30°C.

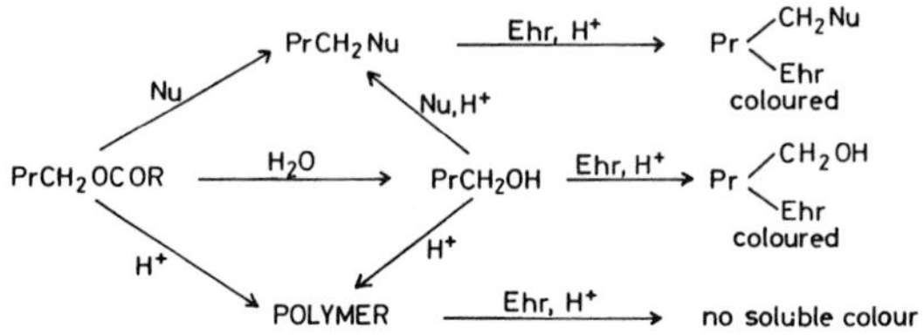

Scheme 5.5.   Pr, Pyrrole; Nu, nucleophile; Ehr, 4-dimethylaminobenzaldehyde.

colour from aqueous dehydroretronecine had ϵ_{max} 13,000; in the presence of 500-fold ascorbic acid the ϵ_{max} was 71,000. Scheme 5.5 summarizes the reactions which can occur. The enhancement of Ehrlich colour thus afforded by nucleophiles in aqueous solution has several practical applications: (1) it identifies pyrrolic derivatives which possess alkylating reactivity; (2) in analytical applications, it allows maximum Ehrlich colour to be obtained from a reactive pyrrole (Mattocks and White, 1970); (3) it can be used as evidence that a nucleophile is capable of being alkylated by a reactive pyrrole (Mattocks and Bird, 1983b).

B. Electrophilic Reactivity

1. Mechanism

Alcohols and esters linked through one carbon atom to the pyrrole ring are made labile by conjugation with the nitrogen. Loss of the oxygen function, which is catalysed by H^+ (Culvenor et al., 1970b), leaves a positively charged pyrrole nucleus which can react with nucleophiles, i.e. act as an alkylating agent (Scheme 5.6). Both the 7- and 9-oxygen functions in (347) can be similarly activated; thus through successive reactions it can act as a bifunctional alkylating agent. The proposed S_N1 character of this reaction has been confirmed in some instances by the high degree of racemization at C-7 when dehydro-alkaloids react with alcohols or amines (Mattocks, 1969a; Culvenor et al., 1970b).

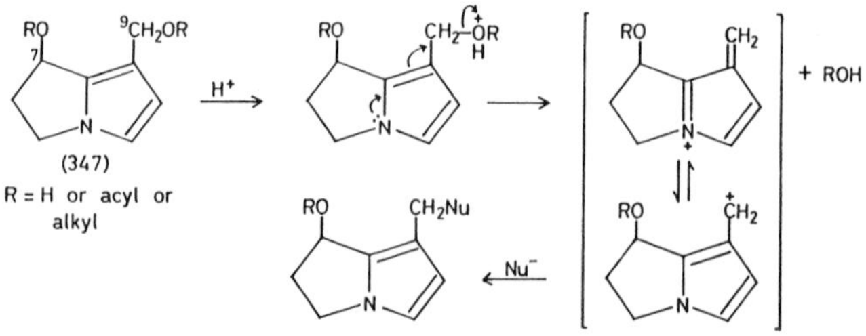

Scheme 5.6.

2. Alkylation by Dehydro-PAs and Dehydro-necines

a. Alkylation of Alcohols. Dehydro-PAs (348) react with methanol at room temperature, with or without the presence of sodium methoxide, to give the racemic diether (349) (Mattocks, 1969a; Culvenor *et al.*, 1970b). Dehydroheliotridine (272) reacts very rapidly with methanol at both hydroxyls in the presence of acid; with methanol alone it reacts slowly and only the 7-hydroxyl is replaced. Monoesters (350) such as dehydroheliotrine react with methanol to give a diether (349). This is because the liberated heliotric acid catalyses alkylation at the less reactive 7-position; if sodium bicarbonate is added to neutralize this acid, the main product is the 9-ether (351) (Culvenor *et al.*, 1970b).

Hydroxymethyl pyrroles behave in a similar way. Thus the dialcohol (279) alone fails to react with methanol (60 h), but with a trace of acid the diether (352) is formed within 2 min, and the acetate (280) without acid gives the same product.

b. Alkylation of Thiols. Dehydroretronecine and hydroxymethyl-*N*-methylpyrroles fail to react with neat thiols (ethyl or benzylmercaptan or 2-mercaptoethanol; up to 90 min; 100°C), but thioethers (353) are readily formed in the presence of water and a trace of acid. The acetate (280) shows little reaction with neat ethanethiol; it reacts slowly with methanethiol if water is present. The carbamate ester (354) refluxed with EtSH in aqueous ethanol gives the thioether (355), not the corresponding *O*-ether.

Dehydroretronecine (201), incubated with cysteine in phosphate buffer at pH 5.7, reacts at the 7-position to give (356); an analogous product is given with glutathione at pH 8 (Robertson *et al.*, 1977).

c. Alkylation of Amines. Culvenor *et al.* (1970b) showed that dehydrosenecionine (357) reacts with ammonia and methylamines to give 7,9-diamines (358). The products are partially or completely racemized at the 7-position. Dehydroheliotrine (350) reacts with amines only at its esterified (9-) position. For example with triethylamine, the quaternary product (359) is formed. A dehydro-PA can also alkylate its parent alkaloid. Dehydroheliotrine (350) reacts with heliotrine (49) to give a quaternary product, isolable as its chloride (360) (Culvenor and Smith, 1969; Culvenor *et al.*, 1970b).

d. Alkylation of Nucleosides and Nucleic Acids. Dehydroheliotridine (DHH) interacts with DNA under mildly acid conditions, as shown by UV spectra; the results are consistent with acid catalysed alkylation of the DNA having taken place (Black and Jago, 1970). Further evidence for the binding of DHH with DNA was given by Curtain and Edgar (1976) using radioactive labelling. White and Mattocks (1972) found evidence that dehydromonocrotaline and dehydroretrorsine can cause cross-linkage of DNA *in vitro;* unlike the much less reactive DHH, these pyrroles are active over a wide pH range (5.5–9). The synthetic bifunctional pyrrole carbamate (361) has the same action as dehydromonocrotaline, whereas the monofunctional pyrrole (354) is ineffective.

Robertson (1982) showed that deoxyguanosine is alkylated by dehydroretronecine at PH 7.4. Two products were identified: the 7α adduct (362) of dehydroretronecine with the N-2 position of deoxyguanosine, and the corresponding 7β adduct [see also Wickramanayake *et al.* (1985).]

e. Alkylation of Enols and Phenols. Dehydroretronecine (DHR) (as well as other pyrrolic alkylating agents) interacts reversibly with ascorbic acid. The reaction product, formed using a large excess of ascorbic acid, has not been purified; it is acidic, and can be isolated on anion exchange resin (Mattocks and Bird, 1983b). Ascorbic acid is an ambident nucleophile, i.e. its anion can exist in either of the two forms (363) or (364); thus it can be alkylated at either C or O (Buncel *et al.*, 1965; Edgar, 1974). The site of reaction of DHR is not known, but it may well be C-2, since the chemically closely related compound 3-hydroxymethyl indole reacts with ascorbic acid at this position to give ascorbigen (365) (Kiss and Neukom, 1966). DHR reacts, apparently in a similar way, with other enolic compounds including oxalacetic acid (366) and dihydroxyfumaric acid (367).

(361) R = CH$_2$OCONHEt
(354) R = H

(362)

(363)   (364)   (365)

(366) R = H
(367) R = OH

Some phenolic compounds, including resorcinol, pyrogallol and 1-naphthol, but not phenol itself, can interact with DHR in acid solution (Mattocks and Bird, 1983b). The nature of the reaction products is again not known, but phenols also can behave as ambident nucleophiles and, depending on the solvent, either O- or C-alkylation is possible, an aqueous system favouring the latter (Kornblum *et al.*, 1963).

f. Other Alkylation Reactions. Mattocks and Bird (1983b) devised a test to detect the interaction of DHR with a large excess of aqueous substrate at pH 5: this depended on the relatively high resistance of the alkylation products to acid-catalysed polymerization, compared with DHR itself; reaction products were not isolated. Comparison of results with a variety of compounds suggests which groups are most labile to interact with DHR. Thus, many amino acids fail to react; reactions of citrulline, histidine, and tryptophan are presumably at the urea group, the imidazole ring and the indole N, respectively. Pyridines, including nicotinic acid and amides, react at the heterocyclic N; pyridinium salts do not react. Thymidine and uridine do not react; cytidine shows some reaction, presumably at the 4-NH_2. Guanine and adenine derivatives react, as do the pyridine nucleotides NAD and NADP, probably at the adenine moiety. Whereas thiols react with DHR, thioethers like methionine and S-methyl-cysteine do not. The highly nucleophilic sulphite, thiosulphate and azide ions react strongly, but cyanide and thiocyanate do not. DHR also intereacts with some ureides, including barbituric and parabanic acids, but not with uric acid or hydantoin. It reacts with protein (albumin) but not with carbohydrate hydroxyls (inositol).

Sun *et al.* (1977) also demonstrated an interaction between DHR and albumin at pH 4.2, but not at pH 7.

3. Hydrolysis of Dehydro-PAs

The dehydro-alkaloids and similar pyrrolic esters hydrolyse extremely rapidly to corresponding alcohols in neutral or basic aqueous solution (polymers being formed in acid solution). Thus the initially high reactivity of dehydroretrorsine with 4-*p*-nitrobenzylpyridine is lost less than 1 min after it has been added to aqueous buffer at pH 7.2, leaving a persistent, lower alkylating reactivity due presumably to the hydrolysis product (dehydroretronecine) (Mattocks, 1969a). Similarly, the ability of dehydromonocrotaline to cross-link DNA is lost very quickly after it is exposed to water, owing to hydrolysis (White and Mattocks, 1972).

4. Kinetics of Alkylation Reactions

a. Reactivity of Pyrrolic Alkylating Agents. Alkylating reactivities have been assessed by the rates of formation of a coloured product with 4-*p*-nitro-

TABLE 5.2 Relative Rates of Alkylation
of 4-*p*-Nitrobenzylpyridine
by Some Dihydropyrrolizine Derivatives[a]

Compound	k (min^{-1})[b]	$t_{1/2}$ (min)
Dehydrosupinidine (345)	0.85	0.82
Dehydroretronecine (201)	0.033	20.8
Diacetyl dehydroretronecine	ca. 17.4[c]	ca. 0.04[c]
Dehydromonocrotaline (276)	0.76	0.91
Dehydroretrorsine (336)	ca. 1.7; 0.031[d]	ca. 0.4; 22[d]

[a] Approximately first-order reaction using 230-fold excess of 4-*p*-nitrobenzylpyridine in aqueous acetone at 50°C (method of Mattocks, 1978a).

[b] Pseudo first-order rate constant.

[c] Rate curve indicates biphasic reaction. Figures represent initial (fast) reaction.

[d] Approximate values deduced from biphasic rate curve.

benzylpyridine (NBP). Some dihydropyrrolizines are compared in Table 5.2, and synthetic pyrrole derivatives in Table 5.3 (cf. Mattocks, 1978a). Karchesy and Deinzer (1981) have also compared the alkylating activities of some dehydroretronecine derivatives under somewhat similar conditions. These results show the following. (1) Pyrrolic esters are more reactive than corresponding alcohols. (2) The reactivity of the oxygen function depends on its location in the molecule; e.g. in N-methylpyrroles, 2-CH_2OH is more reactive than 3-CH_2OH. Thus when two alkylating groups are present they may differ considerably in reactivity: in this case the initial reaction rate reflects mainly the more reactive centre. (3) Reactivity of esters depends on the structure of the acid moiety; thus acetyl esters are more reactive than corresponding pivalyl esters. (4) Reactivity is also affected by substituents on the pyrrole ring; it is enhanced by electron-donating substituents such as Me and MeO, and is lower in N-phenyl than in N-methyl pyrroles. The pyrrolidine ring in dehydrosupinidine (Table 5.2) has a similar influence to that of the 2-Me substituent in 3-hydroxymethyl-1,2-dimethylpyrrole (Table 5.3). (5) Pyrrole-like alkylating activity is also shown by analogous indole and 4-dimethylaminobenzyl esters, which are activated by a similar mechanism. (6) Besides alcohols and esters, pyrrolic ethers (e.g. methyl-ether and TMS-ether) also have alkylating reactivity (Table 5.3; see also Karchesy and Deinzer, 1981).

b. Kinetics under Pseudo First-order Conditions. Alkylation of NBP has been studied using a large excess of the reagent in aqueous acetone (Mattocks, 1969a, 1978a; Karchesy and Deinzer, 1981). Our recent results with pyrrolic

TABLE 5.3 Relative Rates of Alkylation of 4-*p*-Nitrobenzylpyridine by Some Pyrroles and Related Compounds[a]

Structure	Position(s) of -CH$_2$OR	R	Other substituents	k (min^{-1})	$t_{1/2}$ (min)
(i)	2	H		0.12	5.9
(i)	2	Ac		7.5	0.09
(i)	2	Cb		3.97	0.17
(i)	3	H		0.014	49.5
(i)	3	Ac		—	ca. 4
(i)	3	Cb		0.14	4.8
(i)	3	H	2-Me	0.74	0.94
(i)	3	Cb	2-Me	11.4	0.06
(i)	3	H	2-CH$_2$CH$_2$OH	0.31	2.2
(i)	3	Cb	2-CH$_2$CH$_2$OCb	0.80	0.87
(i)	3	Piv	2-CH$_2$CH$_2$OPiv	0.33	2.1
(i)	2,3	H		0.1	6.9
(i)	2,3	Ac		2.03	0.34
(i)	2,3	Cb		1.24	0.56
(i)	2,3	Piv		0.44	1.58
(i)	2,3	Me		0.01	49.5
(i)	2,3	SiMe$_3$		0.09	8.0
(i)	2,3	H	4-MeO	0.24	2.9
(i)	2,3	H	5-Me	1.68	0.41
(i)	3,4	H		0.03	20.6
(i)	2,5	H		0.14	5.0
(i)	3,5	H		0.06	10.8
(ii)	2,3	H		0.4	1.73
(ii)	2,3	Ac		7.5	0.09
(ii)	2,3	Piv		0.78	0.89
(iii)	—	H		0.14	4.9
(iii)	—	Cb		2.92	0.24
(iv)	—	Ac		1.25	0.56

[a] See Table 5.2, footnote (*a*).
[b] Cb, *N*-Ethylcarbamoyl (CONHEt); Piv, pivaloyl (COBu^t).

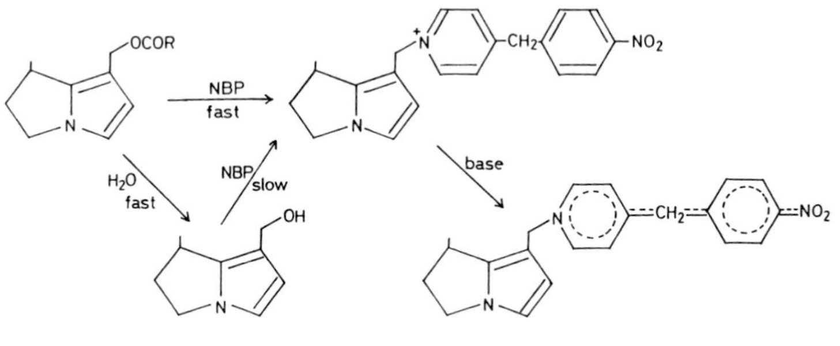

Scheme 5.7.

alcohols and many of their esters (Tables 5.2 and 5.3) have generally shown simple first-order kinetics, as expected. However, in some instances there is a biphasic response, with evidence of a very fast and a much slower reaction occurring simultaneously: this is usually seen with the more reactive pyrrolic esters, such as dehydroretrorsine. The fast alkylation has been attributed to reaction of the pyrrolic ester with NBP, and the slower alkylation to reaction of the corresponding pyrrolic alcohol, formed by hydrolysis of the ester (Mattocks, 1969a): see Scheme 5.7. However, this has been disputed by Karchesy and Deinzer (1981), who obtained biphasic rate curves using some pyrrolic alcohols as well as esters.

c. Reversibility of Alkylation Reactions. Alkylations by pyrrolic esters are not reversible, in that the ester cannot be re-formed. However, the alkylation product may dissociate, the bound nucleophile being replaced by another (e.g. OH$^-$ in aqueous solution). The driving force for reversibility is the tendency of the pyrrolic nucleus to form a resonance-stabilized carbonium ion. The stability of the pyrrole-nucleophile complex depends also on the nature of the nucleophile: thus S-derivatives are relatively stable, whereas O- and quaternary N-derivatives are less stable and can themselves act as alkylating agents. This is demonstrated by their ability to react with 4-p-nitrobenzylpyridine.

To illustrate reversal of alkylation, the dissociation of a compound from a reactive pyrrole and a pyridine derivative is shown in Scheme 5.8. Alkylations of pyridine nitrogen are known to be reversible (Brown and Cahn, 1955), and pyridinium salts are capable of alkylating activity (Brewster and Eliel, 1953): here this action is greatly enhanced by the pyrrole moiety. The pyrrolic cation can react with other nucleophiles (Nu$^-$) including OH$^-$; thus in aqueous solution the pyrrole–pyridinium compound can hydrolyse to the hydroxy-pyrrole; pyrrolic polymer can also be formed.

The reversible reaction of dehydroretronecine (DHR) with nicotinamide under

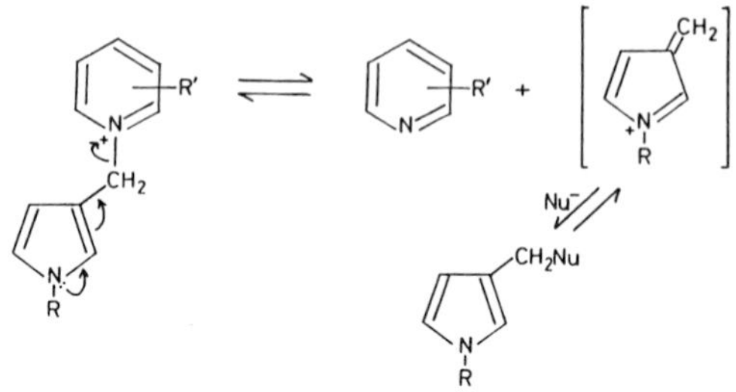

Scheme 5.8.

physiological conditions might play a part in forming a reservoir of 'stabilized' pyrrolic alkylating agent which could subsequently interact with other tissue constituents (Mattocks and Bird, 1983b). A similar role has been proposed by Sun *et al.* (1977) for the reversible complex formed between DHR and albumin. Curtain and Edgar (1976) demonstrated that the binding of dehydroheliotridine with DNA is reversible, the binding being low at pH levels above 7; they pointed out that although this binding is covalent, it might be preceded by noncovalent binding, e.g. through intercalation with the DNA.

A comparison of equilibria between dehydroretronecine and some nucleophiles (Table 5.4) shows that binding is often weak; the strongest observed is

TABLE 5.4 Equilibrium (Stability) Constants
for Alkylation of Some Nucleophiles
by Dehydroretronecine (DHR) at 37°C in Aqueous
Solution, pH 5 under Comparable Conditions[a]

Nucleophile (Nu)	K (approx.)[a,b]
Ascorbic acid	0.072
Nicotinamide	0.059
Adenine	0.17
Guanosine 5'-monophosphate	0.033
Tryptophan	0.06
Azide ($N_3{}^-$)	0.013
Thiosulphate ($S_2O_3{}^{2-}$)	0.5

[a] Our unpublished results.
[b] For reaction:

DHR + Nu $\rightleftharpoons$ product

$$K = \frac{c_{\text{product}}}{c_{\text{DHR}} c_{\text{Nu}}}$$

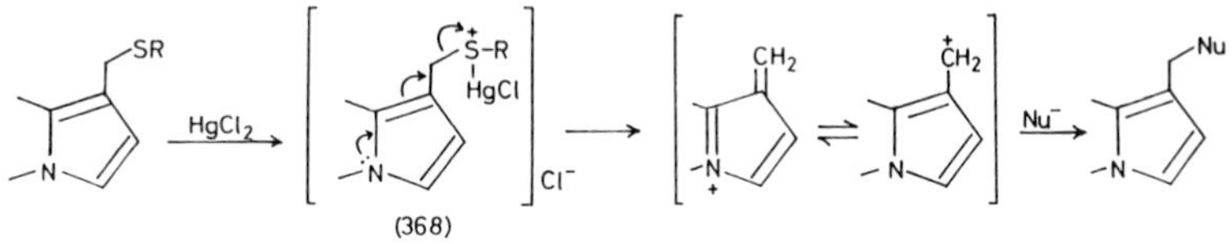

Scheme 5.9.

that with adenine and with the thiosulphate ion, which is highly nucleophilic (Swain and Scott, 1953). A consequence when binding is weak is that a high proportion of the pyrrole remains unreacted unless a large excess of the nucleophile is present; isolation of the product can be difficult.

5. Cleavage of S-Alkylation Products by Metal Salts

Pyrrolic thioethers are relatively stable, as shown by their failure to alkylate NBP. However, after being treated with mercuric chloride they are highly reactive towards NBP. A suggested mechanism is shown in Scheme 5.9: the initially formed cation (368) can readily lead to a pyrrolic carbonium ion, which can alkylate nucleophiles (Nu^-) and be hydrolysed by aqueous alkali to the pyrrolic alcohol. Silver salts have a similar action to mercuric salts on pyrrolic thioethers.

C. Polymerization

Dehydro-necines and similar pyrrolic alcohols are moderately stable in alkaline aqueous solution, but in acidic solution they rapidly form insoluble red polymers. Their esters (dehydro-PAs) are still less stable, being polymerized by a trace of moisture, which liberates the acid moiety by hydrolysis. The rates of polymerization of pyrrolic alcohols in aqueous solution are directly proportional to the H^+ concentration (Fig. 5.2). Thus at pH 5 and 25°C the half-times for polymerization of 10^{-3} M dehydroretronecine (DHR), 2,3-bishydroxymethyl-1-methylpyrrole, and 2-hydroxymethyl-1-methylpyrrole are 63, 28 and 6.8 min, respectively; these times are inversely related to the pyrrole concentration, so that $t_{1/2}$ for 0.1 M DHR at pH 5 is 0.63 min. The rate of polymerization of DHR at pH 5 and 37°C is about four times that at 25°C.

The exceptional susceptibility of these pyrroles to polymerization is due to their possession of both nucleophilic and electrophilic centres. An indication of the tendency of a pyrrolic alcohol to polymerize is given by considering both its nucleophilic reactivity, as measured by its rate of reaction with Ehrlich reagent, and its electrophilic reactivity towards NBP. Thus the relative rates of polymerization of different pyrroles may correlate poorly with either of these two

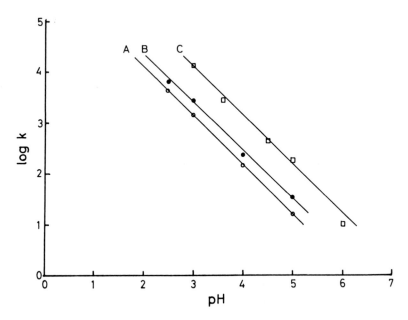

Fig. 5.2. Relationship between pH and rates of polymerization of pyrrolic alcohols, in aqueous solution at 25°C. A, Dehydroretronecine; B, 2,3-bishydroxymethyl-1-methylpyrrole; C, 2-hydroxymethyl-1-methylpyrrole. Here k is the second-order rate constant (l mol^{-1}min^{-1}).

TABLE 5.5 Comparison of Relative Rates of Polymerization
of Some Pyrrolic Alcohols with Their Nucleophilic and Electrophilic Reactivities
towards Ehrlich Reagent and 4-p-Nitrobenzylpyridine (NBP)

Compound	Relative[a] Ehrlich reaction rate[b] (i)	Relative[a] NBP reaction rate[c] (ii)	(i) × (ii)	Relative[a] rate of polymerization
Dehydroretronecine	1.0	1.0	1.0	1.0
2,3-Bishydroxymethyl-1-methylpyrrole	0.71	3.03	2.15	2.25
2-Hydroxymethyl-1-methylpyrrole	1.73	3.64	6.3	9.3

[a] Relative to DHR = 1.
[b] From results in Table 5.1.
[c] From results in Tables 5.2 and 5.3.

reactivities, yet show quite good correlation with the product of these reactivities (Table 5.5).

Protonation of the pyrrolic N would be expected to inhibit both the alkylating reactivity and polymerization; however, assuming the pK_a to be similar to that of pyrrole itself (about 0.4), this inhibition would occur only at an extremely low pH.

6

Metabolism, Distribution and Excretion of Pyrrolizidine Alkaloids

I. METABOLISM AND THE CYTOTOXICITY OF PAs

There are close links between the cytotoxic actions of PAs and their metabolism in animals. Hence, routes of metabolism will be discussed in this chapter with particular emphasis on their relationships to PA toxicity.

Evidence suggesting that it is metabolites, not the alkaloids themselves, which are responsible for toxic effects, includes the following (Mattocks, 1972a)

1. PAs are not locally toxic at the site of injection, or when they are applied to the skin.

2. The alkaloids do not injure some organisms, e.g. cinnabar moth larvae, even though these are able to accumulate relatively large amounts of PAs in their tissues (Aplin *et al.*, 1968).

3. The main organ damaged is usually the liver, regardless of the route of administration of the alkaloid; PAs are known to be metabolized in the liver.

4. The susceptibility of animals to PA intoxication is considerably influenced by treatments which modify the activity of the hepatic drug metabolizing enzymes.

5. Some animals are much more resistant than others to PA hepatotoxicity. Thus, guinea pigs can withstand 20 times the dose of retrorsine that is an LD_{50} for male rats; stimulation of their liver microsomal enzymes by pretreatment with phenobarbitone increases the susceptibility of guinea pigs fourfold (White *et al.*, 1973). Newborn rats given retrorsine within 1 h of birth are more resistant to the chronic hepatotoxicity of the alkaloid than are rats aged 1 day or more when their hepatic microsomal enzyme activity is much greater (Mattocks and White, 1973).

6. PAs are chemically rather unreactive. It is unlikely that they would be able to react with cell constituents under physiological conditions. On the other hand, metabolites known to be formed from PAs in the liver are highly reactive and more cytotoxic than their parent alkaloids.

Once ingested by an animal, a PA can follow various pathways of disposal, and its eventual biological effects will depend on the balance of these. Of the alkaloid which is not excreted unchanged, some may be activated by conversion to toxic metabolites, and some detoxified to harmless metabolites which can then be eliminated from the body. In laboratory animals, the principal known routes of metabolism are ester hydrolysis, conversion to N-oxides, and dehydrogenation to pyrrolic derivatives. The first two appear to be detoxication pathways. The last is the only one known to be associated with cytotoxicity. It is also the most studied, because the pyrroles are easily detected and measured, so that perhaps too little attention has been given to other routes of metabolism. Nevertheless, the properties of pyrrolic metabolites are consistent with the known cytotoxic action of PAs, and it appears improbable, though not impossible, that other metabolic routes of PA activation may remain to be discovered.

II. METABOLIC PATHWAYS OF HEPATOTOXIC PAs IN ANIMAL TISSUES

A. Hydrolysis

Hydrolysis of a toxic pyrrolizidine ester leads to the necine and necic acid moieties, neither of which are hepatotoxic; thus hydrolysis amounts to detoxication. Whereas the chemical hydrolysis and associated reactions of various PAs have been extensively studied (Bull *et al.*, 1968), less is known of the biological hydrolysis of these compounds. There are three kinds of evidence relating to the hydrolysis of PAs and analogous compounds by mammalian enzymes: (1) indirect, in that metabolic activation and hepatotoxicity to hydrolysis-prone PAs is enhanced in animals whose esterase activity is inhibited; (2) direct measurements of hydrolysis by esterase preparations *in vitro;* and (3) identification of hydrolysis products formed *in vivo* or *in vitro*.

1. Effects of Esterase Inhibition

Semi-synthetic PAs which are esters of retronecine with a simple acid moiety are easily hydrolysed compared with natural PAs, and they probably owe their relatively low hepatotoxicity to hydrolytic detoxication (Mattocks, 1982b). Thus, the hepatotoxicity of diacetyl retronecine is only expressed in rats which have also been given an esterase inhibitor, which makes a larger proportion of the dose available for conversion to toxic metabolites (Mattocks, 1970). Table 6.1 shows the effects of predosing rats with an esterase inhibitor (TOCP) on their ability to convert PAs and analogues into toxic pyrrolic metabolites. The enhancement of the latter shows the extent to which a compound is subject to

TABLE 6.1 Pyrrolic Metabolite Levels in Livers of Male Rats
2 h after Equimolar i.p. Doses of PAs and Analogues[a]

| | | Relative pyrrole level[b] | | |
| | | In normal rats | In rats predosed with TOCP[c] | Predosed/ normal |
Compound	Number			
Retronecine ditiglate	(241)	1.19	2.47	2.1
Heliotridine ditiglate	(370)	0.53	2.06	3.9
Synthanecine A ditiglate	(371)	0.07	0.78	11.1
Synthanecine A diacetate	(372)	0.045	0.34	7.6
Synthanecine A bis(*N*-ethylcarbamate)	(373)	3.5	—	—
Synthanecine A bis(diethylphosphate)	(374)	2.5	—	—
Retronecine bis(*N*-ethylcarbamate)	(249)	2.91	—	—
Monocrotaline	(74)	1.89	3.47	1.8
Heliotrine	(49)	0.64	0.62	1.0

[a] Data from Mattocks (1978b, 1981a).
[b] Arbitrary units: extrapolated from colourimetric estimations on 0.5-g liver samples.
[c] TOCP, Tri-orthocresylphosphate, 0.5 ml/kg given p.o. 1 day before the alkaloid.

hydrolytic detoxication in normal rats. For example, a comparison of tiglate esters shows how the amino alcohol moiety can affect hydrolysis. In retronecine ditiglate (241), the ester groups provide mutual steric hindrance; in heliotridine ditiglate (370), they are further apart and hence less hindered; in the synthanecine ditiglate (371) they are free to rotate and much more accessible. Accordingly these three compounds show progressively less conversion to pyrroles in normal rats and greater enhancement after esterase inhibition (Mattocks, 1981a). The results show the natural PA monocrotaline to be less prone to hydrolysis *in vivo,* while heliotrine is hydrolysed too slowly to significantly reduce its conversion to pyrroles. This is consistent also with the very low chemical rate of hydrolysis of heliotrine: less than 1/600 that of monocrotaline (Bull *et al.,* 1968).

(241) R = Tig
(249) R = CONHEt

(370)

(Tig =)

(371) R = Tig
(372) R = Ac
(373) R = CONHEt
(374) R = PO(OEt)₂

Esters such as carbamates and phosphates which are themselves inhibitory to esterases (Baron *et al.,* 1966; Aldridge and Reiner, 1972), are converted to relatively high levels of pyrrolic metabolites in rats (Table 6.1). Thus retronecine bis(*N*-ethylcarbamate) (249) gives over twice the liver pyrrole level compared with the tiglate (241), while the synthanecine carbamate (373) and phosphate (374) esters give 36–60 times more pyrrole than corresponding acetate and tiglate esters (Mattocks, 1978b, 1981a).

2. Hydrolysis Rates in Vitro

Mattocks (1982b) compared the rates of hydrolysis of various retronecine diesters (Table 6.2). Steric hindrance around the ester groups is a major factor inhibiting hydrolysis, whether catalysed by base or by esterases in a rat liver homogenate. As might be expected, the more hindered natural PAs are more resistant to hydrolysis than simpler, semi-synthetic esters. However, in a series of retronecine esters with unbranched aliphatic acids, the rate of enzymic hydrolysis increases with chain length up to a maximum for the valeryl ester.

3. Identification of Hydrolysis Products

The semi-synthetic PA retronecine di-isovalerate (306) is hydrolysed *in vitro* by a rat liver esterase preparation first at the chemically more sensitive allylic primary ester group to give retronecine 7-isovalerate (375) (Mattocks, 1982b).

TABLE 6.2 Comparison of Rates of Base- and Enzyme-Catalysed Hydrolysis of Some Retronecine Diesters[a]

Retronecine diester RCOO—/—OCOR	R	Basic hydrolysis		Enzymic hydrolysis rate (μmol/min/g liver)[d]
		$t_{1/2}$ (h)[b]	Relative rate[c]	
Acetate	Me-	0.21	240	2.4
Propionate	MeCH$_2$-	0.23	217	8.8
Butyrate	Me(CH$_2$)$_2$	0.94	53	16.5
Valerate	Me(CH$_2$)$_3$-	0.75	67	20
Hexanoate	Me(CH$_2$)$_4$-	1.4	36	3.9
Isobutyrate	Me$_2$CH-	1.2	42	8.5
Isovalerate	Me$_2$CHCH$_2$-	2.4	21	4.3
Cyclopentanecarboxylate	CH$_2$(CH$_2$)$_3$CH-	0.9	56	17.7
Benzoate	C$_6$H$_5$-	2.5	20	0.6
Tiglate	MeCH=CH(Me)-	10	5	—
t-Butylacetate	Me$_3$CCH$_2$-	16	3.1	vs[e]
Pivalate	Me$_3$C-	18	2.8	0.5
Senecioate	Me$_2$C=CH-	38	1.3	0.35
2,2-Dimethylvalerate	MeCH$_2$C(Me$_2$)-	50	1.0	—
Natural PAs				
Retrorsine		61	0.8	vs
Monocrotaline		35	1.4	vs
Indicine		—	vs	—

[a] Adapted from Mattocks (1982b).
[b] Approximate half-life of one ester group at 23°C, initial conc. 0.02 M, pH 12.
[c] Pseudo first-order rates relative to 2,2-dimethylvalerate = 1.0.
[d] Anaerobic rates in rat liver homogenate at pH 7.5, 37°C.
[e] vs, very slow.

(306) R = CO—/—Me
(375) R = H

When a non-hepatotoxic dose (300 mg/kg i.p.) of synthanecine A ditiglate (371) was given to male rats, a 24-h urine sample yielded pyrrolic metabolites (3%), pyrroline N-oxides (13%) and pyrroline bases (91%). The pyrroline fraction contained no diester and only a trace of monoester: i.e. the majority of the ester had been hydrolysed (Mattocks, 1978b). In contrast, when the highly toxic N-ethylcarbamate (373) was given to rats (40 mg/kg), the urine yielded pyrrolic metabolites (7.5%), *N*-oxides (1.8%) and pyrrolines (23.8%), mainly in the form of unhydrolysed carbamate. These results illustrate the inverse relationship that exists between hepatotoxicity and ester hydrolysis *in vivo*.

B. Dehydrogenation and *N*-Oxidation

1. Introduction

Two major kinds of metabolites are formed by enzymic oxidation in animals: *N*-oxides and pyrrolic derivatives (Scheme 6.1). These will be discussed together, because they are formed by similar enzymes and usually produced at the same time.

PAs are metabolized to *N*-oxides in rats (Bull *et al.*, 1968; Mattocks, 1968a) and in sheep (Jago *et al.*, 1969). The reaction is brought about by the hepatic microsomal system (Mattocks and White, 1971a). The reverse reaction, reduction of *N*-oxides to tertiary bases, which is easily brought about by chemical reducing agents, may also occur in mammalian liver (Mattocks, 1971c; Powis and Wincentsen, 1980), but this appears to be a relatively minor pathway, perhaps because the *N*-oxides once formed are less lipophilic than their parent alkaloids and readily excreted. However, substantial reduction of PA *N*-oxides can occur in the gut (Mattocks, 1971c).

Hepatotoxic PAs are also metabolized in animals to pyrrolic derivatives (dehydro-pyrrolizidine alkaloids; dihydropyrrolizines), which are readily detectable by the strong magenta colour they give with a modified Ehrlich reagent (4-dimethylaminobenzaldehyde). This colour reaction is given by the urine, liver, and sometimes the lung and other tissues of rats a short time after they have been given various PAs; it is also given by rat liver slices which have been incubated with PAs (Mattocks, 1968a). Like the *N*-oxides, pyrrolic metabolites are also formed from PAs by the action of microsomal enzymes in the liver: the process amounts to dehydrogenation of the unsaturated (pyrroline) ring of the necine

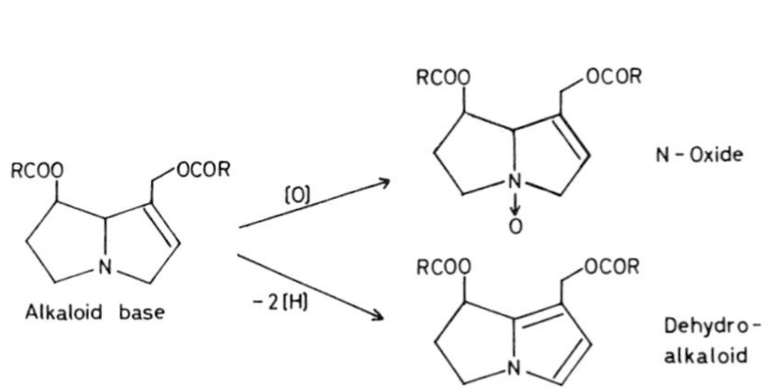

Scheme 6.1.

moiety, and is analogous to the dehydrogenation of these alkaloids, which is readily brought about by chemical oxidizing agents (Mattocks, 1969a; Culvenor *et al.*, 1970b,c).

2. *Conversion of PAs to Pyrroles and* N-*Oxides in* Vitro*: Enzyme Induction and Inhibition*

Experiments with subcellular fractions of rat liver show that conversion of PAs to both pyrrolic and *N*-oxide metabolites is catalysed by enzymes having properties characteristic of mixed-function oxidases, with a requirement for oxygen and reduced NADP (Mattocks and White, 1971a). The conversion of retrorsine to pyrroles is not significantly altered by pretreating rats with the microsomal enzyme inducer 3-methylcholanthrene, but it is increased threefold by pretreatment with phenobarbitone and fourfold by DDT. Both inducers cause a smaller increase in *N*-oxide production; this difference is also seen in the effect of phenobarbitone pretreatment on the metabolism of retrorsine by microsomes from mice and guinea pigs (White *et al.*, 1973) (see Table 6.3) and of monocrotaline by rat liver microsomes (Chesney *et al.*, 1974b). The capacity of liver microsomes to produce both metabolites is inhibited in the presence of carbon

TABLE 6.3 Activities of Microsomal Enzymes
from Male Animals of Various Species Converting Retrorsine
to Pyrrolic and *N*-Oxide Metabolites *in Vitro*[a]

Animal species	Specific enzyme activities (nmol/min/mg microsomal protein)		Approximate ratio, pyrrole/N-oxide
	For conversion to		
	pyrroles	N-oxides	
Rat	3.20	2.27	1.4
Mouse	2.65	1.58	1.7
Hamster	2.70	0.94	2.9
Guinea pig	0.31	1.08	0.3
Fowl	0.35	0.14	2.5
Quail	0.30	0.44	0.7
Sheep	0.46	0.70	0.7
After pretreatment with phenobarbitone			
Rat	9.9	5.3	1.9
Mouse	11.0	3.1	3.5
Guinea pig	1.12	2.1	0.5

[a] Data from Mattocks and White (1971a); White *et al.* (1973).

monoxide or SKF 525A. It is higher in microsomes from male rats than from female rats, and reduced in rats fed a protein-free diet or acutely intoxicated with retrorsine (Mattocks and White, 1971a). Pretreatment of rats with SKF 525A or chloramphenicol greatly inhibits their microsome activity for conversion of monocrotaline to pyrroles but not to N-oxide (Chesney et al., 1974b). These and other experiments have shown that there is no consistent relationship between rates of pyrrole and N-oxide formation by liver microsomes, and these metabolites are evidently formed via independent pathways (White and Mattocks, 1971; Mattocks and Bird, 1983a).

Guengerich (1977) showed that rabbit and rat liver microsome preparations incorporating different purified forms of cytochrome P-450 convert lasiocarpine to pyrrolic metabolites at different rates; rates of metabolism of a number of other toxic substrates were also compared.

Shull et al. (1977) have found that changes in dietary selenium status do not alter the ability of rat liver microsomes to metabolize PAs.

3. Effects of Age and Sex

The rate of conversion of retrorsine to pyrrolic metabolites by liver microsomes or liver tissue slices taken from rats immediately after birth is extremely low; it rises rapidly with age, so that in the liver of 5-day-old rats of either sex it

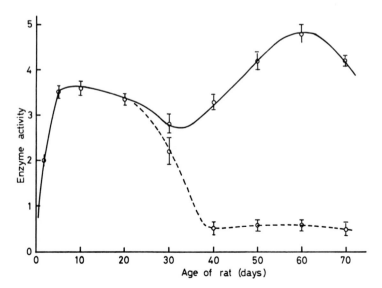

Fig. 6.1. Enzyme activity (nmol/min/mg microsomal protein) associated with the conversion of retrorsine to pyrrolic metabolites by liver microsomes *in vitro* from rats of various ages. Solid line, male rats; dashed line, female rats. From Mattocks and White (1973).

is nearly as high as in adult males (Mattocks and White, 1973). However, as rats become sexually mature, a difference emerges: the metabolic activity towards retrorsine remains high in male rats, but in females it steadily falls between 20 and 40 days of age, until it is about one-eighth that in males (Fig. 6.1). A similar, but not identical age related pattern is seen in the microsomal conversion of retrorsine to N-oxide.

4. Species Differences

Liver microsomal enzyme activities associated with the conversion of retrorsine into pyrrolic and N-oxide metabolites have been measured by White et al. (1973) in a number of animal species (Table 6.3). There are wide variations between species, both in microsome activities and in the ratio of pyrrole to N-oxide production. N-Oxide production from retrorsine is relatively high in the guinea pig, a species that is particularly resistant to PA toxicity; Chesney and Allen (1973a) obtained similar results with monocrotaline. Shull et al. (1976b) also compared rates of pyrrole production from monocrotaline and from mixed alkaloids of Senecio jacobaea, by liver microsome preparations from male animals of different species. The rates were in the decreasing order: hamster > rabbit > mouse > rat > beef steer > beef bull > wether lamb > chicken > Japanese quail; unconfirmed results also suggested that horse microsomes have very high activity. According to Hendricks et al. (1981), trout liver microsomes also are able to convert S. jacobaea alkaloids to pyrrolic metabolites, at a similar rate to rat liver microsomes.

5. Identity and Route of Formation of Pyrrolic Metabolites

Jago et al. (1970) showed that dehydroheliotridine (DHH) (272) is the main extractable pyrrolic metabolite formed when heliotrine (49) or lasiocarpine is metabolized in vitro by rat liver microsomes. The most likely route of metabolism is dehydrogenation of the alkaloid to give the pyrrolic ester, dehydroheliotrine (350). Unlike the parent alkaloid, this would be capable of hydrolysing rapidly in an aqueous environment to DHH, as well as behaving as an alkylating agent (Scheme 6.2), but would be too labile to be isolable. The identification of a quaternary product formed by alkylation of heliotrine by dehydroheliotrine (see Chapter 5) is good evidence that the latter is a primary metabolite: DHH itself is not capable of alkylating heliotrine under the reaction conditions used. An alternative possibility, that the alkaloid is first hydrolysed to heliotridine (184) and that this is then dehydrogenated to DHH, is improbable, since the latter is formed at a much lower rate from heliotridine than from heliotrine or lasiocarpine (Jago et al., 1970).

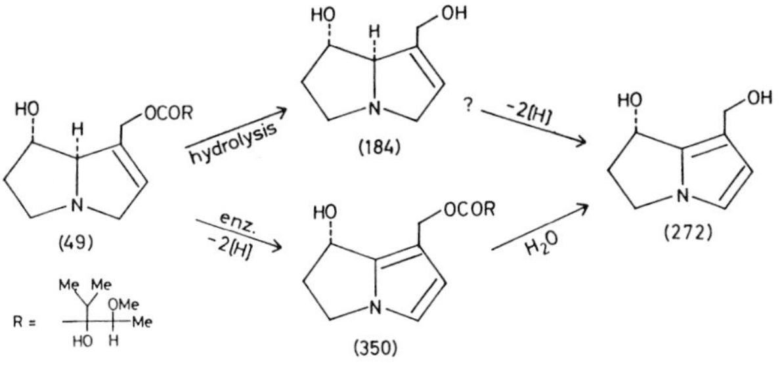

Scheme 6.2.

Apparently in a similar way, the enantiomer of DHH, dehydroretronecine (DHR) (201) is formed by the metabolism of the retronecine-based PA monocrotaline: Hsu *et al.* (1973b) isolated DHR from the liver, blood and urine of rats given this alkaloid. If we can extrapolate from the above results, it seems probable that, in general, hepatotoxic PAs are dehydrogenated by microsomal enzymes in the liver to dehydro alkaloids (dihydropyrrolizine esters). These primary pyrrolic metabolites are short-lived, and can hydrolyse rapidly to somewhat more stable dehydro-necines (dihydropyrrolizine alcohols), of which DHH and DHR are examples. Recent evidence (Kedzierski and Buhler, 1985) suggests that both DHH and DHR formed this way are really identical racemic mixtures of these two enantiomers.

6. *Mechanisms of Pyrrole and* N-*Oxide Formation*

It is well known that N-oxidation of a variety of tertiary amines is catalysed by hepatic microsomal enzymes, with a requirement for NADPH and oxygen: the conversion of PAs to *N*-oxides appears to be typical of this type of metabolism. However, microsomal N-oxidation may differ from other types of microsomal oxidative drug metabolism (Bickel, 1969), and differences between routes of pyrrole and *N*-oxide formation from PAs have been alluded to above. The detailed mechanism of PA N-oxidation awaits further investigation.

Pyrrolic derivatives of unsaturated PAs are easily made chemically from their *N*-oxides by what amounts to removal of the elements of water (Mattocks, 1967c, 1969a), and it might be supposed that pyrroles are similarly formed from the *N*-oxides *in vivo*. However, PA *N*-oxides are not metabolized to pyrroles by hepatic microsomal enzyme preparations *in vitro* (Jago *et al.*, 1970; Mattocks and White, 1971a), and in rats an i.p. dose of retrorsine *N*-oxide is converted to much less pyrrole than a similar dose of the parent PA (Mattocks, 1972b): thus the available evidence favours separate pathways of pyrrole and *N*-oxide formation, the latter being a detoxication pathway (Mattocks, 1971c; Mattocks and

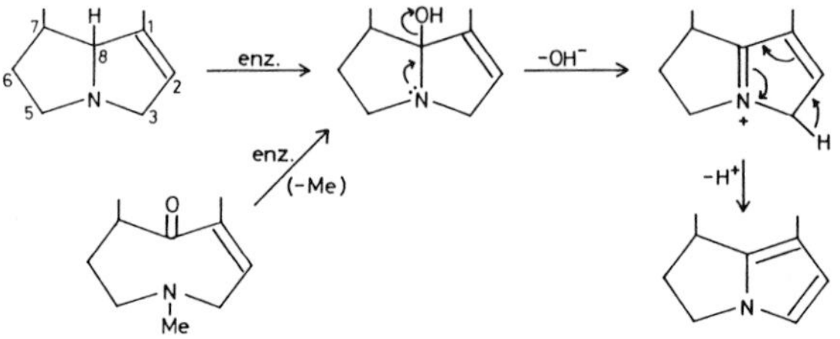

Scheme 6.3.

White, 1971a; Chesney *et al.*, 1974b; Mattocks and Bird, 1983a). It has been suggested (Mattocks, 1968c) that pyrrolic metabolites might be formed by a mechanism similar to microsomal dealkylation, involving hydroxylation of a carbon adjacent to the nitrogen (McMahon, 1966). Dehydrogenation would be expected to result from such hydroxylation at either C-3 (Mattocks and White, 1971a) or C-8 (Mattocks and Bird, 1983a) on the unsaturated necine moiety; the most recent results favour the latter pathway (Scheme 6.3).

Thus, enzymic hydroxylation of a PA at C-8 leads to the chemically unstable carbinolamine which decomposes spontaneously, losing first OH^-, then H^+, giving the dihydropyrrolizine. Otonecine-based PAs could be dehydrogenated by a similar route, after N-demethylation (Mattocks and White, 1971a).

The activation of a PA to a toxic pyrrole may be viewed as a 'metabolic mistake'. The microsomal system attempts to convert the alkaloid to a more water soluble derivative which can be more readily excreted, but the chemical instability of this intermediate leads instead to the formation of a more lipophilic and highly toxic metabolite.

7. *Effect of Molecular Structure on the Formation of* N-*Oxide and Pyrrolic Metabolites*

Rates of metabolism of different PAs by rat liver microsomes *in vitro* vary widely, dependent on the structure and physical properties of the alkaloid (Mattocks and White, 1971a; Mattocks and Bird, 1983a). The most lipophilic PAs are metabolized at the fastest rates (Table 6.4). There are also considerable differences between the proportions of pyrroles and N-oxides formed from PAs with different ester groups, and these appear to be due to the different degrees of steric hindrance caused by the acid moieties at the sites of these metabolic reactions (Mattocks and Bird, 1983a). The acid moieties can exert greater steric hindrance at C-8, the putative site of oxidation leading to pyrroles, than at the nitrogen (cf. Scheme 6.3). Accordingly, PAs which are 'open' diesters, able to cause the most hindrance at C-8, give the highest proportion of N-oxide compared with pyrrole, whereas monoesters, and also macrocyclic diesters in which movement of the acid moiety is restricted, give relatively more pyrrole.

TABLE 6.4 Relative Activities of Rat Liver Microsomal Enzymes
Converting Various PAs to Pyrrolic and N-Oxide Metabolites *in Vitro*[a]

Alkaloid[b]	Enzyme activity for conversion to pyrroles plus N-oxides (nmol/min/mg microsomal protein)	N-oxide as % of total activity	Approx. partition coefficient[c]
Retronecine esters			
Retrorsine (MD)	7.75	15	1.28
Senecionine (MD)	9.54	12	6.3
Monocrotaline (MD)	2.62	19	0.24
Dimethyldidehydrocrotalanine (MD)	10.56	31	3.3
Retronecine diacetate (D)	3.03	47	0.5
Retronecine ditiglate (D)	10.44	43	13.4
Retronecine 9-heliotrate (M)	1.45	21	low[d]
Indicine (M)	0.325	12	0.13
Heliotridine esters			
Lasiocarpine (D)	19.39	49	11.9
Heliotrine (M)	1.70	14	0.33
Heliotridine ditiglate (D)	8.27	35	high[e]
Supinidine ester			
Supinine (M)	0.86	13	0.16
Crotanecine ester			
Anacrotine (MD)	9.05	14	2.1

[a] Adapted from Mattocks and Bird (1983a).

[b] Type of ester shown in parentheses: M, monoester; D, 'open' diester; MD, macrocyclic diester.

[c] For system octan-1-ol: aqueous 0.1 M phosphate buffer, pH 7.4, at 23°C.

[d] Not measured, but cf. indicine and heliotrine.

[e] Not measured, but cf. retronecine ditiglate.

A comparison of retronecine esters, e.g. the heliotrate and ditiglate, with corresponding heliotridine esters (heliotrine and heliotridine ditiglate) shows that the different orientations at C-7 have relatively little effect on the balance between metabolic activation to pyrroles and detoxication by way of N-oxidation (see Table 6.4). Other factors, including ester hydrolysis, appear to be more important in determining relative toxicity to retronecine and heliotridine esters *in vivo* (Mattocks, 1981a).

8. Microsomal Oxidation of PAs in Tissues Other than Liver

The liver appears to be the only important site of metabolic activation of PAs. Only a few *in vitro* experiments have been reported with other tissue preparations, all derived from lung. Rat lung microsomes failed to convert retrorsine to pyrrole (Mattocks and White, 1971a); Guengerich (1977) found negligible conversion of lasiocarpine to pyrrole by microsomes from rabbit lung, compared with liver. The PAs lasiocarpine, retrorsine and fulvine were converted to pyrrolic metabolites by human embryo liver tissue slices, but not by lung slices (Armstrong and Zuckerman, 1970), and rat lung slices incubated with monocrotaline gave very little pyrrole compared with liver slices (Hilliker *et al.*, 1983a).

The highest levels of pyrrolic metabolites are normally found in the livers of animals given hepatotoxic PAs. Smaller amounts of pyrroles often appear in extrahepatic tissues such as lungs and kidneys, but there is reason to believe these have come from the liver; for example they are increased by phenobarbitone pretreatment, which only stimulates the liver microsomal system.

Lanfranconi and Huxtable (1984) have demonstrated that monocrotaline is converted to pyrrolic metabolites in isolated perfused rat liver. Such metabolites subsequently perfused through isolated lungs can cause endothelial cell damage, as evidenced by decreased serotonin transport.

C. Hydroxylation

The actions of hepatic mixed-function oxidase enzymes on PAs may lead to other metabolites besides pyrroles and *N*-oxides; such alternative routes of metabolism have been little studied. Eastman and Segall (1982) have demonstrated the hydroxylation of the acid moiety of senecionine (92) when this alkaloid, labelled with ^{14}C, was incubated with liver microsomes from female mice. The product, 19-hydroxysenecionine, represents a small proportion (6.8%) of the recovered radioactivity. Hydrolysis accounts for 13%, *N*-oxidation 7.8%, and unchanged senecionine, 55%.

D. Dealkylation

Jago *et al.* (1969) found that in sheep, heliotrine (49) is demethylated in the acid moiety to heliotridine trachelanthate (376); this is a detoxication pathway, since the product is about half as toxic as heliotrine (Scheme 6.4).

Although not proven, the N-demethylation of otonecine-based PAs is a probable route leading to the conversion of the latter to pyrrolic metabolites, as already mentioned (Section I,B,6).

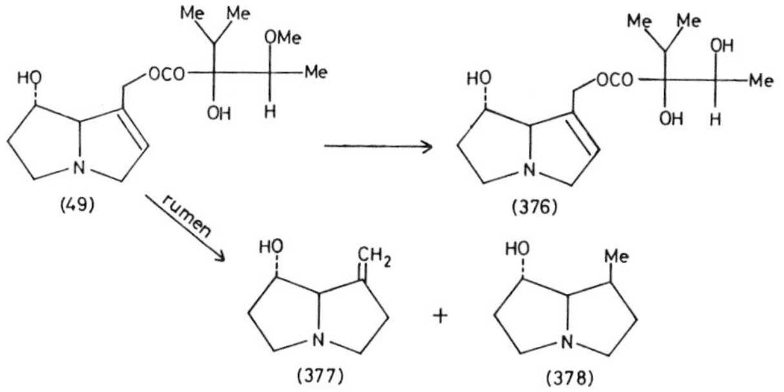

Scheme 6.4.

E. Detoxication in the Sheep Rumen

In ovine rumen fluid, heliotrine is converted to the non-toxic bases (377) and (378) (Dick *et al.*, 1963; Lanigan and Smith, 1970): see Scheme 6.4. However, Swick *et al.* (1983a) found that the alkaloids in *Senecio jacobaea* are not detoxified in this way when the plant is incubated with sheep rumen fluid *in vitro;* thus rumen detoxication does not account for the resistance of sheep to this plant.

F. Epoxidation

Epoxidation of the double bond in the necine moiety of hepatotoxic PAs has been suggested (Bull *et al.*, 1968; Schoental, 1970) as being a possible route of metabolic activation, since epoxides are often highly reactive and toxic. However, no evidence has been forthcoming that PAs are thus epoxidized in animals. Moreover, Culvenor *et al.* (1969a, 1971a) have shown that the chemically prepared α and β epoxides of monocrotaline are not hepatotoxic to rats, and we have not found them to be metabolized to pyrroles by rat liver microsomes (Mattocks and White, 1971a).

G. Other Metabolites

The 1-aldehyde (278) and the 7-methyl ether derivative (379) of dehydroretronecine are detectable after the *in vitro* metabolism of senecionine by mouse liver microsomes (Segall *et al.*, 1984).

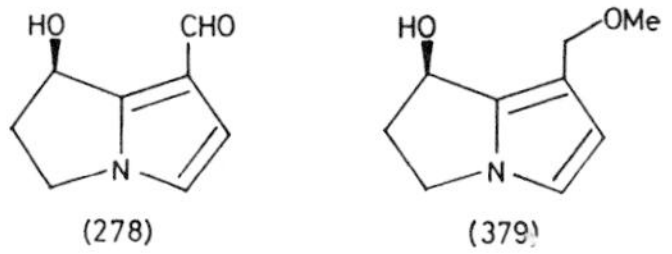

III. METABOLISM OF PA *N*-OXIDES

As mentioned earlier (Section I,B,6), the *N*-oxides of PAs are not converted to pyrrolic metabolites by microsomal enzymes; they are more water soluble and much less toxic than their parent PAs (Mattocks, 1971c). Their main route of metabolism in animals appears to be reduction to the corresponding *tert*-bases, and any hepatotoxic effects result from subsequent metabolic activation of the latter.

Lanigan (1970) showed that PA *N*-oxides are reduced to the bases in sheep rumen fluid. Retrorsine *N*-oxide incubated with rat gut contents is reduced to retrorsine (Mattocks, 1971c): the reaction proceeds best under anaerobic conditions, and the evidence suggests it may be brought about by enzymes in the intestinal flora, rather than from the gut itself. Pyrrolic metabolites accumulate much more slowly in the liver of rats when they are given retrorsine *N*-oxide i.p. than when it it is given by stomach tube, confirming that reduction to base in the gut is a key step in metabolic activation (Mattocks, 1972b). Powis *et al.* (1979a) showed that gut flora plays the major role in the reduction of indicine *N*-oxide to indicine in rabbits. Although given i.v., the *N*-oxide was thought to find its way into the gut by passive diffusion. Antibiotic treatment, which considerably lowers the bacterial content of the gut, causes a large decrease in the amount of *N*-oxide reduced.

Liver enzymes are also, to a more limited extent, able to reduce PA *N*-oxides. Mattocks (1968a) found that retrorsine *N*-oxide is converted to pyrrolic metabolites when incubated with rat liver slices, and it is likely that reduction to retrorsine base is the first step in this process (Jago *et al.*, 1970). Indicine *N*-oxide is reduced to indicine under anaerobic conditions by hepatic microsomal fractions from rabbits (Powis *et al.*, 1979a) and from rats (Powis and Wincentsen, 1980), in the presence of either NADH or NADPH. The reduction is completely inhibited by carbon monoxide, showing that cytochrome P-450 is involved; also, whereas NADH is only about 15% as effective as NADPH in cytochrome P-450 catalysed oxidations, it is 80% as effective in bringing about indicine reduction (Powis and Wincentsen, 1980). The reduction of indicine *N*-oxide in rats and by isolated rat hepatocytes, rat hepatic microsomes, and murine leukemia cells is stimulated by an iron(III)–EDTA complex (Powis *et al.*, 1982).

IV. METABOLISM OF NON-HEPATOTOXIC PAs AND NECINES

A. Esters of Saturated Necines

Pyrrolic metabolites, different from those formed from hepatotoxic PAs, are formed in rats from PAs having a saturated necine moiety such as rosmarinine (Mattocks, 1968a) and platyphylline (Culvenor *et al.*, 1969a). Mattocks and White (1971a,b) investigated the structure of the pyrrolic metabolite formed when the non-hepatotoxic alkaloid rosmarinine (181) is given to rats or incubated with rat liver slices or microsome preparations. Whereas in hepatotoxic PAs the right-hand (pyrroline) ring of the necine moiety is converted to a pyrrole, in rosmarinine it is the left-hand ring which is dehydrogenated, with loss of the associated secondary ester group.

The remaining ester group in the product (380) is not chemically activated by conjugation with the N, so the product cannot act as an alkylating agent and is not cytotoxic.

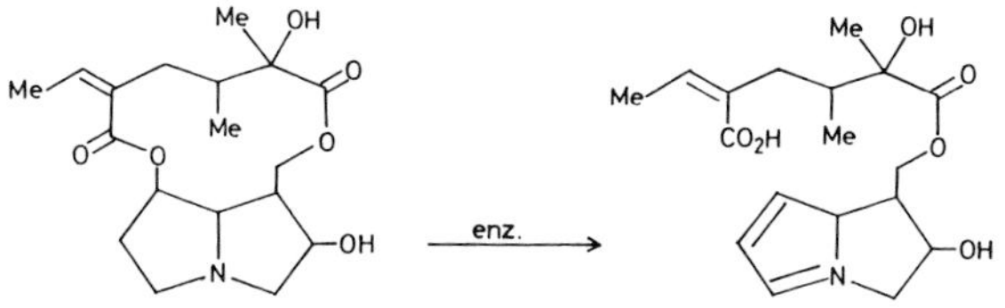

(181) enz. → (380)

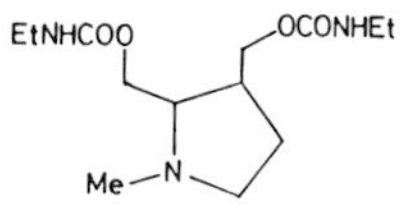

(381) 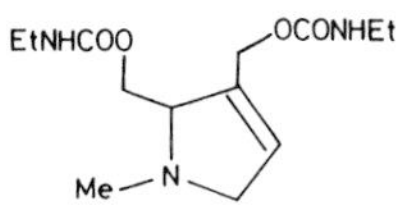   (373)

The synthetic compound synthanecine B bis(N-ethylcarbamate) (381), which has only one ring, equivalent to the right-hand ring of rosmarinine or platyphyl-line, is converted to only traces of a pyrrole in the liver of rats, in contrast with its unsaturated analogue synthanecine A bis(N-ethylcarbamate), (373), which gives large amounts of toxic pyrrole under similar circumstances (Mattocks, 1975).

B. Necines

Microsomal metabolism of necines is probably limited by the very high water solubility of these compounds. Jago et al. (1970) were not able to detect pyrrolic metabolites when heliotridine (184) was incubated with rat liver microsomal preparations. However, there is a low level of conversion of retronecine (183) to pyrroles by rabbit liver microsomes (Powis et al., 1979a) and in the liver of rats given this compound (Mattocks, 1981a). Anhydroretronecine (326), which is more lipophilic than retronecine but is probably quickly hydrolysed to the latter, also gives very little pyrrole in vitro; retronecine 9-methyl ether (327) which is very water soluble, gives only a little more pyrrole (Mattocks, 1981a) (see Table 6.6).

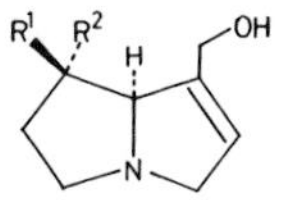

(183) R^1 = OH, R^2 = H
(184) R^1 = H, R^2 = OH

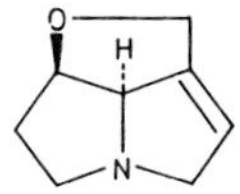

(326)

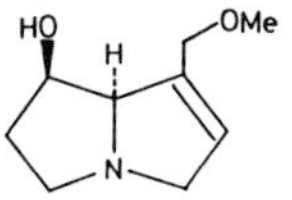

(327)

V. PYRROLES AS TOXIC METABOLITES

A. Pyrrole Formation and Measurement *in Vivo*

The cytotoxicity of pyrrolic metabolites from PAs is associated with their chemical reactivity, which enables them to alkylate tissue constituents *in vivo*. Pyrrolic metabolites are easily measured in tissues colourimetrically, using the Ehrlich reaction (Mattocks and White, 1970). This method gives approximate estimates of total pyrroles, whether primary metabolites (pyrrolic esters), their hydrolysis products (pyrrolic alcohols), or products formed by reaction of these compounds with tissue constituents. Toxic effects are likely to be related to only a small proportion of the latter. Furthermore, the amounts of Ehrlich colour produced from different pyrrole alkylation products are likely to differ considerably: e.g. the Ehrlich colour from N-alkylated products is much weaker than that from O- or S-alkylated products. Nevertheless, total pyrroles measured this way appear to be roughly proportional to those involved in toxic reactions, in that they often show excellent correlation with observed pathological effects, and have given valuable information on the relationship between PA metabolism and toxicity.

B. Tissue Pyrrole Levels in Relation to Alkaloid Structure, Dose and Toxicity

When animals are given a hepatotoxic PA, pyrrolic metabolites accumulate rapidly in the liver. Thus, in rats given retrorsine i.p., liver pyrrole levels reach a maximum after about 20 min (Mattocks, 1972b). Some excess metabolite is lost within the first hour, after which the pyrrole level falls only slowly during 24 h and is still detectable after 2 days (Fig. 6.2). After oral administration, the picture is similar but the alkaloid is absorbed more slowly so the early peak of metabolism is absent. In general, liver pyrroles are best measured after 2 h, when they have reached a fairly stable level. There is a roughly linear relationship between the dose of alkaloid and the liver pyrrole concentration at this time. This is illustrated in male rats given retrorsine (Fig. 6.3): the 2-h pyrrole levels are proportional to the dose up to well above the LD_{50} (ca. 35 mg/kg) of this alkaloid.

In rats given the non-hepatotoxic PA rosmarinine, the initial peak of pyrrolic metabolite disappears from the liver much more quickly than the pyrrole from a toxic PA: this is presumably because the chemically unreactive rosmarinine metabolite is less strongly bound to liver tissue (Mattocks, 1972b).

Different PAs having a wide range of acute hepatotoxicity lead to widely different liver pyrrole levels when given to rats (Mattocks, 1972b): see Table

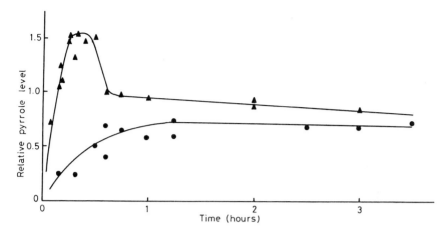

Fig. 6.2. Relative pyrrole metabolite levels in livers of male rats given retrorsine (50 mg/kg) and killed after various times. Alkaloid given by mouth, ●; given i.p., ▲. From Mattocks (1972b).

6.5. The liver pyrroles are roughly related to the LD_{50}: thus there is only a relatively small variation between liver pyrroles in rats given LD_{50} doses of different alkaloids. Liver pyrroles are similarly associated with the hepatotoxicity of synthetic PA analogues (synthanecine carbamates) (Mattocks, 1975, 1978b): see also Table 8.1.

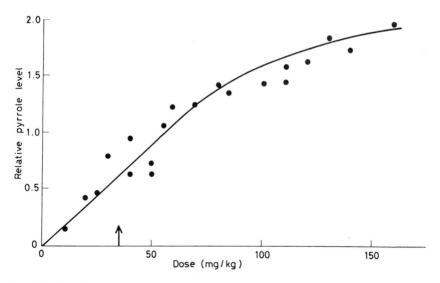

Fig. 6.3. Relative pyrrole metabolite levels in livers of male rats 2 h after various i.p. doses of retrorsine. The arrow marks the acute LD_{50} level. From Mattocks (1972b).

TABLE 6.5 Acute Toxicity Compared with Liver Pyrrole Levels
2 h after i.p. Administration of Some PAs and Synthetic Analogues to Rats[a]

Alkaloid	Sex	Approximate acute LD_{50} (mg/kg)	Relative liver pyrrole level[b] per dose of	
			100 mg/kg	LD_{50}
Retrorsine	M	34	1.88	0.64
Retrorsine	F	150	0.54	0.81
Monocrotaline	M	109	0.64	0.70
Monocrotaline	F	230	0.42	0.96
Heliotrine	M	280	0.27	0.76
Lasiocarpine	M	77	0.87	0.67
Senecionine	M	50	1.85	0.92
Indicine	M	>1000	0.09	>0.9
Triacetylindicine	M	164	0.38	0.62
Synthanecine A bis(N-ethylcarbamate)	M	44	1.23	0.54
Synthanecine C bis(N-ethylcarbamate)	M	88	0.48	0.42
Synthanecine D bis(N-ethylcarbamate)	M	50	1.2	0.60

[a] Adapted from Mattocks (1972b, 1978b).
[b] Arbitrary units representing absorbance of Ehrlich colour from 0.5 g of liver tissue.

It is important to distinguish between the rate of enzymic conversion of a PA to pyrrolic metabolites, and level of pyrroles bound to the liver 2 h after dosing. Toxicity is better related to the latter than to the former (Mattocks, 1973). High rates of metabolism do not necessarily lead to high levels of tissue binding: for example, microsomes convert lasiocarpine to pyrroles *in vitro* at a higher rate than retrorsine (see Table 6.4), but retrorsine gives a much higher liver pyrrole level than lasiocarpine *in vivo* (Table 6.5). There can be various reasons for this; thus primary pyrrolic metabolites may differ in their ability to react with tissue constituents or in their rates of detoxication to less reactive pyrroles.

The relationship between liver pyrroles and hepatotoxicity means that measurements of the former can provide a good indication of potential toxicity of alkaloids which are available in amounts too limited for a full toxicity test (Mattocks, 1981a).

The most important features of a PA which favour the accumulation of pyrrolic metabolites in the liver when it is given to an animal, are resistance to ester hydrolysis; lipophilic character, which enhances access to microsomal enzymes; and a conformation which favours microsomal dehydrogenation in preference to N-oxidation (Mattocks, 1981a). These points have already been touched on in relation to the metabolism of PAs *in vitro*. In the living animal, there may be complex relationships between these factors influencing the formation of pyrrolic and alternative metabolites, and no one factor alone can account for the effects seen. Table 6.6 provides some illustrations. The more water soluble non-esters give relatively little pyrrole, as do monoesters when compared with corresponding diesters. Nevertheless, the highly lipophilic synthanecine A ditiglate (371) gives little pyrrole because of its susceptibility to hydrolysis. Conversion of retronecine ditiglate to the much more water soluble tetrahydroxy ester (242) lowers the pyrrole level fivefold; acetylation of the hydroxy groups to give (382) restores the pyrrole level produced. Similarly, more pyrrole is given by triacetyl indicine than by indicine (see Table 6.5). However, the steric effect of the acetyl

TABLE 6.6 Pyrrolic Metabolite Levels in Livers of Male Rats Given Various PAs and Derivatives[a]

Compound	Structure	Relative liver pyrrole level[b]
Non-esters		
Anhydroretronecine	(326)	0.02
Retronecine	(183)	0.013
Retronecine 9-methyl ether	(327)	0.06
Retronecine monoesters		
9-Tiglate	(240)	0.075
9-(*N*-Ethylcarbamate)	(250)	0.28
7-(*N*-Ethylcarbamate)-9-methyl ether		0.50
Retronecine diesters		
Ditiglate	(241)	1.19
Tetrahydroxy diester	(242)	0.22
Tetra-acetoxy diester	(382)	1.50
7-(*N*-Ethylcarbamate)-9-tiglate		1.76
Bis-*N*-ethylcarbamate	(249)	2.90
Cyclic diesters		
Monocrotaline	(74)	1.89
Retrorsine	(85)	6.60
Other diester		
Synthanecine A ditiglate	(371)	0.07

[a] Adapted from Mattocks (1981a).

[b] Absorbance of Ehrlich colour which would be obtained from 0.5 g of liver of rats, 2 h after an i.p. dose of 1 mmol/kg of alkaloid.

Me Me   Me Me
H—|—|—COO   OCO—|—|—H
  OR OR   OR OR   (242) R = H
                                          (382) R = Ac

groups may also play a part, by inhibiting hydrolysis of the necine ester. Hydrolysis gives products which are less lipophilic, so less susceptible to microsomal oxidation, but it has another important effect on the action of PAs *in vivo*: the pyrrolic metabolite must retain an ester function to have the high chemical reactivity which leads to rapid tissue binding and acute hepatotoxicity. A PA with much steric hindrance around the ester groups is more resistant to hydrolysis; however, it might also be hindered from access to microsomal dehydrogenation, so that a large proportion of *N*-oxide rather than pyrrole may be formed. The carbamate esters are much more effective than many carboxylate esters in producing high liver pyrrole levels, and this may be not only because they inhibit esterases but also because they provide relatively little steric hindrance at the top of the necine moiety, where enzymic oxidation leading to dehydrogenation probably occurs. The particularly high level of liver pyrrole produced by some macrocyclic diesters, such as retrorsine, may similarly be due to a combination of resistance to enzymic hydrolysis coupled with good access to the necine moiety provided by restricted movement of the acid moiety.

C. Factors Affecting the Formation of Pyrroles in the Liver of Animals *in Vivo*

1. Sex

There is a large difference between male and female rats in the level of liver pyrroles from comparable doses of retrorsine and of monocrotaline; in females this is much lower (Mattocks, 1972b) (see Table 6.5). These differences are related to different activities of microsomal mixed function oxidase enzymes between the sexes. Such differences do not occur in mice given retrorsine (White *et al.*, 1973).

2. Age

The way in which liver microsomal enzyme activity converting retrorsine to pyrroles changes in rats from birth to sexual maturity has been described earlier (see Fig. 6.1). Corresponding changes take place in the levels of pyrrolic metabolites which can be measured in liver tissues of rats of various ages given retrorsine (Mattocks and White, 1973). The liver pyrroles in very young rats are

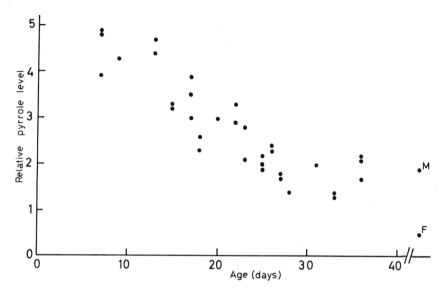

Fig. 6.4.   Relative pyrrole metabolite levels in livers of young rats of various ages, 2 h after similar i.p. doses of retrorsine. M, F: Mean levels in male and female rats respectively, aged over 35 days. From Mattocks and White (1973).

about twice those in adult males, falling to the level of the latter between about 17 and 27 days of age; after 30 days there is a further drop in the females to adult levels less than one-third those in males (Fig. 6.4). The higher pyrrole level in young rats goes some way towards explaining their higher susceptibility (about three times) than adult males to the hepatotoxicity of PAs (Jago, 1970); a further factor in this is the probable higher sensitivity of developing tissues to the pyrrole metabolites.

3. Pretreatment of Animals

Some pretreatments can affect the metabolic formation or disposal of pyrrolic metabolites in animals subsequently given PAs. Changes brought about in hepatic microsomal enzyme activity can result in raised or lowered liver levels of pyrrolic metabolites from PAs, and corresponding changes in susceptibility of the animal to the toxicity of these alkaloids. Alternatively, pretreatments can increase the animal's resistance by enhancing the detoxication of active metabolites. The effects in rats of some pretreatments are shown in Table 6.7. Pretreatment of rats with phenobarbitone stimulates the hepatic microsomal enzymes so that their ability to convert PAs to pyrroles *in vitro* is greatly enhanced (Jago,

TABLE 6.7 Relative Liver Pyrrole Levels and Acute (4 Day) Toxicity of PAs Given i.p. to Rats Following Various Pretreatments[a]

Alkaloid	Pretreatment	Sex	Acute LD_{50} (approx.) mg/kg	Liver pyrroles[b] per dose of 100 mg/kg	LD_{50}
Retrorsine	none	M	34	1.88	0.64
Retrorsine	phenobarbitone[c]	M	65	1.67	1.10
Retrorsine	none	F	150	0.54	0.81
Retrorsine	phenobarbitone[c]	F	90	1.11	1.00
Retrorsine	sucrose diet[d]	M	120	1.32	1.58
Retrorsine	SKF 525A[e]	M	55	1.21	0.67
Monocrotaline	none	F	230	0.42	0.96
Monocrotaline	phenobarbitone[c]	F	75	0.99	0.74

[a] Data from Mattocks (1972b).

[b] Absorbance of Ehrlich colour from 0.5-g liver samples 2 h after giving alkaloid (Mattocks and White, 1970).

[c] 0.1% in drinking water for 1 week.

[d] Rats fed only sucrose for 4 days.

[e] 70 mg/kg i.p., 1 h before alkaloid.

1971; Mattocks and White, 1971a; cf. Table 6.3). However, the situation *in vivo* is more complex. In most cases, phenobarbitone pretreatment causes an increase in pyrrolic metabolites in the liver 2 h after giving the PA, and a corresponding enhancement of the alkaloid's acute hepatotoxicity. This effect is particularly striking in the guinea pig, whose normally low liver pyrrole level after a dose of retrorsine and its susceptibility to this alkaloid are thus greatly increased (White *et al.*, 1973). However, sometimes the reverse occurs: the liver pyrroles are lower and toxicity is reduced. Evidently when the rate of metabolism of a PA in the normal rat is low, phenobarbitone increases the animal's susceptibility to the alkaloid; however, when the normal rate of metabolism is high, as with lasiocarpine (Jago, 1970), and with retrorsine in male rats (Mattocks, 1972b), phenobarbitone reduces susceptibility. The reason for this is not clear, but possibly the pretreatment accelerates the conversion of labile pyrrolic metabolites to less acutely toxic secondary pyrroles. Some support for this comes from the fact that a much higher liver pyrrole level is associated with an LD_{50} dose of retrorsine in male rats given phenobarbitone than in normal male rats (Table 6.7). However, the metabolic products formed in phenobarbitone treated rats are probably not chemically different from those formed in normal rats: thus, rat liver microsomes from both normal and phenobarbitone pretreated rats form the same secondary metabolite, dehydroheliotridine (Jago *et al.,* 1970), both from lasiocarpine whose toxicity is decreased and from heliotrine whose toxicity is increased by phenobarbitone.

An effect of phenobarbitone apparently unrelated to its stimulating effect on microsomal enzymes is seen if it is given i.p. to rats shortly before a dose of retrorsine: this results in liver levels of pyrrolic metabolites much greater than in control rats (Mattocks, 1972b). The enhancement is transitory, being greatest if retrorsine is given 6 h after the phenobarbitone, and it seems to occur only while the metabolism of phenobarbitone itself is taking place.

A sucrose diet given to rats also affords considerable protection against the acute hepatotoxicity of retrorsine. This again is not due to a diminution of pyrrole production: an acute toxic dose of retrorsine in sucrose treated rats leads to well over twice the liver pyrrole level seen in normal rats given an equitoxic dose. Although the liver is thus protected, the lungs remain susceptible to toxic metabolite escaping from the liver, so that lung damage is commonly seen in sucrose-fed rats given high doses of retrorsine that would be fatally hepatotoxic in rats given a normal diet (Mattocks, 1972b). Newberne *et al.* (1971) found that rats fed a lipotrope-deficient diet have reduced susceptibility to monocrotaline, associated with diminished microsomal enzyme activity and lowered levels of pyrrolic metabolites bound to the liver.

Predosing rats with the microsomal enzyme inhibitor SKF 525A also causes decreased liver pyrrole levels and a corresponding decrease in the hepatotoxicity of PAs (Mattocks, 1972b).

Miranda *et al.* (1981a) showed that a diet containing ethoxyquin, which protects mice against the acute hepatotoxicity of monocrotaline, does not reduce the level of pyrrolic metabolites from this alkaloid in the liver. The protection was attributed to detoxication by reaction of pyrroles with an enhanced liver gluthathione (GSH) concentration. However, this seems to contradict the results of White (1976), who showed that an increase in hepatic GSH, while protecting rats against the toxicity of retrorsine, is also associated with a large reduction in the relative liver pyrrole level, while a decrease in liver GSH has the opposite of these actions (see Table 7.4).

D. Pyrrolic Metabolites in Tissues Other than Liver

The conversion of PAs to pyrroles in laboratory animals occurs almost entirely in the liver. In animals given some PAs, pyrrolic metabolites can be found in other tissues, chiefly the lungs and kidneys, although the highest levels are initially in the liver. Table 6.8 gives some examples. Pyrroles in the lungs probably arise from the binding to lung tissue of reactive primary metabolites originating in the liver, and they reflect the ability of the alkaloids to cause lung

TABLE 6.8 Relative Levels of Pyrrolic Metabolities in Organs of Male Rats 2 h after Being Given PAs or Analogous Compounds[a]

		Relative liver pyrrole level[b] per 1-nmol/kg dose	Approx. pyrrole level (% of liver level) in	
Alkaloid	Pretreatment		Lungs	Kidneys
Monocrotaline	none	1.9	33%	54%
Monocrotaline	phenobarbitone[c]	3.1	45%	74%
Retrorsine	none	6.6	3%	54%
Synthanecine A bis(diethyl phosphate)	none	2.5	0%	—
Synthanecine A bis(N-ethylcarbamate)	none	3.5	24%	—
Synthanecine A bis(NN-diethylcarbamate)	none	1.8	75%	—

[a] Data from Mattocks (1972b, 1975) and unpublished results.
[b] Absorbance of Ehrlich colour from 0.5-g liver samples (Mattocks and White, 1970).
[c] Given as 0.1% in drinking water for 1 week prior to experiment.

damage. Thus retrorsine rarely damages normal rat lungs, whereas mono-crotaline does so frequently. The amount of pyrrole from monocrotaline is increased in the lungs and kidneys as well as the liver of rats when they have been treated with phenobarbitone, which stimulates pyrrole production only in the liver (Mattocks, 1972b).

Kidney damage from PAs is relatively uncommon in the rat, and the pyrroles detectable in kidneys are probably largely water soluble secondary metabolites in transit for urinary excretion, which is known to occur within a short time after giving PAs (Mattocks, 1968a).

The level of pyrroles in some extrahepatic tissues, particularly the lungs, is probably a reflection of the chemical stability (half-life) of the reactive primary metabolite, controlling the distance this can travel from the liver before breaking down. Accordingly, synthanecine A bis-N,N-diethyl carbamate (383), which leads to exceptionally large amounts of pyrrole in the lungs, causes more widespread damage than the other compounds, with cytotoxic effects in the gut, spleen and bone marrow as well as in the liver and lung (Mattocks, 1975). Synthanecine A bis-diethylphosphate (374), on the other hand, yields a primary metabolite which is probably very labile, and hence cannot reach the lungs; even

(373) R = CONHEt
(374) R = PO(OEt)$_2$
(383) R = CONEt$_2$

(384)

(385)

in the liver it is not acutely toxic, but causes only chronic damage consistent with the action of pyrrolic alcohol formed by hydrolysis.

E. Possible Further Metabolism of Pyrroles

Guengerich and Mitchell (1980) showed that the model compounds 1,3,4-trimethypyrrole (384) and 3,4-bishydroxymethyl-1-methylpyrrole (385) can be oxidized by hepatic microsomal fractions containing cytochrome P-450, in the presence of NADPH, to metabolites which are able to bind to proteins and nucleic acids. High doses (up to 1000 mg/kg) of these compounds given to rats cause some mild liver necrosis, and the authors suggest that such further metabolism of pyrrolic metabolites from PAs might contribute to the total activation of these alkaloids. Some support for this come from the results of Mattocks and Driver (1983), who found centrilobular liver damage in some rats given highly reactive pneumotoxic pyrrolic esters via a tail vein. These compounds mainly damage the lung, and any which survive to reach the liver would be expected to produce lesions in the periportal regions (cf. Butler et al., 1970). Hence the centrilobular damage might be due to metabolites formed in the liver from the injected compounds or (more probably) from their decomposition products.

Nevertheless, it has been pointed out (Mattocks and Bird, 1983a) that the putative primary pyrrolic metabolites from PAs are themselves sufficiently reactive and cytotoxic to account for the cell damage produced by these alkaloids.

VI. TISSUE DISTRIBUTION AND EXCRETION

A. Early Experiments

Hayashi (1966) gave monocrotaline to rats and recovered 50–70% of it in the urine. The excreted material was said to be 'unchanged' but this was not proven, because a non-specific method of analysis for bases was used. After an injection of tritiated monocrotaline, 30% of the radioactivity was recovered in the bile of rats, but analysis showed very little of this to be basic alkaloid. Alkaloid base was not chemically detectable 24 h after the injection, but persistent radioactivity was found in the liver (only) after 72 h. These results are understandable with present knowledge of PA metabolism. Pyrrolic metabolites and N-oxides, neither of which react as bases, are formed rapidly in the liver. Mattocks (1968a) showed that pyrrole and N-oxide metabolites are excreted, along with unchanged alkaloid, when retrorsine is given to rats (Table 6.9). Jago et al. (1969) gave slow i.v. infusions of heliotrine to sheep: excretion and metabolism of the alkaloid commenced rapidly. A total of 10–15% of metabolites retaining a heliotridine nucleus, together with unchanged heliotrine, was recovered, 80% of this within the first 2 h, mostly in the urine: less than 2% of these excreted

TABLE 6.9 Alkaloid and Metabolites
in the Urine of Rats Injected i.p.
with Retrorsine (60 mg/kg)[a]

Time after dosing (h)	Recovery (% of dose) in urine of		
	Alkaloid	N-oxide	Pyrroles
0–3.5	8.6	10.8	4.1
3.7–7	0.9	1.2	3.0
7–24	1.1	1.3	5.3
24–48	0.1	0.2	1.8
Totals	10.7	13.5	14.2

[a] From Mattocks (1968a). Reprinted by permission from *Nature* **217,** 723–728. Copyright © 1968 Macmillan Journals Limited.

compounds were in the bile. However, the metabolites measured were products of N-oxidation, demethylation and hydrolysis: pyrroles were not measured.

B. Biliary Excretion

White (1977) found that about 25% of an i.v. dose (40 mg/kg) of retrorsine given to rats is excreted in the bile as pyrrolic metabolites, together with about 4% of unchanged retrorsine.

The excretion is greatest during the first hour after dosing and negligible by 7 h. After a similar dose of tritium-labelled synthanecine A bis(N-ethylcarbamate) (386), 25% of the radioactivity appears in the bile within 4 h, but only 5% represents pyrroles. In rats given 10 mg/kg of tritiated 2,3-bishydroxymethyl-1-methylpyrrole (297), a putative pyrrolic metabolite of (386), 17% of the radioactivity is excreted in the bile but only 3% appears still to be pyrrole. (However, the pyrrole measurements in all these experiments must be interpreted with caution, since they were made using the non-specific Ehrlich reaction, and the molar absorptivity used to calculate results may not have been correct for all the pyrrolic metabolites present.) A large proportion of the excreted radioactivity, including most of the pyrrole, is in a highly polar form. White concluded that in rats, retrorsine and the synthetic analogue are converted to pyrroles, which can

(386)   (297)

be rapidly excreted by way of the bile as more polar derivatives which are no longer cytotoxic and, at least from the synthanecine, are largely either unreactive or only weakly reactive to Ehrlich reagent.

An investigation of enterohepatic circulation showed that this does not play an important role in the elimination of the synthanecine carbamate or its metabolites from the rat.

C. Distribution and Excretion

1. Of Radioactive PAs and Analogues

The tissue distribution of radioactivity from a compound having PA-like hepatotoxicity was first studied in male rats by Mattocks and White (1976) using [^{3}H]synthanecine A bis(N-ethylcarbamate) (386). The highest concentrations appear in the liver, where the compound is metabolically activated, and in the lungs, kidneys and spleen. Radioactivity is lost much more slowly from the last three than from other tissues. In the spleen this can be attributed to the large reservoir of bound radioactivity in the erythrocytes. The Ehrlich reaction shows that 68% of the radioactivity in the liver 2–6 h after dosing represents pyrrolic metabolites. About 69% of the dose is eliminated via the urine and faeces during the first day; the expired air contains negligible radioactivity. After the first day there is practically no loss of radioactivity from the lungs during 8 weeks.

The similar semisynthetic PA retronecine bis(N-ethylcarbamate) (387), shows a fairly similar pattern of tissue distribution to the above when it is given as an equitoxic dose (120 mg/kg) to rats (Mattocks, 1977b). Both compounds lead to relatively substantial binding in the liver and lungs, where major toxic effects are seen.

EtNHCOO   C^{3}H$_2$OCONHEt

(387)

Eastman et al. (1982) studied the distribution of uniformly ^{14}C-labelled senecionine and seneciphylline in lactating mice. At 16 h after an i.p. injection of senecionine (82 mg/kg), about 96% of the radioactivity has been recovered, mainly in the urine (75%) and faeces (14%). There is very little in expired CO_2 (0.21%) and in the milk (0.04%). A relatively high level in liver (1.92%) reflects the hepatotoxicity of this PA, but in comparison with the radioactivity from the synthanecine carbamate (Table 6.9), a much smaller proportion appears in the kidneys (0.05%), blood (0.32%) and lungs (0.03%). The latter is somewhat surprising in view of the apparent pneumotoxic activity of senecionine (Culvenor

et al., 1976b). Some of the radioactivity in liver from senecionine or seneciphylline is bound to proteins, RNA and DNA.

A mixture of senecionine and seneciphylline, tritiated in either the acid or the
necine moiety, was given orally to lactating rats by Luthy *et al.* (1983b); after 5
h, a much higher proportion of the base moiety remained in the blood than the
acid moiety, indicating the formation of protein-bound necine. During 3 h about
0.02% of unchanged alkaloids were excreted in the milk along with 0.08%
unidentified, water-soluble metabolites containing the necine moiety. The highest tissue levels (6 h) were in the liver and lungs.

2. Of Radioactive Pyrrolic Metabolites and Analogues

a. [^{3}H]Dehydroretronecine (DHR). The distribution of [^{3}H]DHR (292) has
been measured in male rats given an s.c. dose of 65 mg/kg (Hsu *et al.*, 1974;
Shumaker *et al.*, 1976a). Some of the results are condensed in Table 6.10. There

TABLE 6.10 Distribution of Radioactivity
from Tritium-labelled Dehydroretronecine (292)
Given s.c. to Young Male Rats[a,b]

	Approx. specific activity (dpm $\times$ 10^{-2})[c]		
Organ	6 h	24 h	7 d
Liver	67	29	5
Kidney	58	34	9
Testis	58	45	6
Lung	38	12	3
Spleen	47	32	6
Brain	41	33	5
Thymus	39	15	3
Small intestine	29	15	3
Heart	24	9	2
Stomach (glandular)	307	251	138
Stomach (squamous)	44	28	3
Blood	46	41	7

[a] Data from Hsu *et al.* (1974).

[b] Dose 65 mg/kg to rats weighing about 60 g: approximately 8.58 $\times$ 10^8 dpm/rat.

[c] Results rounded to nearest 100 for ease of
comparison.

HO   C³H₂OH

(292)

is a rapid general distribution of radioactivity throughout the body tissues. Unlike PAs and analogous compounds which are activated in the liver, DHR is not preferentially bound in this organ. However, there is a very large amount in the glandular part of the stomach. Acid, which is secreted by this tissue, is known to catalyse both the polymerization and the alkylating reactivity of DHR (see Chapter 5), and thus it seems likely that the high activity in this organ represents both polymeric DHR and products from alkylation of tissue constituents (cf. White and Mattocks, 1976). Hsu *et al.* (1975) showed that DHR binds more readily to cellular macromolecules under acidic conditions, and Black and Jago (1970) similarly found that the interaction with DNA of its enantiomer dehydroheliotridine is catalysed by acid. Investigating the tissue distribution of [³]DHR in rhesus monkeys, Hsu *et al.* (1976) again found the highest tritium level in the gastric mucosa (although it is much less than in rats); over 20% of this radioactivity is bound to the mucosal protein and under 1% to nucleic acids. Excretion occurs via the urine (up to 25% in 24 h) and the bile.

b. [³H]-2,3-Bishydroxymethyl-1-methylpyrrole (BHMP). The distribution in rats of BHMP (297), a putative secondary metabolite of synthanecine A esters and analogous to DHR in chemical properties, has been studied by White and Mattocks (1976). After an i.v. dose of 20 mg/kg, almost 69% is excreted during the first day, most of this (61%) in the urine. As with DHR, the radioactivity is distributed fairly uniformly throughout the body except that it is high in the liver and extremely high in the glandular stomach (Table 6.11). To test the effect of increasing the stomach pH, a rat was given a large amount of sodium bicarbonate prior to an i.v. dose of BHMP: this caused a considerable fall in the radioactivity in the glandular stomach (see Table 6.11). The strong reaction with gastric tissue only occurs when the pyrrole reaches the latter via the bloodstream. When BHMP is given by stomach tube there is no high level of binding to stomach tissue; instead, a large proportion of the radioactivity is localized in the stomach contents, presumably as polymer and other decomposition products. The amounts reaching other tissues are also reduced.

When the tissue concentrations of radioactivity from equimolar doses of BHMP and synthanecine A bis(*N*-ethylcarbamate) (386) are compared, the carbamate leads to much stronger binding in liver and lung, to which it is highly toxic, and it is not concentrated in gastric tissue. It is concluded that BHMP cannot be a major metabolite present in the blood stream of rats given a toxic dose of the carbamate. In a similar way, a comparison of the distribution of DHR

TABLE 6.11 Radioactivity in Organs of Rats
1 h after a Dose (20 mg/kg) of [³H]BHMP[a]

Organ	Radioactivity (% of dose) for route		
	i.v.	i.v.[b]	p.o.
Liver	6.6	6.8	0.64
Lungs	0.34	0.35	0.05
Kidneys	0.8	1.36	0.17
Spleen	0.24	0.28	0.03
Blood	3.4	4.0	0.58
Stomach (squamous)	0.09	0.08	0.09
Stomach (glandular)	14.28	6.3	0.19
Stomach contents	0.31	0.42	69.1
Caecum	0.13	0.31	0.03
Caecum contents	0.88	1.0	0.08

[a] Adapted from White and Mattocks (1976).
[b] Animal predosed with sodium bicarbonate (500 and 300 mg/kg) i.g., 90 and 5 min before BHMP.

(Table 6.10) with that of senecionine and seneciphylline (Eastman *et al.*, 1982) suggests that DHR is not a major circulating metabolite from these and similar retronecine-based PAs.

c. [³H]Bisacetoxymethyl-1-methylpyrrole (BAMP). This compound (298) is analogous to primary metabolic activation products supposedly formed from toxic PAs. When given to rats via a tail vein it causes lung damage following chemical interaction with lung vascular endothelial tissue; high doses lead to only limited liver necrosis (Mattocks and Driver, 1983). Rats thus injected with a pneumotoxic dose of [³H]BAMP (8 mg/kg) excrete over 60% of the radioactivity during the first day, rising to over 80% by 7 days; about three-quarters of this is in the urine (Mattocks, 1979). High concentrations are initially in the lungs and heart—the first organs reached after injection—but after 1 day the lungs only contain 2% of the total dose, showing that only a small proportion of the latter is responsible for lung damage, whereas a large amount (18%) in blood is con-

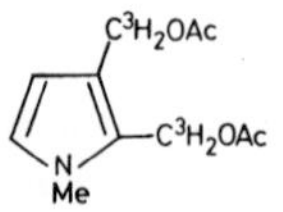

C^3H_2OAc

$-C^3H_2OAc$

Me

(298)

sistent with the high chemical reactivity of BAMP leading to rapid reaction with blood cells. The liver is not appreciably damaged after this dose of BAMP, and the radioactivity in liver, as in most other tissues, is probably due largely to less reactive secondary decomposition products or metabolites rather than to BAMP itself (cf. Mattocks and Driver, 1983).

The high reactivity of BAMP is demonstrated by the fact that most of it is localized at the injection site (90% after 1 day) when it is injected into subcutaneous tissue (Mattocks, 1979).

VII. PAs AND THEIR METABOLITES IN INSECTS

Whereas PAs ingested by mammals are quickly metabolized or excreted, these compounds can persist in some insects for long periods, unchanged or as apparently non-toxic metabolites. Thus, plants containing PAs are eaten by some members of the Lepidoptera, which can store the alkaloids in their bodies without suffering harm.

Cinnabar moth (*Callimorpha jacobaeae* L.) larvae can feed on *Senecio jacobaea*, *S. vulgaris* (Aplin *et al.*, 1968), or *Tussilago farfara* (Culvenor *et al.*, 1976a), and accumulate much higher levels of PAs, including *N*-oxides, than are present in the food plant (Mattocks, 1971b). Most of the PAs in *S. jacobaea*, including senecionine, seneciphylline and integerrimine, and a metabolite which is an ester of retronecine with a modified acid moiety are found in the larvae and transferred to the adult moths (Aplin *et al.*, 1968). Many examples have now been found of PAs in moths and butterflies. The bodies of *Utetheisa* moths (family Arctiidae) from a field of *Heliotropium europaeum* contained heliotrine, heleurine and supinine, but not lasiocarpine or europine, which also occur in the plant (Culvenor and Edgar, 1972). PAs have been found in many butterflies of *Amauris, Danaus* and *Euploea* species (Edgar *et al.*, 1979), and in six species of Arctiid moths reared on *Senecio* or *Crotalaria* plants (Rothschild *et al.*, 1979). Alkaloids of *Senecio spathulatus* are ingested by larvae of the moth *Nyctemera annulata* and are transferred to the adults and their eggs as well as to a parasite of the larvae (Benn *et al.*, 1979). Some butterflies and moths are attracted to withered or dead plants containing PAs, then feed on the plants and accumulate the alkaloids (Edgar and Culvenor, 1974; Schneider *et al.*, 1975; Goss, 1979); in the case of dried *Heliotropium indicum*, the attractant is a volatile decomposition product from the acid moiety of PAs (Pliske *et al.*, 1976).

The accumulated PAs may act as a defence against potential predators, for although their toxicity takes effect too slowly to be associated with ingestion of the prey (Aplin *et al.*, 1968), the PAs might make the insects unpalatable (Edgar *et al.*, 1976, 1979). Brown (1984) has shown that PAs accumulated in some adult Ithomiine butterflies feeding on flowers (mostly of Eupatorieae) or decomposing foliage (Boraginaceae) are able to make the butterflies unpalatable to a

TABLE 6.12 Dihydropyrrolizines Derived from PAs in Moths and Butterflies

Species	Dihydropyrrolizines found	Reference
Arctiid moths		
Utetheisa pulchelloides	(390)	Culvenor and Edgar (1972)
U. lotrix	(389) (390)	
Creatonotos gangis	(390)	Schneider *et al.* (1982)
C. transiens	(390)	
Danaid butterflies		
Daunus gilippus berenice	(388)	Meinwald *et al.* (1969)
African monarch (*Daunus* sp.)	(388)	Meinwald *et al.* (1971)
Daunus and *Euploea* spp.	(388) (389) (390)	Edgar *et al.* (1971);
		Edgar and Culvenor (1974)
Daunus and *Amauris* spp.	(388)	Meinwald *et al.* (1974)
D. chrysippus dorippus	(388)	Schneider *et al.* (1975)

predatory spider (*Nephila clavipes*). Goss (1979) has suggested that PAs might be a nutrient source in some moths.

An intriguing function of PAs in some moths and butterflies is to act as precursors for the biosynthesis of sex-attractants (pheromones). These are relatively volatile dihydropyrrolizine derivaties arising from dehydrogenation of PAs and loss of the acid moiety. Three such compounds which have been identified are 1-methyl-6,7-dihydro-5*H*-pyrrolizine-7-one (388), 1-formyl-6,7-dihydro-5*H*-pyrrolizine (389), and the 7-hydroxy derivative (390). They are found in the hairpencils or analogous scent organs (brushes) of the male insects, and are only produced after PAs have been ingested; some species in which they are found are listed in Table 6.12. The size of the scent organs in *Creatonotos* moths is said to be controlled by dietary PAs, being larger when the moths have access to PAs (Schneider *et al.*, 1982).

The grasshopper *Zonocerus variegatus* is able to sequester and store PAs, among other poisons from plants, probably for defensive purposes (Bernays *et al.*, 1977). *Zonocerus elegans* grasshoppers are attracted to pure PAs (Boppre *et al.*, 1984). Remarkably, male danaid butterflies sometimes feed on the bodies of disabled *Zonocerus* grasshoppers, thus obtaining a secondhand supply of PAs.

7

Toxicology of Pyrrolizidine Alkaloids in Animals

I. INTRODUCTION

There are many different PAs, and this group of compounds can have a variety of toxic actions in various species of laboratory and farm animals. These effects can be broadly classified as *pharmacological,* which can result in rapid death, and *cytotoxic,* leading to later death associated with tissue damage. The latter effects are caused by PAs which are esters of unsaturated necines, and are usually seen primarily in the liver, although the lungs and other tissues may also be affected.

Acute cytotoxicity may result in deaths between 1 day and about a week after ingestion of the alkaloid. Animals surviving a sublethal dose of a cytotoxic PA often show delayed (chronic) toxic effects in the liver and sometimes the lungs and elsewhere; such effects may also result from multiple (chronic) intake of smaller amounts of PAs, as when these alkaloids contaminate the diet. Some PAs have proved to be carcinogenic after such chronic ingestion (see Chapter 11).

Some PAs (e.g. senecionine) are cytotoxic to primary rat hepatocyte cultures (Green *et al.,* 1981; Hayes *et al.,* 1984).

Mutagenic and other DNA-damaging effects of PAs are considered in Chapter 10; embryotoxic and teratogenic actions are considered in Chapter 8.

The toxicity of some PAs and their metabolites to the lungs and heart has been of particular interest to experimental pathologists, and is dealt with in more detail in Chapter 8.

The toxic and pathological actions of PAs have previously been discussed by Bull *et al.* (1968), by McLean (1970), and by Peterson and Culvenor (1983).

II. TOXIC ACTIONS IN LABORATORY ANIMALS

A. Peracute Toxicity

Sufficiently large doses of most PAs can cause rapid death—between a few minutes and a few hours after the alkaloid intake. This type of toxicity, referred to as peracute (or hyperacute), is not related to cytotoxic action. It is associated with pharmacological actions of the compounds and death may be preceded by convulsions or by coma. Peracute toxicity is most often encountered when PAs are absorbed rapidly, as after i.p. or i.v. injections, and it may sometimes be averted by injecting the alkaloid very slowly or by dividing the dose into several lots, separated by short time intervals. The more lipophilic PAs, pyrrolic metabolites, and semisynthetic esters are especially liable to cause peracute death.

Gallagher and Koch (1959) showed that doses of heliotrine or lasiocarpine (but not their N-oxides) larger than the acute LD_{50} values which would cause death associated with liver damage, would kill rats within 30 min with respiratory failure. Different PAs and related compounds have a variety of pharmacological actions (see Chapter 8) which might account for their peracute toxicity.

B. Acute Hepatotoxicity

1. Relative Toxicity of Different PAs

The principle acute toxic action in animals of many PAs is on the liver cells. All the hepatotoxic alkaloids are esters of unsaturated necines such as heliotridine, retronecine, crotanecine, otonecine and supinidine. Different unsaturated PAs can produce similar hepatotoxic effects, but their degree of toxicity may vary widely. Table 7.1 lists LD_{50} values for some of them. Most of these values refer to acute (subacute) deaths, which are associated with severe haemorrhagic necrosis of the liver and occur most frequently within a few days after the animal has received a single dose of the alkaloid. The acute hepatotoxicity of some alkaloids, e.g. supinine, cannot be determined because peracute death

TABLE 7.1 Acute Toxicity Data for Unsaturated PAs: Results from Various Sources

Alkaloid	Animal species	Sex[a]	Route of administration[b]	LD_{50} (mg/kg)	Time range (days)[c]	Reference
7-Angelyl heliotridine	rat	M	i.p.	ca. 260	P	Bull et al. (1968)
Cynaustine	rat	M	i.p.	ca. 260	3	Bull et al. (1968)
Echimidine	rat	M	i.p.	ca. 200	3	Bull et al. (1968)
Echinatine	rat	M	i.p.	ca. 350	3	Bull et al. (1968)
Europine	rat	M	i.p.	>1000	3	Bull et al. (1968)
Heleurine	rat	M	i.p.	ca. 140	3	Bull et al. (1968)
Heliosupine	rat	M	i.p.	ca. 60	3	Bull et al. (1968)
Heliotrine	rat	M	i.p.	296	3	Bull et al. (1958)
Heliotrine	rat	F	i.p.	478	3	Bull et al. (1958)
Heliotrine	rat	U	i.v.	274	7	Harris et al. (1957)
Heliotrine	mouse	U	i.v.	255	7	Harris et al. (1957)
Heliotrine N-oxide	rat	M	i.p.	ca. 5000	3	Bull et al. (1968)
Heliotrine N-oxide	rat	F	i.p.	ca. 2500	3	Bull et al. (1968)
Indicine	rat	M	i.p.	>1000		Schoental (1968a,b)
Indicine, triacetyl	rat	M	i.p.	164		Mattocks (1972b)
Integerrimine	mouse	U	i.v.	78	7	Anonymous (1949)
Jacobine	mouse	U	i.v.	77	7	Harris et al. (1942)
Jacobine	rat	F	i.p.	138	3	Bull et al. (1968)
Jaconine	rat	F	i.p.	168	3	Bull et al. (1968)
Lasiocarpine	rat	U	i.v.	88	5	Rose et al. (1959)
Lasiocarpine	mouse	U	i.v.	85	5	Rose et al. (1959)
Lasiocarpine	hamster	U	i.v.	67.5	5	Rose et al. (1959)

(continued)

TABLE 7.1 (*Continued*)

Alkaloid	Animal species	Sex[a]	Route of administration[b]	LD_{50} (mg/kg)	Time range (days)[c]	Reference
Lasiocarpine	rat	M	i.p.	77	3	Bull et al. (1958)
Lasiocarpine	rat	F	i.p.	79	3	Bull et al. (1958)
Lasiocarpine N-oxide	rat	M	i.p.	547	3	Bull et al. (1958)
Lasiocarpine N-oxide	rat	F	i.p.	181	3	Bull et al. (1958)
Latifoline	rat	M	i.p.	ca. 125	3	Bull et al. (1968)
Monocrotaline	rat	M	i.p.	109	4	Mattocks (1972b)
Monocrotaline	rat	F	i.p.	230	4	Mattocks (1972b)
Monocrotaline	mouse	F	i.p.	259	7	Miranda et al. (1981c)
Retrorsine	rat	U	i.v.	38	7	Anonymous (1949)
Retrorsine	rat	M	i.p.	34	4 or 7	Mattocks (1972b)
Retrorsine	rat	F	i.p.	153	4 or 7	Mattocks (1972b)
Retrorsine	mouse	U	i.v.	59	7	Anonymous (1949)
Retrorsine	mouse	M	i.p.	65	4	White et al. (1973)
Retrorsine	mouse	F	i.p.	69	4	White et al. (1973)
Retrorsine	hamster	M	i.p.	81	4	White et al. (1973)
Retrorsine	guinea pig	M	i.p.	>800	4	White et al. (1973)
Retrorsine	fowl	M	i.p.	85	4	White et al. (1973)

Retrorsine	quail	M	i.p.	279	4	White et al. (1973)
Retrorsine N-oxide	rat	M	p.o.	48	7	Mattocks (1972b)
Retrorsine N-oxide	rat	M	i.p.	250	7	Mattocks (1972b)
Retrorsine N-oxide	mouse	U	i.v.	834	7	Anonymous (1949)
Riddelliine	mouse	U	i.v.	105	7	Anonymous (1949)
Rinderine	rat	M	i.p.	ca. 550	3	Bull et al. 1968
Sceleratine	mouse	U	i.p.	135	7	Anonymous (1949)
Senecionine	rat	M	i.p.	50	7	Mattocks (1972b)
Senecionine	mouse	U	i.v.	64	7	Anonymous (1949)
Senecionine	hamster	U	i.v.	61	7	Anonymous (1949)
Seneciphylline	rat	M	i.p.	77	3	Bull et al. (1968)
Seneciphylline	rat	F	i.p.	83	3	Bull et al. (1968)
Seneciphylline	mouse	U	i.v.	ca. 90	7	Anonymous (1949)
Senkirkine	rat	M	i.p.	220		Hirono et al. (1979)
Spartioidine	mouse	U	i.v.	80	7	Anonymous (1949)
Spectabiline	rat	M	i.p.	ca. 220	3	Bull et al. (1968)
Supinine	rat	M	i.p.	400	P	Bull and Dick (1959)
Symphytine	rat	M	i.p.	130		Hirono et al. (1979)
Usaramine	mouse	U	i.p.	300		Singh et al. (1969)

[a] U, Unspecified.

[b] i.p., Intraperitoneal; i.v., intravenous; p.o., by mouth (via stomach tube).

[c] P (Peracute) means that death was within a few hours of dosing.

intervenes at lower dose levels; nevertheless, sublethal hepatotoxicity is demonstrable in animals surviving after single or repeated doses of supinine (Bull and Dick, 1959; Bull et al., 1968).

From comparisons of PAs tested under similar conditions (e.g. i.p. in male rats), it appears that the most toxic are certain macrocyclic diesters of retronecine, followed by 'open' heliotridine and retronecine diesters, then otonecine diesters, then (least toxic) monoesters of heliotridine, retronecine and supinine—though there are exceptions to this sequence.

Recent toxicity studies of PAs (other than LD_{50} determinations), and of plants or plant extractions containing PAs, are listed in Tables 7.2 and 7.3, respectively.

2. Methods for the Assessment of Toxicity

Acute LD_{50} tests may consume unacceptibly large amounts of an alkaloid, especially if its toxicity is low. Jago (1970) described a method which can be used to determine the relative hepatotoxicity, both acute and chronic, of PAs using small numbers of 14-day-old rats. This very economical in material, not only because of the low weight of the animals, but also because they are more susceptible to the alkaloids: the acute LD_{50} in the young rats is 0.3–0.4 times that of adults. Animals, in small groups, are given single i.p. doses of the test compound at a series of dose levels increasing in geometric progression. Livers of animals which die, and of survivors after 4 weeks, are examined histologically. The method has been used to screen a large number of PAs and derivatives (Culvenor et al., 1976b).

It is plainly desirable that the numbers of animals subjected to toxicity tests should be kept to a minimum. The existing knowledge of relationships between acute heptatotoxicity, molecular structure and metabolism enables an assessment of probable toxicity to be made without the need for animals to be acutely poisoned. Rough estimates of acute hepatotoxicity can often be made from a knowledge of the alkaloid's structure (see Chapter 12). A quantitative assessment can be made by measuring levels of pyrrolic metabolites in the livers of rats given PAs (Mattocks, 1981a). There is an approximately direct relationship between amounts of such metabolites and acute hepatotoxicity (Mattocks, 1972b). A few rats are given different doses of an alkaloid and are killed 2 h later, and liver pyrroles are measured by a simple procedure (Mattocks and White, 1970) to establish the relationship between dose and pyrrole level. In this way, small amounts of PAs can be screened for potential toxicity. The alkaloid structure must be taken into account, because some non-toxic PAs based on saturated necines (e.g. rosmarinine) may give rise to large amounts of a different kind of pyrrolic metabolite in liver tissue (Mattocks and White, 1971b). Thus,

TABLE 7.2 Experimental Studies of Toxic Effects of PAs
in Laboratory Animals (1967–1983)

Alkaloid	Other treatments	Animal species	References
Anacrotine		young rats	Jago (1970)
Fulvine		rats	Persaud et al. (1970)
Heliotrine		rats	Kerr (1969)
Heliotrine		young rats	Jago (1970)
Heliotrine	phenobarbitone	rats	Jago (1971)
Heliotrine	enzyme inducers	rats	Tuchweber et al. (1974)
Hydroxysenkirkine		rats	Schoental (1970)
Hygrophylline		rats	Mattocks (1972b)
Jacobine		rats, hamsters, guinea pigs, gerbils	Cheeke and Pierson-Goeger (1983)
Lasiocarpine		young rats	Jago (1970)
Lasiocarpine	phenobarbitone	rats	Jago (1971)
Lasiocarpine	enzyme inducers; SKF 525A; partial hepatectomy	rats	Tuchweber et al. (1974)
Lasiocarpine	diet; antioxidants; acetylaminofluorene	rats	Rogers and Newberne (1971)
Lasiocarpine	thioacetamide	rats	Reddy et al. (1976)
Monocrotaline		rats	Hayashi and Lalich (1967)
Monocrotaline		young rats	Jago (1970)
Monocrotaline	phenobarbitone	rats	Mattocks (1972b)
Monocrotaline	phenobarbitone; chloramphenicol	rats	Allen et al. (1972)
Monocrotaline	low lipotrope diet	rats	Newberne et al. (1971)
Monocrotaline	light; riboflavin; carotene	rats	Newberne et al. (1974)
Monocrotaline	enzyme inducers; SKF 525A; partial hepatectomy	rats	Tuchweber et al. (1974)
Monocrotaline	cysteine; mercaptoethylamine	rats	Hayashi and Lalich (1968)
Monocrotaline	diet restriction	rats	Hayashi et al. (1979)
Monocrotaline	metabolic inhibitors; cysteine	rats	Eisenstein et al. (1979)
Monocrotaline		rats	Roth et al. (1981)

(continued)

TABLE 7.2 (*Continued*)

Alkaloid	Other treatments	Animal species	References
Monocrotaline	ethoxyquin	mice	Miranda *et al.* (1981a)
Monocrotaline	butylated hydroxy-anisole	mice	Miranda *et al.* (1981c)
Monocrotaline		rats, hamsters, guinea pigs, gerbils	Cheeke and Pierson-Goeger (1983)
Monocrotaline		guinea pigs	Chesney and Allen (1973a); Swick *et al.* (1982a)
Monocrotaline		monkeys	Allen *et al.* (1967)
Monocrotaline		chickens	Allen *et al.* (1970a)
Monocrotaline epoxides		rats	Culvenor *et al.* (1971a)
Otosenine		rats	Culvenor *et al.* (1971a)
Retrorsine	phenobarbitone; SKF 525A; sucrose diet	rats	Mattocks (1972b)
Retrorsine		young rats	Schoental (1970); Mattocks and White (1973)
Retrorsine	cysteine; chlorethanol	rats	White (1976)
Retrorsine	phenobarbitone	mice; guinea pigs; hamsters; chickens; quail	White *et al.* (1973)
Retrorsine		monkeys	van der Watt and Purchase (1970); van der Watt *et al.* (1972)
Rosmarinine		rats	Mattocks (1972b)
Senecionine		young rats	Jago (1970)
Senkirkine		rats	Schoental (1970)
Various		young rats	Culvenor *et al.* (1976b)

TABLE 7.3 Experimental Studies of Toxicity
of PA-Containing Plants and Extracts in Laboratory Animals (1968–1984)

Plant	Other treatments	Animal species	Reference
Cassia auriculata, Crotalaria verrucosa and *Holarrhena antidysenterica*		rats	Arseculeratne *et al.* (1981)
Echium plantagineum		rats	Peterson and Jago (1984)
Senecio abyssinicus		rats	Williams and Schoental (1970)
Senecio glabellus, S. jacobaea and *S. vulgaris*		rats	Goeger *et al.* (1983)
S. jacobaea		rats	Burns (1972); Swick *et al.* (1979)
S. jacobea	diet; *S*-amino acids; DDT	rats	Cheeke and Garman (1974)
S. jacobea	cysteine; methionine	rats	Buckmaster *et al.* (1976)
S. jacobea	copper	rats	Miranda *et al.* (1981d)
S. jacobea (alkaloids)	zinc	rats	Miranda *et al.* (1982a)
S. jacobea		mice	Hooper (1974)
S. jacobea	BHA; ethoxyquin; cysteine	mice	Miranda *et al.* (1982b,c)
S. jacobea		rabbits	Pierson *et al.* (1977)
S. jacobea		quail	Buckmaster *et al.* (1977)
S. jacobea		trout	Hendricks *et al.* (1981)
S. jacobea	cysteine	guinea pigs	Swick *et al.* (1982a)
S. jacobea	cysteine (in chicks)	gerbils, guinea pigs, hamsters, mice, chickens, turkeys	Cheeke and Pierson-Goeger (1983)
S. longilobus (alkaloids)	BHA; ethoxyquin; disulfiram	mice	Kim and Jones (1982)
S. vulgaris		rats	Delaveau *et al.* (1979)
S. vulgaris	cysteine	rats	Buckmaster *et al.* (1976)
Various plants		rats	Schoental and Coady (1968)

metabolite measurements should not rule out the need for a confirmatory toxicity tests; however, this can be limited, and animals and alkaloid can be conserved by using the metabolite results as a guide to suitable dose levels.

3. Non-hepatotoxic PAs and Derivatives

PAs having a saturated necine moiety are not hepatotoxic; among those which have been tested are platyphylline (Rose et al., 1959; Jago, 1970; Culvenor et al., 1976b), rosmarinine (Rose et al., 1959; Mattocks, 1972b); hygrophylline (Mattocks, 1972b), and cynaustraline (Culvenor et al., 1976b). Saturated PAs of the loline group (for list, see Robins, 1982a), laburnine, and similar saturated necines and their esters which occur in the Orchidaceae and other genera (Smith and Culvenor, 1981) are also non-hepatotoxic.

Pyrrolizidine quaternary salts are free of hepatotoxicity—e.g. the methiodides of the hepatotoxic PAs monocrotaline and senecionine (Culvenor et al., 1976b); so also are the epoxides of monocrotaline (Culvenor et al., 1971a).

Unesterified necines such as retronecine and heliotridine also are not hepatotoxic (Bull and Dick, 1959; Schoental and Mattocks, 1960; Peterson et al., 1972; Culvenor et al., 1976b).

C. Factors Affecting Hepatotoxicity

1. Mode of Administration

The majority of LD_{50} data for PAs relates to administration via the i.p. route (Table 7.1). Where i.v. data are available for comparison, it appears that these two routes lead to similar hepatotoxicity. LD_{50} values are not generally available for PAs administered orally—a surprising ommission considering that this is practically the only route of exposure to PAs outside the laboratory. However, other experimental work has been done using this route; a comparison of the results of Schoental and Magee (1959) for rats given several PAs orally, including retrorsine, lasiocarpine, heliotrine and monocrotaline, with i.p. LD_{50} values in the same strain of rats (Mattocks, 1972b), suggests that the oral hepatotoxicity is the same or slightly lower than the i.p. toxicity.

PAs given i.v. or i.p. reach the liver much more rapidly than when given s.c. or orally. The levels of potentially toxic pyrrolic metabolites in the livers of rats given retrorsine are initially much higher after i.p. than after oral administration of similar doses, but they reach comparable levels by 2 h (Mattocks, 1972b), confirming that the subsequent hepatotoxic actions are likely to be similar. Thus, as far as is known, the hepatotoxicity of PAs in rats does not differ greatly, regardless of the route of administration. This may not be true in other species. Pierson *et al.* (1977) found that rabbits are much less susceptible to PAs contained in a plant (*Senecio jacobaea*) in the diet than to injected PAs; this may be due to efficient excretion of the *Senecio* alkaloids (Swick *et al.*, 1982b), rather than restricted intestinal absorption as was first thought.

In contrast with the alkaloid bases, *N*-oxides of PAs can differ greatly in toxicity to the rat according to the route of administration. Because of metabolic activity in the gut, *N*-oxides may be much more hepatotoxic after oral intake than after other routes (Mattocks, 1971c): see Section II,G (below).

2. Species

Similar toxic actions of PAs have been demonstrated in many animal species, but there may be large quantitative differences. The effects of one alkaloid, retrorsine, have been compared in rats, mice, hamsters, guinea pigs, fowl and quail (White *et al.*, 1973)—see Table 7.1. The LD_{50} varies from 34 mg/kg in male rats to 279 mg/kg in quail and over 800 mg/kg in guinea pigs. A major factor determining acute toxicity is the extent to which hepatic microsomal enzymes are able to convert the alkaloid to pyrrolic metabolites (Mattocks, 1972b). In guinea pigs, in which this metabolic activity is particularly low, a dose of 800 mg/kg produces liver necrosis but is insufficient to kill the animals, whereas higher doses cause peracute death (within 1 h). Pretreatment of guinea pigs with phenobarbitone increases the hepatic metabolic activity and the LD_{50} is lowered to 210 mg/kg (White *et al.*, 1973). Guinea pigs are similarly resistant to monocrotaline (Chesney and Allen, 1973a; Swick *et al.*, 1982a) but, curiously, not to jacobine or to mixed alkaloids from *Senecio jacobaea,* which are highly toxic at 100–150 mg/kg (Swick *et al.*, 1982a).

Pierson *et al.* (1977) found that rabbits also are susceptible to injected PAs, but are resistant to chronic feeding with *Senecio jacobaea.* Buckmaster *et al.* (1977) found that quail also are resistant to dietary *Senecio jacobaea,* although they are acutely poisoned by the alkaloids from this plant (LD_{50} 115 mg/kg, i.p.).

Farm animals affected by PAs include horses, pigs, sheep, goats and cattle. There is no information on the acute toxicity of pure PAs in these species: the available literature concerns poisoning following ingestion of PA-containing plants (Table 7.5). According to Bull *et al.* (1968), sheep are more resistant than cattle to PAs from *Heliotropium* or *Senecio* species, because the alkaloids are detoxified by bacterial action in the sheep rumen (Lanigan, 1970, 1971, 1972; Lanigan and Smith, 1970). This has been disputed by Shull *et al.* (1976b), who found that the (chronic) hepatotoxicity of *Senecio jacobaea* (to rats) is lowered by incubation with cattle rumen fluid but not sheep rumen fluid; these authors suggest that the resistance of sheep is due to their lower rate of hepatic metabolism of PAs.

The relative resistance of sheep to PAs is highlighted in recent studies of the toxicity of *Echium plantagineum* (Paterson's curse). Sheep fed high levels of this plant (providing up to 0.13% of PAs in the diet) for long periods suffered practically no liver damage (Culvenor *et al.*, 1984), whereas rats given equivalent amounts of *Echium* were severely poisoned (Peterson and Jago, 1984). Probably rumen detoxication and the degree of liver metabolism both contribute to the low susceptibility of sheep: the activity of sheep liver microsomes for converting *Echium* alkaloids to pyrrolic metabolites is one-fifth to one-seventh that of rat microsomes (Peterson and Jago, 1984).

Fish also are susceptible to PAs: Hendricks *et al.* (1981) have demonstrated chronic hepatotoxicity of *Senecio jacobaea* in trout.

3. Sex

Some animal species show a distinct sex difference in their susceptibility to PA hepatotoxicity. This again is probably related to differences in hepatic microsomal metabolism; the extent of the sex difference varies according to the alkaloid. Male rats are over four times as susceptible as females to the acute toxicity of retrorsine and over twice as susceptible to monocrotaline (Mattocks, 1972b). In contrast, there is no significant sex difference for retrorsine in mice (White *et al.*, 1973). Jago (1971) found male rats to be more sensitive than females to the development of chronic lesions induced by repeated small doses of heliotrine, but females were more susceptible to lasiocarpine.

The metabolic effects of male sex hormones may be a factor influencing PA hepatotoxicity; Ratnoff and Mirick (1949) found that male or female rats fed a protein deficient diet are more susceptible to monocrotaline if they are pretreated with testosterone; the survival time of male rats on this diet is not altered by castration or treatment with estradiol.

4. Age

Young rats are more susceptible than adults to the hepatotoxicity of monocrotaline (Schoental and Head, 1955) and retrorsine (Schoental, 1959). Suckling rats are particularly susceptible, and can be poisoned by PAs in milk from mothers given these alkaloids while the mothers themselves are unaffected (Schoental, 1959). Rats aged 2 weeks are about three times more sensitive than adults to hepatotoxic PAs (Jago, 1970). However according to Jago (1971), although rats aged 1–2 weeks are more susceptible to heliotrine and lasiocarpine than older rats, the sensitivity of both sexes to the effects of chronic administration of these alkaloids increases with age after 2 or 3 months.

Schoental (1970) found rats aged 1–4 days to be more susceptible to the hepatotoxicity of retrorsine, senkirkine and hydroxysenkirkine than rats aged 25–30 days, and suggested that this was not due to activation of the alkaloids by liver microsomes because these enzymes have very low activity in newborn animals. However, Mattocks and White (1973) found that the toxicity of retrorsine is considerably lower in rats given the alkaloid within 1 h after birth than in 1- to 4-day-old rats. The liver levels of pyrrolic metabolites from retrorsine are low only if the alkaloid is given within a few hours after birth; if it is given between the ages of 1 and 20 days, the metabolite levels are much higher than those found in older animals, and are compatible with the higher toxicity.

The higher toxicity of retrorsine in 14-day-old rats compared with adults has been attributed to three factors (Mattocks and White, 1973): the relatively lower liver weight (2.8% of body weight compared with 4.2–4.4% in older rats), which makes the dose of alkaloid to the liver higher; the higher proportion of pyrrolic metabolites bound to liver tissue; and the higher susceptibility of tissues in which the cells are rapidly dividing.

5. Diet

The effects of PAs in animals can be modified by food intake and nutritional status. Schoental and Magee (1957) found that rats fed a low protein diet are more susceptible to the chronic toxicity of an oral dose of lasiocarpine than rats fed normally: they survived for a shorter time, and had very fatty livers. Conversely, rats given riddelliine, retrorsine or isatidine survived longer if fed a high casein diet, and the prolonged survival permitted the development of liver tumours (Schoental and Head, 1957). Similarly, the toxicity of rats to PAs in *Senecio jacobaea* is enhanced when the diet is low in protein, whereas a high protein diet affords some protection (Cheeke and Garman, 1974).

The effects of a low-lipotrope diet have been studied by Newberne (1968), who found that this enhances the hepatotoxicity of lasiocarpine in pregnant rats given an oral dose of the alkaloid, and also in the foetal livers. However, the diet protects young male rats against the acute toxicity of monocrotaline, an explanation for this being that the conversion of the alkaloid into active (pyrrolic) metabolites in the liver is reduced (Newberne *et al.*, 1971, 1974). A low-lipotrope diet also slightly increased the incidence of liver tumours in a group of rats given multiple doses of monocrotaline (Newberne and Rogers, 1973).

The cardiopulmonary toxicity of a single dose of monocrotaline was significantly lowered in rats fed a restricted diet, compared with rats fed *ad libitum*

(Hayashi *et al.*, 1979). The restricted diet consisted of about half the food intake of control rats; since it only commenced after the alkaloid treatment, the decreased toxicity could not be due to altered metabolism of the alkaloid. Rather, it seems to have been associated with the reduced growth rate of the animals: thus, when a group of diet-restricted rats was allowed free access to food, 30 days after monocrotaline treatment, they developed increased signs of toxicity.

Mattocks (1972b) found that the acute hepatotoxicity of retrorsine is diminished more than three-fold in rats fed only with sucrose for 4 days prior to injection of the alkaloid, and then returned to a normal diet. The action of sucrose was attributed partly to decreased conversion of the alkaloid to toxic metabolites in the liver, but also to a protective effect on the liver, since these animals survive with a higher level of pyrrolic metabolites in the liver than is produced by a lethal dose of retrorsine in control rats. The sucrose diet does not similarly protect the lungs against toxic metabolites: lung damage, rare in control rats given retrorsine, is common in sucrose-fed rats given higher doses of this alkaloid; so also are chronic hepatotoxic effects in surviving rats.

Miranda *et al.* (1981d) produced evidence that a diet with a high copper content can increase an animal's susceptibility to PAs. Incorporation of $CuSO_4$ (50 ppm) in the diet of rats increases the hepatotoxicity of the simultaneously fed PA-containing plant *Senecio jacobaea,* as judged by enzyme measurements. The significance of this appears to be that the plant itself can contain a high level of copper, so that PAs might be more hepatotoxic when ingested by eating the plant than would appear from toxicity studies with the pure alkaloids.

6. *Pretreatments with Substances Which Affect Metabolism*

a. Hepatic Microsomal Enzyme Inducers. The hepatotoxicity of PAs in animals is affected by predosing the animals with substances which affect the hepatic microsomal enzyme activity. Thus, pretreatment of rats for up to a week with phenobarbitone [which enhances the rate of conversion of PAs to pyrrolic metabolites by hepatic microsomal enzymes (Mattocks and White, 1971a; Tuchweber *et al.*, 1974)] increases the susceptibility of female rats to retrorsine and of male and female rats to monocrotaline and heliotrine (Mattocks, 1972b; Allen *et al.*, 1972; Jago, 1971; Tuchweber *et al.*, 1974). Similar pretreatment has a protective action against the hepatotoxicity of retrorsine in male rats—the LD_{50} changing from 34 to 67 mg/kg (Mattocks, 1972b)—and of lasiocarpine in both male and female rats (Jago, 1971; Tuchweber *et al.*, 1974). It appears that where the rate of metabolism of a PA in rats is normally low, pretreatment with phenobarbitone causes an increase in susceptibility to the alkaloid; where metabolism is normally fast, phenobarbitone is protective, even though it increases pyrrole production (Mattocks, 1972b).

The susceptibility of guinea pigs to retrorsine is greatly increased by pretreating them with phenobarbitone; however, in mice, although phenobarbitone increases the liver microsomal enzyme activity *in vitro*, it decreases the liver level of bound pyrrolic metabolites from retrorsine, and lowers its hepatotoxicity (White *et al.*, 1973). Swick *et al.* (1982e) found that prolonged phenobarbitone administration induces hepatic mixed function oxidase activity in sheep, but does not alter their susceptibility to *Senecio jacobaea*. The enhanced production of potentially toxic metabolites would be expected to lead to increased toxicity—and this often occurs; however, further metabolism might lead to detoxication of active metabolites, and pretreatment with phenobarbitone or other substances might produce conditions in the cell which protect tissue constituents against toxic action. Thus the *in vivo* action of each PA is likely to be the result of a complex balance of mechanisms.

Other microsomal enzyme inducers which have a similar effect to that of phenobarbitone on PA toxicity are diphenylhydantoin and the steroidal compounds spironolactone, ethylestrenol, and pregnenolone-16α-carbonitrile: all these reduce the hepatotoxicity of lasiocarpine and enhance the toxicity of heliotrine and monocrotaline (Tuchweber *et al.*, 1974). Dietary DDT is said by Cheeke and Garman (1974) to have some protective action against *Senecio jacobaea* toxicity in rats.

b. Inhibitors of Microsomal Enzyme Activity. Pretreatment with SKF 525A (2-diethylaminoethyl-2,2-diphenylvalerate) reduces the susceptibility of male rats to retrorsine: the LD_{50} is increased from 34 to 53 mg/kg and pyrrolic metabolites in the liver are reduced proportionately (Mattocks, 1972b). This compound also protects rats against the lethality of monocrotaline, but it greatly enhances the hepatotoxicity of lasiocarpine (Tuchweber *et al.*, 1974). The microsomal inhibitor chloramphenicol also protects rats against the acute hepatotoxicity of monocrotaline; however, chronic liver lesions (giant hepatocytes) are able to develop in animals thus protected (Allen *et al.*, 1972). The protective action of SKF 525A, and also of metyrapone (2-methyl-1,2-dipyrid-3-yl-1-propanone), has also been demonstrated against the lung damaging effects of monocrotaline (Eisenstein *et al.*, 1979).

Pretreatment with zinc lowers the ability of rat liver microsomes to convert PAs to pyrrolic metabolites and gives some protection against hepatotoxicity (Miranda *et al.*, 1982a).

c. Esterase Inhibitors. Some PAs are susceptible to detoxification by hydrolysis of their ester groups (Mattocks, 1982b). Pretreatment of animals with substances, such as tri-orthocresyl phosphate, which inhibit esterase activity can result in a larger proportion of such alkaloids being converted to potentially toxic metabolites (Mattocks, 1981a). Thus, the hepatotoxicity of the semisynthetic

PA, diacetyl retronecine, is expressed in rats only when they have been given an organophosphate esterase inhibitor (Mattocks, 1970).

7. Pretreatments Able to Affect Liver Thiol Levels

Hayashi and Lalich (1968) showed that mercapto-ethylamine or cysteine, given i.p. to rats shortly before, and again after an s.c. dose of monocrotaline, protects the animals against the toxicity of the alkaloid: survival is prolonged, and lung damage is less severe than in control animals given monocrotaline. Mercapto-ethylamine is more effective than cysteine, but either compound must be given before the alkaloid to afford protection. Eisenstein et al. (1979) found that cysteine fails to protect rats against monocrotaline-induced lung damage if it is given at the same time as the alkaloid. Neither does glutathione, given 0.5 h before monocrotaline.

Other workers have confirmed the protective action of thiols against pyrrolizidine toxicity. Rogers and Newberne (1971) found that mercapto-ethylamine protects rats against hepatotoxicity and mortality due to lasiocarpine, but not against inhibition of cell division by the alkaloid. (This suggests that the thiol might interact with metabolites responsible for necrosis but that different, less reactive metabolites remain which are able to inhibit mitosis.) Buckmaster et al. (1976) partially protected rats against the toxicity of dietary Senecio jacobaea, or of injected PAs from this plant, by the inclusion of 1% cysteine in their diet. Methionine (which does not possess a free thiol group) failed to protect the rats. Dietary cysteine likewise protected mice against the toxicity of monocrotaline, the acute LD_{50} of the alkaloid being raised thereby to 335 mg/kg from a control value of 259 (Miranda et al., 1981c). However, cysteine did not protect either guinea pigs (Swick et al., 1982a) or chicks (Cheeke and Pierson-Goeger, 1983) against the chronic toxicity of dietary Senecio jacobaea.

White (1976) made a careful study of the effect of altered liver thiol concentrations on the hepatotoxicity of retrorsine in male rats. The main findings are summarized in Table 7.4. Pretreatment of rats with cysteine roughly doubles the liver glutathione (GSH) level and halves the toxicity of retrorsine; pretreatment with 2-chlorethanol reduces the liver GSH to a quarter that in controls and doubles the alkaloid's toxicity. (Depletion of liver thiols by predosing diethyl maleate similarly increases the susceptibility of mice to PAs: Kim and Jones, 1982). Liver microsomes from either group of pretreated rats do not differ from controls in their ability to convert retrorsine to pyrrolic metabolites in vitro, but levels of such (potentially toxic) metabolites bound to liver tissue in vivo are lower after cysteine, and higher after chlorethanol pretreatment. Thus the protection afforded by thiols in rat liver is associated with a reduction in the levels of pyrrolic metabolites from the alkaloid which bind to liver tissue, but not with an

TABLE 7.4 Effects of Pretreating Male Rats with Cysteine or Chlorethanol on Liver Thiol Concentrations, and the Acute Hepatotoxicity and Metabolism of Retrorsine in These Rats[a]

Pretreatment (with time before retrorsine injection)	Acute LD_{50} of retrorsine (mg/kg, i.p.)	Approx. liver GSH level (% of control)	Pyrrolic metabolites in liver 2 h after retrorsine (60 mg/kg) (% of control)	Hepatic microsomal enzyme activity converting retrorsine into pyrrolic metabolites in vitro (% of control)
Controls	42	100	100	100
Cysteine, 200 mg/kg (0.5 h)	83	200	60	108[b]
2-Chlorethanol, 30 mg/kg (1 h)	23	25	200	120[b]

[a] Data from White (1976).

[b] Not significantly different from controls.

alteration in the rate of production of such metabolites. These results are consistent with the view that thiol can react with toxic (pyrrolic) metabolites in the liver, thus reducing the amounts available to alkylate vital tissue constituents.

The antioxidants ethoxyquin (6-ethoxy-2,2,4-trimethyl-1,2-dihydroquinoline) and butylated hydroxyanisole (BHA) afford some protection against PA toxicity, if given to animals before the alkaloid. Their action is unlikely to be due to their antioxidant properties—other antioxidants are not protective against PAs (cf. Rogers and Newberne, 1971)—but it may be associated with the increased tissue thiol levels which occur in rodents given these compounds (Batzinger et al., 1978). Miranda et al. (1981a) found that ethoxyquin fed in the diet (0.25%) of female mice for 38 days protects the animals against the lethality and acute hepatotoxicity of monocrotaline, given i.p. on day 10. The acute LD_{50} of monocrotaline is raised from 243 to 364 mg/kg. The ethoxyquin diet leads to elevated liver glutathione levels in the mice, and greatly increased glutathione-S-transferase activity. However it does not affect the concentration of pyrrolic metabolites from monocrotaline in the liver. A similar diet containing ethoxyquin also partially protects mice against the hepatotoxicity of injected PAs from Senecio jacobaea (Miranda et al., 1982b) or from S. longilobus (Kim and Jones, 1982).

A diet containing BHA (up to 0.75%) also protects mice against the acute toxicity of monocrotaline (Miranda et al., 1981c), S. jacobaea alkaloids (Miranda et al., 1982b) and S. longilobus alkaloids (Kim and Jones, 1982), and rats against S. jacobaea alkaloids (Miranda et al., 1982c). It is clear from the published results that these diet additives are not totally protective against PA hepatotoxicity. They can increase resistance to these alkaloids, but toxicity is manifested if the animals receive sufficiently large amounts of alkaloids.

Kim and Jones (1982) found that a diet containing disulfiram [bis-(diethylthiocarbamyl)-disulphide] is able to protect mice against PA hepatotoxicity, to a lesser extent than the antioxidants discussed above. Although disulfiram contains sulphur, this is not in the free thiol form, and the mode of its protective action is not understood.

Rats pretreated with zinc (as $ZnCl_2$) show some resistance to PA hepatotoxicity (Miranda et al., 1982a). This may be associated with lowered microsomal metabolism, but there is also a greatly increased level in the liver metallothionein, a protein rich in thiol groups, which might be able to react with toxic alkaloid metabolites.

D. Pathology of Acute Pyrrolizidine Hepatotoxicity

The various hepatotoxic PAs have similar actions in small animal species; descriptions have been given by many authors, including Davidson (1935), Bull et al. (1958), Barnes et al. (1964), and in reviews by Bull et al. (1968) and McLean (1970). Similar lesions are produced both by otonecine-based PAs (such as senkirkine) and retronecine-based PAs (such as retrorsine) (Schoental, 1970). Death of the animal commonly ensues 1–4 days after a fatal dose of an alkaloid; the liver is firm, congested, with a deep red granular appearance; haemorrhagic ascitic fluid is often present.

Microscopically, the characteristic effect is zonal haemorrhagic necrosis of the liver, which becomes apparent from about 12 h after the alkaloid intake. The zone of necrosis can vary according to the animal species and sometimes its nutritional status, or a chemical pretreatment it may have received. For example, after a dose of retrorsine, liver necrosis in the rat, mouse and guinea pig is centrilobular (shifted to midzonal in mice pretreated with phenobarbitone); in the hamster it is periportal; in fowl, focal; in male quail it is a general diffuse haemorrhagic necrosis (White et al., 1973). Lasiocarpine produced periportal necrosis in a monkey (Rose et al., 1959), but midzonal in rats (Bull et al., 1958). However, in cases of severe poisoning the area of necrosis may extend to almost the entire liver lobule.

In addition to necrosis, the sinusoids may become dilated with blood, and blood 'lagoons' may be present, with compression of surrounding hepatocytes. Barnes et al. (1964) described extensive central and midzonal blood lagoons in rats given fulvine, and narrowing of the lumina of central veins, a condition resembling human acute veno-occlusive disease (see Chapter 9); the veins are occluded by cells which are of uncertain origin, but may arise by proliferation of endothelial tissue. These cells are capable of blocking the vessel, and may (in surviving rats) become replaced by fibrous tissue. Allen *et al.* (1969) have presented a detailed light- and electron-microscopic study of the sequential changes which occur in the hepatic veins of monkeys with monocrotaline-induced veno-occlusive disease.

A suggested sequence of events in acute pyrrolizidine hepatotoxocity is as follows (Barnes *et al.*, 1964): the PA is metabolized to a proximal toxin in hepatic parenchymal cells; this toxin first causes necrosis in these cells; some escapes to damage the endothelium of hepatic veins leading to cell proliferation and veno-occlusion; some of the metabolite may proceed further by way of the bloodstream to damage other organs (particularly the lungs). The conversion of the PA to a toxic metabolite in the liver would explain why this organ is usually the principle site of injury; the zone of necrosis in the liver lobule would correspond to the cells having maximum metabolic activity towards the particular alkaloid in that animal.

Changes in hepatic subcellular particles following acute intoxication by PAs have been reviewed by McLean (1970).

E. Chronic Hepatotoxicity

The progress of chronic pathological effects of hepatotoxic PAs has been described by Schoental and Head (1957), Schoental and Magee (1957, 1959), and Bull and Dick (1959), and reviewed by Bull *et al.* (1968) and McLean (1970). Chronic liver lesions are similar whether they are due to an animal receiving a single sublethal dose of a PA or a succession of smaller doses. Animals receiving a single dose may lack appetite and lose weight for a time (typically from the second day after intake), then grow more slowly than controls. Progressively, the liver becomes small and of a tough consistency; its surface has a granular, mottled or (later) sometimes a nodular appearance, the edges of the lobes sometimes being rounded. Ascites may be present. Later, the liver may be grossly deformed, some lobes being affected more than others: for illustrations see Schoental and Magee (1957, 1959).

Microscopically, by 10 days there may be some post-necrotic fibrosis. The most striking feature of PA hepatototoxicity is the progressive enlargement of parenchymal cells and their nuclei: by 4–8 weeks, extremely large cells ('megaloytces') are abundant. These tend to occupy periportal areas, while there are areas of normal sized cells in the centrilobular zone. The development of these giant hepatocytes is a manifestation of the antimitotic action of the alkaloids (cf. Chapter 10). Examination of their ultrastructure (1 month) shows that they have increased metabolic activity but a lack of organisation (Afzelius and Schoental, 1967). An unusual feature of their nuclei is the presence of multiple centrioles, consistent with the cell making preparation for division, then failing to complete it; thus it continues to undergo division cycles but bypasses mitosis. The giant hepatocytes often show varying degrees of degeneration; however, many persist for the lifetime of the animal (up to 2 years or more in rats) and the liver never returns to normal, even though the animal may appear to be in good health.

Other common chronic features of PA hepatotoxicity include a proliferation of bile ducts, or simply of bile duct cells (Schoental and Magee, 1957); there is often a diffuse infiltration of the parenchyma by elongated cells whose origin is not clear, but which may be biliary cells, or regenerating Kupfer cells (Schoental and Magee, 1959). There may also be varying degrees of fibrosis, with fibrous thickening of central veins. In animals surviving longer, hyperplastic nodules may develop. Formation of these is greatly accelerated by the synergistic action of thioacetamide, given to rats concurrently with the PA lasiocarpine (Reddy *et al.*, 1976; Rao *et al.*, 1983).

F. Toxicity in Organs Other Than the Liver

The liver, where many PAs are converted to toxic metabolites, is the commonest site of injury by these alkaloids. At present there is no evidence that substantial amounts of toxic metabolites are formed from PAs in other tissues, although this might occur in species that have not been investigated. Thus the extent to which an organ is damaged will depend not only on its intrinsic susceptibility to the toxic metabolites, but probably also on the amounts of the latter, formed in the liver, which are able to reach it in active form. The differences between the actions of various PAs on extrahepatic tissues in different animal species can probably be accounted for by such factors as the varying chemical stabilities of their active metabolites, and the rates at which the latter enter the bloodstream from the liver.

1. Lungs

After the liver, the lungs are the most common site of toxic action of PAs. Not all PAs affect the lungs. Among those which do, 11-membered macrocyclic diesters such as fulvine (Barnes *et al.*, 1964) and monocrotaline are particularly active. The latter alkaloid, or plant material (*Crotalaria spectabilis* seed) containing it, has been extensively used to produce experimental lung damage in animals (Kay and Heath, 1969; Lafranconi and Huxtable, 1981). Culvenor *et al.* (1976b) recognised a number of other PAs which are able to produce chronic lung damage in young rats, including crispatine (an isomer of fulvine); senecionine, seneciphylline and usaramine (12-membered macrocyclic retronecine diesters); anacrotine and madurensine (crotanecine esters); and the heliotridine esters, heliosupine, lasiocarpine and rinderine. These workers found that pulmonary lesions are only produced by the same or larger doses of PAs as are needed to produce liver damage.

The development of lung damage can be markedly enhanced by pretreatments which affect the hepatotoxicity of PAs. Thus, lung damage is rare in normal rats given retrorsine, but it is commonly seen in male rats protected from the acute hepatotoxicity of this alkaloid by pretreatment with phenobarbitone, SKF 525A, or a sucrose diet, and then given higher doses of the alkaloid (Mattocks, 1972b).

Early lung changes due to PAs include alveolar oedema and haemorrhage; the principle chronic effects are a progressive proliferation of alevolar walls, and pulmonary arteritis and hypertension. For a more detailed discussion of the lung toxicity of PAs and metabolites, see Chapter 8.

2. Heart

Chronic heart damage often occurs as a secondary result of PA-induced lung damage. Thus, right ventricular hypertrophy develops in rats fed monocrotaline in the diet (Turner and Lalich, 1965) or given a single s.c. injection of the alkaloid (Hayashi et al., 1967). Cor pulmonale has been similarly induced with other PAs, including seneciphylline (Ohtsubo et al., 1977), and with pyrrolic metabolites of PAs (see Chapter 8).

3. Kidneys

There are relatively few reports of kidney damage in animals poisoned with PAs. Ratnoff and Mirick (1949) found necrosis of renal tubules in rats given monocrotaline. Van der Watt and Purchase (1970) found a severe toxic nephritis, with damaged glomeruli, in monkeys given retrorsine; there was also haemorrhage in the adrenal glands. Persaud et al. (1970) reported damage to the epithelium of proximal convoluted tubules in rats given large amounts of fulvine; this was interpreted as a functional adaptation, rather than primary epithelial damage. In pigs fed with Crotalaria retusa seed (main alkaloid, monocrotaline), the main fatal lesion was, unexpectedly, in the kidneys rather than liver (Hooper and Scanlan, 1977); the findings included megalocytosis in the renal tubules and glomeruli, atrophy of glomeruli, and tubular necrosis. McGrath et al. (1975) studied renal lesions in pigs given Crotalaria spectabilis seed daily for 43 days. They found moderate to severe damage to the glomeruli, with haemorrhages and markedly reduced cellularity. There was also necrosis of tubular cells. Glomerular lesions have also been induced in rats by a single i.p. injection of monocrotaline, and studied by light and electron microscopy (Kurozumi et al., 1983). Cell enlargement has been seen in renal tubules of mice fed with Senecio jacobaea by Hooper (1974), who cites a few other instances of this in pigs, sheep and horses. Roth *et al.* (1981) found minor functional changes in the kidneys of rats after chronic administration of monocrotaline.

We have seen severe damage to kidney glomeruli in some rats given the synthetic PA analogues synthanecine A bis-*N*-ethylcarbamate and synthanecine C bis-*N*-ethylcarbamate (Driver and Mattocks, 1984): see Chapter 8.

4. Pancreas

Infrequent enlarged islet cells have been seen in pigs fed *Crotalaria retusa* (Hooper and Scanlan, 1977). The toxicity of fulvine to pancreatic acinar cells in rats has been studied by Putzke and Persaud (1976). Islet cell tumours of the pancreas have been reported in rats given PAs from *Amsinckia intermedia* (Schoental *et al.*, 1970) or monocrotaline (Hayashi *et al.*, 1977).

5. Brain

Hooper (1972) observed spongy degeneration in brains of calves and sheep fed with PAs. The lesions were shown to be due to increased blood ammonia, secondary to the severe liver damage which was also present: a similar lesion was produced in sheep given ammonium acetate intravenously.

6. Other Tissues

Bull *et al.* (1968) have described changes in subcutaneous tissue, and in the epithelium of the duodenum, in rats given large doses of heliotrine.

G. Toxicity of PA *N*-Oxides

Published reports have formerly led to some confusion as to whether the *N*-oxides of hepatotoxic PAs are equally toxic as or less toxic than their parent alkaloids. The matter is of practical importance because PAs frequently exist partly in their oxidized form in plants which may be eaten by livestock. The *N*-oxides of lasiocarpine, monocrotaline and fulvine are said to be as toxic to rats as their parent PAs (Schoental and Magee, 1959; Barnes *et al.*, 1964); in other reports, PA *N*-oxides given i.p. or i.v. have been found very much less toxic than the basic alkaloids (Mattocks, 1971c). The conflict has been resolved upon recognition that PAs exert their toxic actions only after being activated to toxic (pyrrolic) metabolites by hepatic microsomal enzymes; they are also converted to *N*-oxides by these enzymes, but via a separate metabolic pathway (Mattocks and White, 1971a; Jago *et al.*, 1970; Chesney and Allen, 1973a). Hepatic microsomes do not convert PA N-oxides to pyrrolic metabolites (Mattocks and White, 1971a; Jago *et al.*, 1970), and the N-oxides are regarded as detoxication products (Mattocks, 1972a,b). Before they can be activated to toxic metabolites, PA N-oxides must be reduced to the basic alkaloids: in animals this occurs principally in the gut (Mattocks, 1971c; Powis *et al.*, 1979a; Brauchli *et al.*, 1982). Liver enzymes (in the rabbit) are capable of reducing N-oxides, but to a much smaller extent (Powis *et al.*, 1979a).

The foregoing accounts for the large differences in toxicity of some PA N-oxides when they are given to animals by different routes. For example, retrorsine N-oxide given i.p. to male rats has an LD_{50} of 250 mg/kg. When given orally it is five times as toxic (LD_{50} 48 mg/kg): this is because gut enzymes are able to convert it to retrorsine base, which is then absorbed and activated by the liver in the usual way (Mattocks, 1972b).

In general, PA N-oxides absorbed via the gut are likely to have hepatotoxicity similar to that of their parent alkaloids; if injected by other routes, they may be much less toxic.

III. TOXICITY OF PAs IN LIVESTOCK

A. Reported Incidents and Experiments

Table 7.5 lists cases of livestock poisoning and feeding trials with plants containing PAs which have been reported since 1968. For a survey of earlier work, see Bull *et al.* (1968).

Because the toxic effects of PAs are often delayed, animals may have eaten plants containing them for some time and received a fatal dose, before signs of poisoning become apparent.

Extensive toxicity studies have been made using *Senecio jacobaea* (ragwort; tansy ragwort). Although this plant has by no means the highest PA content among *Senecio* species, it is an extremely abundant weed; in the U.K. it occurs widely in pastures and on waste ground, is difficult to eradicate, and is said to cause 'more livestock losses than all other poisonous plants put together' (Forsyth, 1968). Poisoning of cattle by this plant in the northwestern United States is said to be a considerable economic problem (Johnson, 1982). Ragwort, like many other poisonous plants, is not willingly eaten by livestock, but it may be eaten if pastures are in poor condition (as in times of drought), or if it is incorporated in hay. PAs (from *Senecio alpinus*) persist for months when incorporated in hay, but they may be largely degraded when in silage (Candrian *et al.*, 1984b). Ironically, ragwort sprayed with the selective weedkiller 2,4-D (2,4-dichlorophenoxyacetic acid) (which over-stimulates growth) has a higher carbohydrate content 10–14 days later and so might be more palatable to cattle, as well

TABLE 7.5 Poisoning of Livestock by PA-Containing Plants,
and Experimental Studies in Farm Animals (1968–1985)

Plant	Field (F) outbreak or experiment (E)	Animal species	Additional treatments	Reference
Crotalaria juncea	E	cattle		Srungboonmee and Maskasame (1981)
C. mucronata	F,E	sheep		Laws (1968)
C. retusa	F,E	pigs, poultry		Hooper and Scanlan (1977)
C. retusa	E	pigs		Ross (1977)
C. retusa	E	chickens		Ross and Tucker (1977)
C. saltiana	E	calves		Barri and Adam (1981)
C. spectabilis	E	turkeys	selenium, aflatoxin B_1	Burguera *et al.* (1983)
Cynoglossum officinale	F	horses		Knight *et al.* (1984)
Echium plantagineum	E	sheep		Culvenor *et al.* (1984)
Heliotropium europaeum	E	sheep	cobalt	Lanigan and Whittem (1970)
H. europaeum	E	sheep	iodoform	Lanigan *et al.* (1978)
H. europaeum	E	chickens, ducks		Pass *et al.* (1979)
H. europaeum	E	pigs		Jones *et al.* (1981)
H. ovalifolium	E	sheep, goats		Damir *et al.* (1982)
Senecio alpinus	F	cattle		Pohlenz *et al.* (1980)
S. bipinnatisectus	E	calves		Mortimer and White (1975)
S. douglasii var. *longilobus*	E	cattle		Johnson and Molyneux (1984)
S. erraticus	F	horses	methionine	Araya and Gonzalez (1979)
	E	sheep		Araya *et al.* (1983)
S. jacobaea	E	calves		Ford *et al.* (1968); Thorpe and Ford (1968); Mortimer and White (1975)
S. jacobaea	E	cattle		Johnson (1976, 1979); Dickinson *et al.* (1976)

TABLE 7.5 (*Continued*)

Plant	Field (F) outbreak or experiment (E)	Animal species	Additional treatments	Reference
S. jacobaea	E	cattle	diet supplement	Johnson (1982)
S. jacobaea	E	cattle, sheep		Hooper (1972)
S. jacobaea	E	cattle, goats		King and Dickinson (1979)
S. jacobaea	E	goats		Goeger *et al.* (1979, 1982a)
S. jacobaea	E	sheep	phenobarbitone	Swick *et al.* (1982e, 1983b)
S. jacobaea	E	sheep	Cu; Mo	White *et al.* (1984a)
S. jacobaea	E	horses	$CHCl_3$; halo-thane	Gopinath *et al.* (1972)
S. lautus	F,E	cattle		Walker and Kirkland (1981); Kirkland *et al.* (1982)
S. montevidensis; S. pampeanus	E	cattle		Venzano and Vottero (1982)
S. riddellii	E	cattle		Johnson *et al.* (1985b)
S. sanguisorbae	F	sheep		Rosiles and Paasch (1982)
S. spathulatus	E	calves		Mortimer and White (1975)
S. vulgaris	E	horses		Qualls (1980)

as having a higher PA content (Irvine *et al.*, 1977). The toxicity of a plant will vary because of differences in alkaloid levels with season and locality. However, according to Goeger *et al.* (1982a), a chronic lethal dose to cattle or horses is 0.05–0.2 kg ragwort/kg body weight; for goats it is 1.25–4.04 kg/kg body weight.

Senecio toxicity is seen principally in the liver (Forsyth, 1968). The development of hepatic lesions was studied by serial liver biopsies in five calves fed ragwort in their diet (Thorpe and Ford, 1968). The plant treatment was ultimately fatal. All the livers developed veno-occlusive disease, fibrosis and megalocytosis (giant hepatocytes). One animal (given 0.54 kg ragwort/day) had centrilobular necrosis of the liver. Dickinson *et al.* (1976) gave dried ragwort to cows via a rumen cannula, at the rate of 10 g/kg/day for 2 weeks. A PA (jacoline) was found in their milk, but this had no effect on their calves. Liver biopsies showed fibrosis and megalocytosis in the cows. Goeger *et al.* (1982a) fed 11 goats a diet containing 25% dried ragwort; four died, displaying chronic liver damage including megalocytosis and bile-duct proliferation. Swick *et al.* (1982e, 1983b) investigated ragwort intoxication in sheep: prolonged phenobarbitone administration did not alter the animals' susceptibility. Feeding of ragwort does not enhance the accumulation of copper in the liver of sheep (White *et al.*, 1984a), although it does so in rats and in rabbits (Swick *et al.*, 1982b,c,d). In contrast, there are large increases in liver copper levels after intoxication of sheep by other PAs (e.g. those in *Heliotropium europaeum*); this can lead to copper toxicosis (Bostwick, 1982).

Gopinath *et al.* (1972) tested the effects of halogenated anaesthetics on ponies fed with *Senecio jacobaea*. They found that an existing ragwort lesion reduced the animals' resistance to the hepatotoxicty of chloroform, but not of halothane.

Other *Senecio* species have similar toxic actions to those of *S. jacobaea*. Mortimer and White (1975) in New Zealand fed calves with *S. jacobaea*, *S. bipinnatisectus* or *S. spathulatus*: each of these produced liver damage including photosensitisation, jaundice, wasting, liver atrophy and necrosis. Dairy cattle in Switzerland have been poisoned when grazing on pastures containing large amounts of *Senecio alpinus* (Pohlenz *et al.*, 1980). Liver cirrhosis and ascites were seen. The plant contained a high proportion of seneciphylline. *S. erraticus* intoxication of cattle is said to be common in Chile; this plant can also cause severe liver damage in horses (Araya and Gonzalez, 1979). Kirkland *et al.* (1982) described poisoning of cattle by *S. lautus* in Australia; 75 out of a herd of 800 died after grazing contaminated pasture; they had liver and kidney lesions.

Heliotropium species also produce liver damage. Poisoning of sheep by *H. europaeum* is common in Australia: a chronic LD_{50} dose of *Heliotropium* alkaloids (calculated as heliotrine) in sheep is 23–33 g/kg (Lanigan and Whittem, 1970). *H. europaeum* poisoning occurred in pigs when their feed contained peas which were contaminated with Heliotrope seed. The affected animals had blood in the gastrointestinal tract, and damage to the livers, which were small and fibrotic, and to the lungs. Megalocytes were seen in livers and kidneys, and haemorrhages in the hearts. Liver damage has also been observed in chickens and ducks fed wheat contaminated with *H. europaeum* (Pass *et al.*, 1979). In the Sudan, *H. ovalifolium* (chemical constituents unknown) has been found toxic to goats (Damir *et al.*, 1982).

The alkaloids in *Crotalaria* species are especially liable to damage other organs besides the liver. Poisoning of livestock by *Crotalaria spectabilis* (main alkaloid monocrotaline) has been reviewed by Sippel (1964), with emphasis mainly on liver lesions. However, sheep which died after grazing on *Crotalaria mucronata* in Queensland, Australia were mainly affected by acute lung damage (Laws, 1968). Onset of the acute illness was sudden, about a day after eating the plant, and death followed rapidly. The thoracic cavity contained fluid, which often coagulated in air; the lungs were oedematus and congested, and there was some concurrent liver damage. Hooper and Scanlan (1977) and Ross (1977) described chronic poisoning in pigs which had consumed seeds of *Crotalaria retusa* (which contains monocrotaline); most had advanced nephrosis, interstitial pneumonia, or both; liver damage was comparatively minor. Chickens receiving this seed at 0.05% or more in the diet, had hepatic necrosis, megalocytosis and bile duct proliferation; there was also some kidney damage, including glomerular atrophy and megalocytosis in the tubules (Hooper and Scanlan, 1977; Ross and Tucker, 1977).

Crotalaria saltiana, common in the Sudan and recognised as toxic by live-stock owners, produced liver necrosis and haemorrhage, pulmonary haemor-rhage and emphysema, and kidney damage, when fed to calves (Barri and Adam, 1981). The alkaloids of this plant are unknown, but the effects are consistent with those produced by PAs from other *Crotalaria* species.

C. juncea, which is known to contain a low level of PAs including junceine and trichodesmine (Adams and Giantuero, 1956b), and is regarded as a useful fodder plant in Asia, produced no pathological changes when fed to cattle for 2 months (Srungboonmee and Maskasame, 1981). The average overall alkaloid consumption was calculated to be 20 g/100 kg. For comparison this represents less than one-hundredth the amount of *Heliotropium* alkaloids required to pro-duce chronic intoxication in sheep, and suggests that a plant containing a suffi-ciently low level of toxic PAs may still be acceptable for animal feed: another example might be comfrey (*Symphytum* spp.), which in spite of its toxic and carcinogenic actions when fed in large amounts to experimental animals (see Chapter 11), has been in use for many years as a feedstuff for livestock (Hills, 1976) with apparently no evidence of ill effects (L. D. Hills, private communica-tion).

In summary, toxic effects in livestock ingesting plants containing PAs vary considerably with the animal species and the nature of the alkaloids involved. Acute or chronic liver damage are commonly seen; also damage to the lungs, kidneys and sometimes the heart is seen. Gastrointestinal disorders may occur, especially in cattle; neurological disturbances secondary to hepatic disease may be seen in cattle, sheep and horses.

B. Tests for PA Hepatotoxicity

Because signs of PA intoxication might not become evident until after animals have been severely affected, there is a need for tests which can enable veterinari-ans to detect this poisoning in its early stages.

Elevated serum levels of glutamic oxalacetic transaminase can be evidence of

Crotalaria poisoning in livestock (Sippel, 1964). Ford *et al.* (1968) made a detailed study of liver enzymes released into the serum of calves poisoned with *Senecio jacobaea.* Dickinson *et al.* (1976) found that elevated blood sorbitol dehydrogenase activity precedes severe liver lesions due to *S. jacobaea* in cattle, and is valuable for assessing liver function, although it is not specific for PA intoxication. However, it is less sensitive to liver damage than glutamate dehydrogenase (Craig, 1979). Enzyme tests for PA toxicosis in cattle and horses have been described by Craig *et al.* (1978) and Craig (1979): increased serum glutamate dehydrogenase is an early indicator (4–6 weeks) of ragwort intoxication in cattle, but it declines to normal levels later (by 14 weeks). On the other hand, gamma glutamyl transpeptidase increases and remains high in cattle and ponies receiving ragwort, and may be the most useful enzyme indicator of chronic or low level intoxication with this plant.

Rogers *et al.* (1979) have proposed a test based on measurements of plasma amino acid levels in horses. A low ratio (0.5–2.0) of branched chain amino acids (leucine + isoleucine + valine) to levels of phenylalanine + tyrosine is correlated with liver damage, which may be fatal; no clinical signs of poisoning are evident if this ratio is 3.1 or higher (see also Gulick *et al.*, 1980).

C. Attempts at Prophylaxis and Therapy

There is little evidence that anything can be done to prevent the onset of PA toxicosis once animals have ingested these alkaloids. The only apparently successful treatment was by Retief (1962), who reported the recovery of two *Senecio*-poisoned horses after they were treated i.v. with methionine. However, methionine was ineffective in horses poisoned with *Senecio erraticus* (Araya and Gonzalez, 1979). The mechanism of action of methionine is not known. Rogers *et al.* (1979) claim to have successfully treated an elderly horse poisoned by *Senecio vulgaris,* by giving it an oral paste containing the branched-chain amino acids leucine, isoleucine and valine.

Lanigan and Whittem (1970) failed to protect sheep against *Heliotropium europaeum* poisoning by treating them with cobalt pellets. [Vitamin B_{12}, which contains cobalt, mediates the detoxication of heliotrine in sheep rumen fluid (Dick *et al.*, 1963)]. Lanigan *et al.* (1978) found that large daily doses of the antimethanolgenic drug, iodoform, prolongs the lifespan of sheep fed a diet containing 50% of *H. europaeum*. However, this does not seem to be a practicable form of prophylaxis for the farmer. Moreover Swick *et al.* (1983a) have found that *Senecio* alkaloids, unlike those from *Heliotropium* plants (Lanigan and Smith 1970), are resistant to rumen detoxication in sheep. White *et al.* (1984a) found that treatment of sheep with molybdenum does not protect them against hepatic copper toxicosis associated with *Senecio jacobaea* consumption.

Diet supplements containing vitamins A and D, copper, cobalt, organic iodine, nicotinamide or asparagine failed to protect calves against the toxic effects of dietary *Senecio jacobaea* (Johnson, 1982).

The only effective method for protecting animals against the effects of PAs is still to prevent the contamination of pastures and feedstuffs by PA-containing plants, and to remove affected animals without delay from sources of these poisons.

8

Further Toxic and Other Biological Actions of Pyrrolizidine Alkaloids, Analogues and Metabolites

I. TOXICITY OF SEMISYNTHETIC PAs AND SYNTHETIC ANALOGUES

A. Semisynthetic PAs

Some semisynthetic esters of retronecine have chronic hepatotoxicity similar to that of natural PAs when given to rats. Branched chain esters such as retronecine di-isovalerate, ditiglate and disenecioate are thus active at dose levels of 200–500 mg/kg. Similar doses of straight chain esters, such as di-*n*-butyryl retro-

(252) $R^1 = R^3 = R^4 = H$; $R^2 = Me$
(74) $R^1 = R^3 = Me$; $R^2 = R^4 = OH$

(249)

necine, are not hepatotoxic (Schoental and Mattocks, 1960). The semisynthetic macrocyclic diester dimethyldidehydrocrotalanine (252) has hepatotoxicity similar to that of monocrotaline (74), with an acute LD_{50} of about 100 mg/kg in male rats (A. R. Mattocks, unpublished). All these highly lipophilic esters also have a high peracute toxicity, causing convulsions (probably associated with respiratory failure) if large amounts of them are given too quickly. Chain branching in the acid moiety (as in most natural PAs) appears to be necessary for the expression of hepatotoxicity; one reason for this is that the branched esters are more sterically hindered, and thus more able to resist detoxication via ester hydrolysis (Mattocks, 1982b). Accordingly, pretreatment of an animal with an esterase inhibitor leads to increased conversion of hydrolysis-prone pyrrolizidine esters to toxic (pyrrolic) metabolites in the liver (Mattocks, 1981a), and enables even the hepatotoxicity of the simple ester, diacetyl retronecine, to be expressed. Retronecine bis-N-ethylcarbamate (249) is more hepatotoxic than carboxylate esters of retronecine (Mattocks, 1971d) because, in common with other carbamates (Baron et al., 1966), it is resistant to enzymic hydrolysis. This compound produces acute centrilobular necrosis of the liver, with LD_{50} about 140 mg/kg in male rats, and delayed lung damage in a proportion of surviving animals after 3–6 weeks (Mattocks 1977b).

Chemical modification of natural PAs can sometimes cause large changes in toxicity. Acetylation of the hydroxyl groups of indicine to give triacetyl indicine causes an eightfold increase in toxicity: this might be due to several factors, including increased lipophilicity, increased steric hindrance to ester hydrolysis and altered properties of the active metabolite (Mattocks, 1972b). Similar factors might account for the somewhat increased hepatotoxicity of diacetylmonocrotaline compared with monocrotaline (Culvenor et al., 1976b). On the other hand, 7-acetyl heliotrine is no more toxic than heliotrine itself, perhaps because the acetyl group is easily hydrolysed (Culvenor et al., 1976b).

B. Synthetic Analogues

Characteristic pyrrolizidine toxicity is shown by some esters of synthetic necines (synthanecines) which are monocyclic analogues having the pyrroline but not the pyrrolidine ring of the PA nucleus (Mattocks, 1971d). The structural similarity will be evident from a comparison of synthanecine A (2,3-bishydrox-ymethyl-1-methyl-3-pyrroline) (235) and retronecine (183) (see Chapter 3, Section II,C). To be hepatotoxic, these esters must be sufficiently resistant to enzymic hydrolysis. Some synthanecine A esters and carbamates are compared in Table 8.1. The hindered ester (391) resembles a natural PA in that it has a branched chain acid moiety. The macrocyclic diester (392), analogous to a macrocyclic PA, is much more toxic. The bis-N-ethylcarbamate (373) has twice the hepatotoxic activity of the natural PA, monocrotaline, and its effects in rats resemble those of the latter: it causes acute haemorrhagic centrilobular necrosis of the liver; it has a powerful antimitotic action, producing chronic hepatic megalocytosis typical of that seen after PA intoxication; and a large dose of it causes acute lung oedema (Mattocks, 1971d). Rats of both sexes are similarly susceptible to this compound; pretreatment with phenobarbitone decreases their susceptibility, raising the LD_{50} to 68 mg/kg in male rats (Driver and Mattocks, 1984).

The hepatotoxicity of synthanecine esters, like that of natural PAs, is associated with the formation of pyrrolic metabolites in the liver (Mattocks, 1975). The extent to which these esters affect extrahepatic tissues varies according to the nature of the acid (or carbamate) moiety, and is probably related to the proportion of pyrrolic metabolite which can escape from the liver to reach other organs:

TABLE 8.1 Hepatotoxicity of Some Synthanecine A Esters and Carbamates Given i.p. to Male Rats[a]

Structure	Type of ester	R^1	R^2	Approx. liver pyrrole level 2 h after 100 mg/kg[b]	Approx. LD_{50} (mg/kg)
(391)	Acyl diester	Ac	—COCMe(Et)(OAc)	0.24	164
(392)	Macrocyclic diester	—COCH$_2$CMe$_2$CH$_2$CO—		2.1	30
(373)	Carbamate	both —CONHEt		1.26	44
(383)	Carbamate	both —CONEt$_2$		0.57	75
(372)	Phosphate	both —PO(OEt)$_2$		0.8	not acutely cytotoxic

[a] Data from Mattocks (1975), Driver and Mattocks (1984), and A. R. Mattocks unpublished results.

[b] Relative values based on absorbance of Ehrlich colour from 0.5-g tissue samples.

TABLE 8.2 Toxic Actions of Some Synthetic PA Analogues, Given i.p. to Male Rats[a]: All Are N-Ethylcarbamates (R = OCONHEt)

(373) (381) (393) (394) (395) (396)

R = OCONHEt

Structure	Compound	Acute LD$_{50}$[b] (mg/kg)	Induction of		
			Giant hepatocytes (3 weeks)	Kidney damage[f]	Lung damage[g]
(373)	Synthanecine A bis-N-ethylcarbamate	44	+	+	+
(381)	Synthanecine B bis-N-ethylcarbamate	NT(R)[c,d]	0	0	0
(393)	Synthanecine C bis-N-ethylcarbamate	89	+	+	+
(394)	Synthanecine D bis-N-ethylcarbamate	120	0	0	0
(395)	Synthanecine E N-ethylcarbamate	NT(R)[c]	0	0	0
(396)	3-Hydroxymethyl-1-methyl-3-pyrroline-N-ethylcarbamate	NT[e]	0	0	0

[a] Data from Driver and Mattocks (1984) and Mattocks (1978b).
[b] For deaths up to 7 days, associated with liver damage.
[c] NT(R): not hepatotoxic even after repeated large doses.
[d] Peracute LD$_{50}$ about 500 mg/kg.
[e] NT: no acute hepatotoxicity seen after single dose.
[f] Glomerular lesions.
[g] Chronic proliferative lesions.

this in turn depends on such factors as the chemical stability and physical proper-'
ties of the metabolite. Thus, the level of pyrrolic metabolites in the lungs of rats
given (373) is about 30% that in the liver, and this compound frequently causes
lung damage; the diethylcarbamate (383) leads to a much higher lung pyrrole
level, amounting to 75% of that in liver, and it damages not only the lungs but
also other tissues including the gut, spleen and bone marrow, reflecting the wide
distribution of toxic metabolites (Mattocks, 1975). On the other hand, no pyrroles are found in the lungs of rats given the phosphate ester (372), and this
compound only affects the liver: thus it causes the chronic development of giant
hepatocytes without being acutely hepatotoxic, even when given in doses which
lead to high liver levels of pyrrolic metabolites. A possible explanation for this is
that its primary pyrrolic metabolite is too unstable to react with cell constituents
before being hydrolysed to a more stable secondary metabolite (pyrrolic alcohol)
which has antimitotic action but no acute toxicity (Driver and Mattocks, 1984).

A comparison of N-ethylcarbamate esters of various synthetic necines (Table
8.2) shows that the saturated compound (381), like PAs which are esters of
saturated necines, is not hepatotoxic. The carbamates of synthanecines A and C,
both capable of being metabolized to bifunctional pyrrolic alkylating agents, can
cause both acute and chronic liver damage as well as injury to the lungs and
kidneys. The synthanecine D carbamate (394), which can be converted to a
mono- but not a bifunctional alkylating agent, is able to cause acute liver necrosis
but has no antimitotic action, failing to produce giant hepatocytes.

Experiments with synthanecine esters have extended structure-activity studies
beyond those possible using natural PAs.

II. TOXIC ACTIONS OF PYRROLIC METABOLITES OF PAs
AND ANALOGOUS COMPOUNDS

A. Dehydro-alkaloids and Other Pyrrolic Esters

Primary toxic metabolites formed from hepatotoxic PAs in mammalian liver
are dihydropyrrolizine esters, often called 'pyrrolic esters' or 'pyrroles' (Mattocks, 1968a). These can also be prepared chemically from PAs (see Chapter 3):
thus dehydromonocrotaline (DHM) is derived from monocrotaline. Acyl or carbamate esters of hydroxymethyl pyrroles are synthetic analogues having chemical and biological properties similar to those of these pyrrolic metabolites (Mattocks, 1971d). Unlike their parent alkaloids, the pyrrolic esters are chemically
reactive and highly unstable in aqueous solution; they must therefore be dissolved in a water miscible non-aqueous solvent for administration to animals.
For this purpose, N,N-dimethylformamide is usually used. This solvent is itself
hepatotoxic in various animal species (Ungar et al., 1976); however, it is satis-

factory in rats if its dose level is kept below 500 mg/kg (Mattocks and Driver, 1983).

(276)

Table 8.3 lists experiments in which dehydro-pyrrolizidine alkaloids and analogous pyrrolic esters have been given to animals. The toxic effects of dehydromonocrotaline (276) are typical. Given by stomach tube this compound is inactive, because it is immediately decomposed or polymerized in the acid aqueous environment of the stomach. Given i.p., it produces severe local necrosis and peritonitis. To reach a target organ, it must be injected into the blood stream flowing directly to that organ. The action of the pyrrole is largely restricted to the first tissue it encounters; what is not absorbed by that tissue appears to be largely inactivated in the circulation before it can act on other organs. Thus when DHM is given to rats via a tail vein, the principle damage is to the lungs (Butler et al., 1970; Plestina and Stoner, 1972). The delayed lung damage resulting from a dose of 5 mg/kg of DHM closely resembles that produced by about 10 times the dose of the PA, fulvine (Barnes et al., 1964): this is consistent with the idea that PA lung damage is caused by pyrrolic ester metabolites from the liver reaching the lungs (Butler et al., 1970). Larger doses of DHM given via the tail vein (up to 30 mg/kg) cause acute lung damage associated with congestion, oedema and large amounts of pleural effusions, and most animals die within 48 h (Plestina and Stoner, 1972). Similar acute and chronic lung damage is produced in rats by comparable i.v. doses of other dihydropyrrolizine esters including the semi-synthetic compounds diacetyl dehydroretronecine and disenecioyl dehydroretronecine: this shows that the complex acid moiety of the natural PA is not essential for cytotoxic action (Mattocks, 1970). Moreover, synthetic hydroxymethylpyrrole esters have similar pneumotoxicity, showing that the essential toxic nucleus is an ester made chemically reactive by its attachment to a pyrrole (or similar) structure (Mattocks, 1971d; Mattocks and Driver, 1983). Some heart damage may also develop after tail-vein injection of pyrrolic esters (Shumaker et al., 1977; Mattocks and Driver, 1983). The cardiopulmonary toxicity of PAs and pyrroles is further discussed later in this chapter (Section III).

Centrilobular necrosis of the liver occurs in a proportion of rats injected via a tail vein with some primarily pneumotoxic esters of dehydroretronecine and analogous pyrroles (Mattocks and Driver, 1983). This action might be due to a proportion of the injected pyrrole which has survived passage through the lungs,

TABLE 8.3 Studies on Effects in Rats of Dehydropyrrolizidine Alkaloids and Similar Chemically Reactive Pyrrolic Esters

Compound	Route and site of injection	Affected tissue or objective of experiment	References
Dehydromonocrotaline; dehydroretrorsine	i.v. (tail and mesenteric veins)	lung; liver	Butler et al. (1970)
Dehydromonocrotaline	i.v. (tail)	lung (ultra-structure)	Butler (1970); Hurley and Jago (1975)
Dehydromonocrotaline	i.v. (tail)	lung	Plestina and Stoner (1972)
Dehydromonocrotaline	i.v. (mesenteric vein)	liver	Newberne et al. (1971); Hsu et al. (1973a)
Dehydromonocrotaline	i.v. (tail)	lung, heart	Chesney et al. (1974a)
Dehydromonocrotaline	s.c.	skin	Hooson and Grasso (1976)
Dehydromonocrotaline	s.c. (scrotum)	cremaster (vascular lesions)	Hurley and Jago (1976)
Dehydromonocrotaline	i.v.	lung	Lalich et al. (1977)
Dehydromonocrotaline	i.v. (tail)	lung (slices in vitro)	Hilliker et al. (1983a)
Dehydromonocrotaline	i.v.	lung	Kido et al. (1981)
Dehydromonocrotaline	i.v. (tail)	lung, heart ECG	Bruner et al. (1983)
Dehydromonocrotaline	i.v. (tail)	platelet aggregation	Hilliker et al. (1983b)
Dehydromonocrotaline	i.v.	pulmonary hypertension; lung 5HT uptake	Hilliker et al. (1983c)
Dehydromonocrotaline	i.v. (tail)	lungs: protection by drugs or diet restriction	Hilliker and Roth (1984); Ganey et al. (1985)
Acetyldehydroretronecine (mixture)	i.p. and i.v. (tail)	lung	Mattocks (1968a)
Diacetyldehydroretronecine	i.v. (tail and mesenteric vein)	lung; liver	Mattocks (1970)
Disenecioyl dehydroretronecine	i.v. (mesenteric vein)	liver	Shumaker et al. (1976c)
Disenecioyl dehydroretronecine	i.v.	lung, heart	Shumaker et al. (1977)

TABLE 8.3 (*Continued*)

Compound	Route and site of injection	Affected tissue or objective of experiment	References
2-(N-Ethylcarbamoyloxy-methyl)-1-methylpyrrole; 2,3-bis-(N-ethylcarbamoyl-oxymethyl)-1-methylpyrrole	into carotid or femoral artery	lung (vascular lesions)	Plestina *et al.* (1977)
[^{3}H]-2,3-Bisacetoxymethyl-1-methylpyrrole	i.v. (tail)	lung (and tissue distribution)	Mattocks (1971d, 1979)
14 Acetates and N-ethylcarba-mates of dehydroretronecine, hydroxymethyl pyrroles, and analogues	i.v. (tail)	lungs, heart, liver	Mattocks and Driver (1983)

but if so, periportal necrosis would be expected (see below). The centrilobular distribution of damage suggests that decomposition products from the injected pyrrole may be further converted to hepatotoxic metabolites in the liver (cf. Guengerich and Mitchell, 1980).

To act directly on the liver, pyrrolic esters must be injected into a mesenteric vein (Butler *et al.*, 1970; Newberne *et al.*, 1971; Hsu *et al.*, 1973a; Chesney and Allen, 1973a; Shumaker *et al.*, 1976c). It is not possible by this means to reproduce exactly the situation which exists when a PA is converted within the liver cell to an active metabolite which then has immediate access to cell constituents. An injected pyrrole must first encounter blood, vascular endothelial tissue and the exterior of the liver cell membrane before it can arrive at the point where a metabolic product would begin its existence. Nevertheless, injected pyrrolic esters are much more toxic than PAs and their actions on the liver bear many similarities to those of the latter. A mesenteric i.v. injection of 15 mg/kg de-hydroretrorsine (retrorsine pyrrole) in rats leads to extensive necrosis in the left lobes of the liver within a few days, with infarcts radiating from portal tracts; there is thrombosis and necrosis of the walls of the portal veins, whereas the central veins appear normal (Butler *et al.*, 1970). After several weeks, enlarged parenchymal cells are present in damaged lobes; these often have greatly abnormal nuclei but lack mitoses. Hyperplastic nodules may also be seen, with uniform small cells and abundant mitoses. After 3 weeks or more, the lobular pattern of the liver is distorted, the left side being atrophic while the right side is hypertrophic or has a greatly nodular appearance. The lungs and other organs are unaffected when the pyrrole is given by this route. Liver damage is produced when dehydromonocrotaline (15 mg/kg) is given in a similar way to guinea pigs, even though these animals are highly resistant to the hepatotoxic effects of the parent alkaloid, monocrotaline (Chesney and Allen, 1973a). The semisynthetic compound disenecioyl dehydroretronecine has a similar action on the liver of rats (Shumaker *et al.*, 1976c).

To summarize, mesenteric vein injections of highly active pyrrolic esters first damage the vascular system into which they are introduced; then the first liver cells which they reach, causing necrosis and mitotic inhibition; and later repair processes. They are rapidly absorbed or inactivated so that more distant tissues, and even the central areas of liver lobules, are hardly affected.

The acute cytotoxicity of dehydromonocrotaline has been investigated *in vitro* by Johnson (1981) using cultured mouse fibroblasts. Initially, the plasma membrane ATPase is inhibited; subsequently there is swelling and disruption of cytoplasmic organelles including mitochondria. Cell death is associated with rupture of plasma membranes and consequent loss of cytoplasmic components from the cell.

B. Dehydro-necines and Pyrrolic Alcohols

Dehydro-PAs (dihydropyrrolizine esters) are rapidly hydrolysed in an aqueous environment to the corresponding alcohols (dehydro-necines), which may therefore be present as 'secondary' metabolites in animals given hepatotoxic PAs. Heliotridine-based PAs can thus give rise to dehydroheliotridine (DHH) (272) (Jago *et al.*, 1970), and retronecine-based PAs (such as monocrotaline) give dehydroretronecine (201) (Hsu *et al.*, 1973b). These pyrrolic alcohols have chemical alkylating ability similar in character to that of their esters, but they are much less reactive (see Chapter 5), and they can thus survive for a longer time in the body and circulate to tissues which may not be reached by the pyrrolic esters. Major toxic effects of DHH are in many respects similar to the effects of ionizing radiation, and are associated with its powerful antimitotic activity. DHH differs from its parent PAs heliotrine and lasiocarpine in that it affects a wider range of tissues, has lower hepatotoxicity, and does not produce lung lesions. The LD_{50} (7 days) of DHH in mice of either sex is approximately 250 mg/kg (Percy and Pierce, 1971). Peterson *et al.* (1972) made a detailed study of the pathological effects of DHH in rats aged 14 days. It particularly affects developing tissues

(201) $R^1 = OH$, $R^2 = H$
(272) $R^1 = H$, $R^2 = OH$

with a high turnover of cells. An aqueous i.p. dose of 61 mg/kg produces no clinical signs of toxicity, but 92 mg/kg kills most animals within 10 days; it is less acutely toxic if dimethyl sulphoxide is used as solvent. In the acute phase, growth is arrested, there is fur loss, which may be considerable, and there are tooth defects. There is atrophy of hair follicles, gut mucosa, spleen, thymus, bone marrow and testis. Necrosis is only seen in the epithelium of the intra-lobular ducts of salivary glands; there is no liver necrosis other than in isolated cells. Chronic effects in surviving rats are milder than those of the parent PAs, and consist of a moderate megalocytosis in hepatic parenchymal cells and in renal tubular epithelium. Chronic liver damage occurs within a shorter time (4 weeks) if two injections of DHH (61 mg/kg) are combined with small doses of carbon tetrachloride. [A similar stimulation of megalocytosis by a combination of a mitotic inhibitor and a hepatotoxin occurs when synthetic pyrrolic alcohols analogous to DHH are given to rats, followed by dimethylnitrosamine (Mat-tocks, 1981b)]. DHH also has a long lasting inhibitory action on mitosis in the liver of adult rats (Samuel and Jago, 1975); it is teratogenic when given to pregnant rats (Peterson and Jago, 1980), it has immunosuppressive activity in mice (Percy and Pierce, 1971), and it causes premature aging and increased tumour incidence when given intermittently to rats (Peterson et al., 1983).

Dehydroretronecine (DHR), which is enantiomeric with DHH, has effects similar to those of DHH when injected s.c. into young rats (Hsu et al., 1973b; Shumaker et al., 1976a). DHR severely depresses the hepatic mitotic index, causes atrophy of the thymus, spleen and testis, and induces chronic hepatic megalocytosis. Some of the effects of DHH observed by Peterson et al., includ-ing alopecia, tooth defects and salivary gland necrosis, were not produced by Hsu et al. (1973b) using DHR; this might have been because the latter workers used older rats in which tissue proliferation had slowed down. Ulceration of the glandular stomach, not seen after DHH, was often observed in rats given DHR: however, much larger amounts of the latter were used—up to six weekly doses of 90 mg/kg. A very high level of radioactivity is found in the glandular stomach of rats given tritiated DHR subcutaneously (Hsu et al., 1974; Shumaker et al., 1976a), showing that DHR, the reactivity of which is enhanced by acid (see Chapter 5), selectively reacts with, or is decomposed in, this acidic tissue.

When DHR (65 mg/kg) is given s.c. to rhesus monkeys, it causes no gross lesions up to 24 h (Hsu et al., 1976). Microscopically there is damage to the endothelium of hepatic veins and to the gut mucosa. Daily s.c. administration of DHR (4 mg/kg) to rats for 2 weeks inhibits their growth and causes right ventricular hypertrophy, a condition similar to that produced by a low level of the PA monocrotaline in the drinking water (Huxtable et al., 1978). After a large dose of DHR (100 mg/kg) to rats, serotonin metabolism in the lungs is increased, suggesting that there is an increase in cell permeability.

Synthetic pyrrolic alcohols, such as 2,3-bis-hydroxymethyl-1-methylpyrrole

Fig. 8.1.   Male rat, small (weight 29 g) and almost hairless, 17 days after being given 2,3-bishydroxymethyl-4-methoxy-1-methylpyrrole (100 mg/kg, i.p.) at age 10 days. Photographed with control rat of the same age (weight 106 g).

(279)

(BHMP) (279), which have chemical properties analogous to those of DHH and DHR (Mattocks, 1974) have similar toxic actions in young rats (Mattocks, 1975), causing growth arrest and fur loss (Fig. 8.1), cytotoxic effects in the thymus, spleen, bone marrow and gut mucosa, and defective tooth development (Fig. 8.2). As with DHR, an injection of tritiated BHMP leads to a high level of binding in the glandular stomach (White and Mattocks, 1976).

III. TOXICITY OF PAs AND PYRROLES TOWARDS THE LUNGS AND HEART

The lungs of animals are injured by many, but not all, PAs which are also hepatotoxic (see Chapter 7, Section II,F). There have been many experimental investigations of the development, mechanisms, physiology, pharmacology and morphology of PA-induced lung and heart damage. For these, monocrotaline or

Fig. 8.2.   Male rat, 70 days after being given 2,3-bishydroxymethyl-1-methylpyrrole (200 mg/kg, i.v.) at age 23 days. Tooth defects include one overgrown and one short upper incisor; the lower incisors are broken and loose.

its source plant *Crotalaria spectabilis* has often been used; see Table 8.4. For reviews, see Kay and Heath (1969), McLean (1970), and Lafranconi and Huxtable (1981).

A. Acute Toxicity of PAs in the Lung

Profound changes are probably initiated in the lungs of animals very quickly after PA intake, but their consequences are usually only apparent a long time afterwards. In the first week after receiving a single oral dose of fulvine or its *N*-oxide, rats have damaged livers but display relatively little lung injury. A clear pleural effusion may be present, and microscopically a perivascular exudate may be seen around lung arteries and veins (Barnes *et al.*, 1964). Occasionally, parts of the lungs are congested, with collapsed alveoli. Large injections of monocrotaline in rats cause interstitial alveolar oedea in the first day, which later increases in severity (Valdivia *et al.*, 1967a); early alterations in the alveolar wall ultra-structure have been studied (Valdivia *et al.*, 1967b). A modest pulmonary oedema occurs in dogs within 2 h after an i.v. injection of monocrotaline (Miller *et al.*, 1978). In rats dying a few days after very large i.p. doses of monocrotaline (in excess of the acute hepatotoxic LD_{50}), pulmonary oedema and pleural effu-

TABLE 8.4 Experimental Studies on Toxicity of PAs and PA-containing Plant Materials to the Lungs and Heart[a]

Plant or alkaloid	Animal species	Method of dosing[b]	Heart or lung effects studied	References
Crotalaria laburnoides	rat	D	pulmonary vascular disease; RVH	Heath *et al.* (1975)
C. spectabilis (seed)	rat	D	mast cells in lung	Takeoka *et al.* (1962); Kay *et al.* (1967)
C. spectabilis (seed)	rat	D	blood 5HT after PH	Kay *et al.* (1968)
C. spectabilis (seed)	rat	D	EM of lungs	Kay *et al.* (1969)
C. spectabilis (seed)	rat	D	pulmonary vascular occlusions; endothelial lysis	Allen and Carstens (1970)
C. spectabilis (seed)	rat	D	PH; RVH; arterial changes	Hislop and Reid (1974)
C. spectabilis (seed)	rat	D	EM of vascular smooth muscle cells in PH	Smith and Heath (1978)
C. spectabilis (seed)	rat	D	development of pulmonary arterial changes	Meyrick and Reid (1979); Meyrick *et al.* (1980)
C. spectabilis (seed)	rat	D	hyperplasia and hypertrophy in alveolar and pulmonary arterial cells	Meyrick and Reid (1982)
Senecio jacobaea	rat	D	pulmonary vascular disease	Burns (1972)
Fulvine	rat	S	toxicity	Barnes *et al.* (1964)
Fulvine	rat (SPF)	S	toxicity not due to lung infections	Schoental (1966)
	rat	S	RVH and EM of pulmonary vasculature	Wagenvoort *et al.* (1974a,b)
Monocrotaline	rat	D	PH; 5HT level: effect of *p*-chlorophenylalanine	Carrillo and Aviado (1969)
Monocrotaline	rat	D	removal of 5HT and noradrenaline by perfused lung	Gillis *et al.* (1978)
Monocrotaline	rat	D	biochemistry of pulmonary endothelium	Huxtable *et al.* (1978); Huxtable (1979b)
Monocrotaline	rat	D	PH; RVH; LDH activity; 5-HT clearance	Roth *et al.* (1981)
Monocrotaline	rat	D	ACE activity in lung	Lafranconi and Hux-

TABLE 8.4 (*Continued*)

Plant or alkaloid	Animal species	Method of dosing[b]	Heart or lung effects studied	References
				table (1983); Huxtable and Lafranconi (1984)
Monocrotaline	rat	D	comparison of heart and lung effects	Lafranconi *et al.* (1984)
Monocrotaline	rat	D	effects on pulmonary endothelial function	Molteni *et al.* (1984)
Monocrotaline	rat	S	interstitial pulmonary oedema	Valdivia *et al.* (1967a)
Monocrotaline	rat	S	alterations in alveoli	Valdivia *et al.* (1967b)
Monocrotaline	rat	S	cor pulmonale	Hayashi *et al.* (1967)
Monocrotaline	rat	S	pulmonary arteritis	Hayashi and Lalich (1967)
Monocrotaline	rat	S	effect of zoxazolamine (etc.) on RVH development	Kay *et al.* (1976)
Monocrotaline	rat	S	protective action of diet reduction	Hayashi *et al.* (1979)
Monocrotaline	rat	S	protection by inhibition of hepatic metabolism	Eisenstein *et al.* (1979)
Monocrotaline	rat	S	lipid peroxides in interstitial pneumonitis	Tsugi (1979)
Monocrotaline	rat	S	collagen synthesis in pulmonary arteries	Kameji *et al.* (1980)
Monocrotaline	rat	S	PH; RVH; small vessel thickening	Ghodsi and Will (1981); Will (1981)
Monocrotaline	rat	S	pathophysiological development of PH	Kido (1981)
Monocrotaline	rat	S	alveolar-capillary membrane permeability	Kido *et al.* (1981)
Monocrotaline	rat	S	LDH activity	Roth (1981)
Monocrotaline	rat	S	PH; RVH; 5HT clearance	Hilliker *et al.* (1982)
Monocrotaline	rat	S	lipid peroxides and GAG in interstitial pneumonitis	Uchiyama *et al.* (1982)

(*continued*)

TABLE 8.4 (*Continued*)

Plant or alkaloid	Animal species	Method of dosing[b]	Heart or lung effects studied	References
Monocrotaline	rat	S	membrane properties of smooth muscle walls	Suzuki and Twarog (1982)
Monocrotaline	rat	S	ACE activity in pulmonary disease	Kay *et al.* (1982); Keane *et al.* (1982); Keane and Kay (1984); Hayashi *et al.* (1984)
Monocrotaline	rat	S	RVH; lung vessel leak	Sugita *et al.* (1983a)
Monocrotaline	rat	S	abnormal alveolar cells	Sugita *et al.* (1983b)
Monocrotaline	rat	S	5-HT and paraquat uptake by lung slices	Hilliker *et al.* (1983a)
Monocrotaline	rat	S	effect of chemical sympathectomy and serotonin inhibition on RVH	Tucker *et al.* (1983)
Monocrotaline	rat	S	norepinephrine removal by perfused lungs	Hilliker *et al.* (1984)
Monocrotaline	rat	S	protection by methylprednisolone	Langleben and Reid (1985)
Monocrotaline	rat	S	alveolar inflammation; arachidonate metabolism	Stenmark *et al.* (1985)
Monocrotaline	rat	S	Effects in neonates and infants	Todd *et al.* (1985)
Monocrotaline	monkey	M	PH, cor pulmonale and endocardial fibrosis	Chesney and Allen (1973b,c)
Monocrotaline	dog	S	acute pulmonary oedema	Miller *et al.* (1978)
Seneciphylline	rat	S	PH, cor pulmonale	Ohtsubo *et al.* (1977)

[a] Abbreviations used: ACE, angiotension converting enzyme; EM, electron microscopy; GAG, glycosaminoglycans; 5HT, 5-hydroxytryptamine; LDH, lactate dehydrogenase; PH, pulmonary hypertension; RVH, right ventricular hypertrophy.

[b] Methods of dosing: S, single injection (usually s.c. or i.p.); M, multiple doses; D, given in diet (food or drinking water).

sions are common (Mattocks, 1972b). Large doses of the synthetic analogue, 2,3-bis-*N*-ethylcarbamoyloxymethyl-1-methylpyrrolline (synthanecine A bis-*N*-ethylcarbamate) have similar action (Mattocks, 1971d). Retrorsine, which rarely affect the lungs in normal rats, can cause severe lung damage in rats pretreated in ways which protect them against the acute hepatotoxicity of this alkaloid, e.g. with phenobarbitone or with a sucrose diet (Mattocks, 1972b).

B. Chronic Cardiopulmonary Toxicity of PAs

1. Pathological Changes

A single large dose of fulvine to rats causes a wave of deaths around 4–6 weeks later, preceded by respiratory distress (Barnes *et al.*, 1964). There is extensive pleural effusion, and microscopy of the lungs shows congestion, thickening of alveolar walls, intra-alveolar oedema, and a marked perivascular exudate. Similar effects have been reported by many workers after single or multiple dosing of rats with monocrotaline or *Crotalaria spectabilis* seed (Table 8.4). A large single dose of monocrotaline can lead to necrotising pulmonary arteritis and pulmonary hypertension, in association with right ventricular hypertrophy (Hayashi and Lalich, 1967). Some very large cells may be present in the thickened alveolar walls and also within the alveoli; the latter were described by Kay *et al.* (1969) as enlarged granular pneumocytes, but more recent evidence (Sugita *et al.*, 1983b) suggests that they are abnormal macrophages. There is a proliferation of mast cells (Takeoka *et al.*, 1962; Kay *et al.*, 1967, 1969).

Most experimental work has been done with monocrotaline, but other PAs can act similarly, e.g. seneciphyline (Ohtsubo *et al.*, 1977). Some PAs, e.g. rinderine, cause an intravascular accumulation of mononuclear cells,which may lead to venous occlusion (Culvenor *et al.*, 1976b). Monocrotaline-induced pulmonary hypertension and cor pulmonale in *Macaca* monkeys has been cited as a model of human cardiopulmonary disease (Chesney and Allen, 1973b).

2. Development of Cardiopulmonary Lesions in Animals Given Monocrotaline

Monocrotaline has no immediate effects on pulmonary arterial pressure (Gillis *et al.*, 1978): development of pulmonary hypertension is progressive, first appearing about 2 weeks after a single large dose of the alkaloid (Hilliker *et al.*, 1982). Meyrick *et al.* (1980) followed the development of pulmonary hypertension in rats fed *Crotalaria spectabilis* in the diet, and found that pulmonary arterial pressure increases from 14 days after commencement of the diet. The initial damage is apparently to endothelial cells of small vessels [cf. the endo-thelial damage seen in hepatic veins of monkeys given monocrotaline (Allen *et al.*, 1969)]. Changes in alveolar walls appear much later; later still comes reduc-tion of the lumen of small vessels. There is an increase in pulmonary arterial muscle (Hislop and Reid, 1974), which begins to develop even in normally non-muscular arteries from as early as 3 days after commencement of the diet (Meyrick and Reid, 1979). The hyperplasia and hypertrophy of pulmonary ar-terial cells has been demonstrated by increased uptake of tritiated thymidine (Meyrick and Reid, 1982). The lungs from monocrotaline-treated rats are heavier and have a higher protein content than control lungs (Gillis *et al.*, 1978).

After a single dose of monocrotaline, there is leakage from the microvascula-ture of rat lungs within the first week (Sugita *et al.*, 1983a). The initial injury leads to a progressive thickening of medial walls of small pulmonary arteries and veins; the right heart responds to this vessel restriction by increasing its force of contraction, thus leading to increased pulmonary pressure and right ventricular hypertrophy (Ghodsi and Will, 1981; Will, 1981). The latter has been well demonstrated by Chesney and Allen (1973c) in young *Macaca* monkeys given monocrotaline.

3. Biochemical Changes in the Lungs

Rat mast cells are rich in 5-hydroxy-tryptamine (5HT): it has been suggested (Turner and Lalich, 1965) that the release of 5HT from proliferated mast cells might be a cause of pulmonary hypertension in rats given monocrotaline; howev-er, lung 5HT levels in rats fed a diet containing *Crotalaria spectabilis* seed did not differ from levels in control rat lungs (Kay *et al.*, 1968). Kay and Heath (1969) consider that the mast cells are not concerned with pulmonary hyperten-sion but are secondary to the exudative lung lesions associated with congestive heart failure. Nevertheless, Carrillo and Aviado (1969) found that treatment with *p*-chlorophenylalanine, which prevents hypertension due to 5HT by inhibiting the enzyme synthesizing 5HT, reduces monocrotaline-induced pulmonary hyper-tension in rats. However, *p*-chlorophenylalanine does not protect against *acute* pulmonary oedema produced by the PA metabolite, dehydromonocrotaline (Plestina and Stoner, 1972).

A number of biochemical studies have been aimed at investigating the re-laionship between pulmonary endothelial damage and the development of pul-monary hypertension. A normal function of pulmonary endothelial cells is to remove and metabolize circulating vasoactive substances and drugs; this activity may be impaired in animals suffering from PA-induced lung injury. Gillis *et al.* (1978) found that isolated lungs from monocrotaline-fed rats remove and metab-olize 50% less 5HT than control lungs. Monoamine oxidase activity is not less than in controls; thus the reduced 5HT metabolism is probably due to impaired delivery of substrate to the enzyme. This is not a direct action of monocrotaline, since perfusion of control lungs with the alkaloid does not affect 5HT removal. Pulmonary removal of noradrenaline (norepinephrine) is also decreased after monocrotaline treatment. This has been confirmed by Hilliker et al. (1984). Hilliker et al. (1983a) found that 5HT uptake is decreased 35% in lung slices from rats treated 14 days earlier with monocrotaline or its pyrrolic metabolite dehydromonocrotaline, but not in lung slices treated with pyrrolic metabolite formed from monocrotaline in vitro by liver slices in the aqueous medium. They concluded that either the pyrrole does not react directly with lung tissue in vitro, or that the injury is slow to develop; however, it is doubtful whether much of the highly reactive primary metabolite would have reached the lung slices intact.

Kay et al. (1982) and Keane et al. (1982) found that the specific activity of angiotension converting enzyme (ACE) in the pulmonary endothelium is reduced in female rats from 10 days after a single injection of monocrotaline. This was concluded to be not a cause of pulmonary hypertension but a result of it, possibly a protective mechanism to limit the elevation of pulmonary arterial pressure. However in young male rats, given monocrotaline for up to 3 weeks in their drinking water, although the specific activity of ACE in the lung is reduced by 64%, this is only due to a large increase (52%) in total lung protein, and the total ACE in the lung is the same as in control rats (Lafranconi and Huxtable, 1983). In similarly treated rats, Huxtable et al. (1978) showed that levels per lung of ACE, 5'-nucleotidase and monoamine oxidase are unaltered. Norepinephrine transport is unaffected [in contrast with Gillis et al. (1978) as cited above], but 5HT transport is specifically impaired (Huxtable et al., 1978). These rats have pulmonary hypertension without inflammatory changes. Differences between the conclusions of these two groups of workers have been discussed in correspondence (Keane and Kay, 1984; Huxtable and Lafranconi, 1984).

Molteni et al. (1984) demonstrated changes in the pulmonary endothelium of male rats given monocrotaline for up to 12 weeks in drinking water. These changes, indicative of endothelial cell injury, accompanied and sometimes preceded the development of pulmonary hypertension. After an early increase, ACE activity fell to about 55% of control lever by 6 weeks; plasminogen activator was 59% of control at 6 weeks, whereas prostacyclin production increased. Similar lung changes occur following treatment with ionizing radiation or bleomycin.

Kameji et al. (1980) found increased collagen synthesis in pulmonary arteries of rats with monocrotaline-induced hypertension; they suggest that this is a consequence of the increased blood pressure.

Elevated lactate dehydrogenase activity in lung lavage fluid has been found a useful indicator of lung injury caused by monocrotaline as well as other chemicals (Roth, 1981).

C. Metabolites Responsible for Lung Damage

Lung damage is not caused by direct action of PAs themselves, but by the interaction with lung tissue of pyrrolic ester metabolites formed in the liver. There is good evidence for this. Pyrrolic metabolites are formed in significant amounts from PAs in liver, but not lung tissue (Mattocks and White, 1971a). The toxicity of PAs to the lungs is influenced by pretreating animals in ways which mainly affect drug metabolizing enzymes in the liver, not the lung. Thus, metabolic inhibition by SKF 525A or metyrapone protects rats against monocrotaline pneumotoxicity (Eisenstein *et al.*, 1979), and prior stimulation of hepatic metabolism with phenobarbitone increases the lung damage and the lung level of pyrrolic metabolites when rats are given monocrotaline (Mattocks, 1972b). Pyrrolic esters injected into rats have much greater pneumotoxic activity than corresponding doses of their parent PAs (Butler *et al.*, 1970).

Direct evidence for pulmonary toxicity of metabolites formed in the liver has been provided by Lafranconi and Huxtable (1984). Pyrrolic metabolites are formed when monocrotaline is perfused through an isolated rat liver, and subsequent passage of these through isolated lungs injures the pulmonary endothelium, as indicated by a reduction in serotonin transport. The latter is not affected by monocrotaline itself. In contrast, pyrrolic alcohol metabolites such as dehydroretronecine and dehydroheliotridine have relatively low toxicity, and single large doses of these do not affect the lungs (Mattocks, 1970; Peterson *et al.*, 1972; Hsu *et al.*, 1973b). Multiple treatment of rats with dehydroretronecine leads to development of right ventricular hypertrophy similar to that produced by less than half the equivalent level of monocrotaline, but its effects on pulmonary endothelial cell metabolism are different (Huxtable *et al.*, 1978): this suggests that the pyrrolic alcohol is not the major metabolite responsible for the pneumotoxicity of monocrotaline.

The primary site of PA-induced lung damage is apparently in the pulmonary vasculature; this is consistent with a reactive toxin arriving by way of the bloodstream, rather than being produced in the lungs, which would be expected to lead to early injury of other pulmonary cell types (Boyd, 1980). Nevertheless, the possibility cannot be excluded that further metabolites might be formed in lung cells either from PAs or from the pyrrolic derivatives, which might also contribute towards pneumotoxicity.

The chemical stability of the pyrrolic ester metabolites might be a factor affecting the extent to which different PAs affect the lungs compared with the liver. The more reactive the metabolite, the less of it would survive the journey from liver to lung. The pyrrolic metabolite from monocrotaline is less reactive than that from retrorsine (Mattocks, 1969a), and this may be the reason why the former alkaloid (and similar *Crotalaria* alkaloids) is the more effective in damaging the lungs.

D. Toxic Actions of Pyrrolic Esters in the Lungs and Heart

The toxicity of pyrrolic esters has been discussed in general terms earlier in this chapter (Section II,B), and experimental work using rats is listed in Table 8.3. In addition, Raczniak *et al.* (1979) have studied dehydromonocrotaline-induced pulmonary fibrosis in dogs. Two aspects of pyrrole ester toxicity will be further discussed here: the mechanisms by which pathological lesions develop in lungs and heart, and the relation of pneumotoxicity to the molecular structure and chemical properties of the pyrrole.

1. Mechanisms

Since pyrrolic ester metabolites are believed to be the principle cause of PA-induced pneumotoxicity, such pyrroles would be expected to have actions similar to those of PAs when injected into animals. This is generally true, but there are a few differences. The pyrrolic esters are so reactive that they only act on the lung when they are introduced directly into the bloodstream leading to that organ. The pyrrole, injected as a single, relatively large pulse, probably does its damage during its first passage through the lungs (Plestina and Stoner, 1972; Raczniak *et al.*, 1979), whereas the metabolite from a PA is fed gradually to the lungs during a relatively long time; thus by injecting the pyrrole it is easy to produce acute lung lesions, which are relatively uncommon after giving the PAs themselves.

a. Acute Lung Damage. Large doses of some pyrrolic esters, e.g. the acetate and *N*-ethylcarbamate of 2-hydroxymethyl-1-methylpyrrole, given to rats via a tail vein, can produce lung damage very rapidly, with congestion and oedema developing within a few minutes after the injection (Mattocks and Driver, 1983). The mechanism of this action has not been studied, but it seems probable that it is a response to the massive interaction of an overwhelming dose of the reactive pyrrole with lung tissue. More commonly, injected PA metabolites such as dehydromonocrotaline (DHM), although taken up rapidly by the lungs, do not produce obvious pathological changes for some hours (Plestina and Stoner, 1972). After a large i.v. dose (30 mg/kg) of DHM to Wistar rats, the first change to be observed is an increase in the amount of residual blood in the lungs, leading to a significant increase in lung weight by 6 h. An increasing amount of fluid passes into the interstitial spaces of the lung, and when this is too great to be drained away by the lymphatics, fluid enters the pleural cavity. There is an increase in vascular permeability in the capillaries, but this is only detectable by EM study of colloidal carbon 'labelling' after 9 h, when the illness is well advanced. Respiratory difficulty appears after about 15 h and increases rapidly, leading to death of all animals within 48 h (Plestina and Stoner, 1972). In a different strain of (hooded) rats (Hurley and Jago, 1975), or using synthetic pyrrole carbamates instead of DHM (Plestina *et al.*, 1977), carbon 'labelling' appears in venules as well as in capillaries. These studies have shown that relatively large doses of DHM (as well as some if not all of the analogous pneumotoxic pyrrolic esters) act primarily upon the vascular endothelial cells in the lung; this leads to fatal pulmonary oedema and pleural effusion. The mechanism of the oedema is uncertain. Hurley and Jago (1975) believe that it is due to increased permeability of small vessels following the endothelial damage; vascular permeability is similarly increased in the rat cremaster by local application of DHM (Hurley and Jago, 1976). This is disputed by Plestina *et al.* (1977), who hold that this lesion is insufficient, or appears too late, to account for the early effects, and that most of the fluid is formed by "an exaggeration of the normal process of lymph production" (Plestina and Stoner, 1972).

b. Chronic Lung Damage. From an electron-microscopic study of the lungs of rats up to 4 weeks after a single injection of DHM into the tail vein, Butler (1970) concluded that the initial lesion is in the capillary endothelium, while changes in interstitial and epithelial cells are secondary. After 1–2 weeks, endothelial cell nuclei are large and bizarre, and cytoplasm is abundant. Alveolar walls are greatly thickened by 3–4 weeks, with all cell types being involved; the capillaries are still very abnormal at this time. Chesney *et al.* (1974a) found that within 4 weeks, many pulmonary vessels become occluded by fibrin and platelet thrombi, by thickening of the endothelium and by hypertrophy of smooth muscle cells; this leads to impaired pulmonary blood flow and pulmonary hypertension. A comparable sequence of events occurs in beagle dogs given DHM (3 mg/kg) via a pulmonary artery; the pulmonary hypertension leads to increased work for the right ventricle, resulting in right ventricular hypertrophy and finally in congestive heart failure (Raczniak *et al.*, 1979). In general, the development of chronic lung damage after the injection of pyrrolic esters resembles that from monocrotaline (Section B,I, above), and the mechanism appears to be essentially the same in each case. Nevertheless the pyrrolic esters differ slightly in that they do not produce medial hypertrophy of the pulmonary arteries: thus the latter, seen after PA intoxication, may be due to a different mechanism, such as the action of pyrrolic alcohol metabolites (Shumaker *et al.*, 1977), or to the more prolonged action of the PA compared with that of the injected pyrrole.

Hilliker and Roth (1984) have demonstrated that several drugs afford some protection of rats against DHM-induced pulmonary hypertension. These are hydrallazine, a vasodilator which also inhibits platelet aggregation; dexamethazone, an anti-inflammatory agent which may decrease vasoconstrictor prostaglandin; and sulphinpyrazone, which decreases platelet prostaglandin synthesis and increases platelet survival time. These actions support the view that platelets and vasoconstrictors are involved in the induction of pulmonary hypertension by DHM.

2. Pneumotoxicity and Molecular Structure

For a PA metabolite to have acute pneumotoxicity, the presence of an ester group appears to be essential; but the often complex acid moiety of the natural alkaloid is not necessary: effects similar to those of the pyrrolic metabolites from retrorsine or monocrotaline are produced by a simpler diacetyl ester (397) of dehydroretronecine (Mattocks, 1970). Neither is the intact dihydropyrrolizine moiety essential: a simpler monocyclic pyrrole is sufficient. Thus, both acute and chronic lung damage, identical to that from DHM, are produced in rats by a tail vein injection of 2,3-bisacetoxymethyl-1-methylpyrrole, (280) (Mattocks, 1971d, 1979). The analogous N-ethylcarbamate ester (361) and that of 2-hydroxymethyl-1-methylpyrrole (354) also cause acute pulmonary oedema in rats: this toxicity is not due to the carbamate moiety—ethyl N-ethylcarbamate is inactive (Plestina et al., 1977).

Mattocks and Driver (1983) have made a detailed comparison of the pneumotoxic actions of pyrrolic and analogous esters given i.v. to rats. Some of the results are summarized and compared with those from DHM (276), in Table 8.5: they suggest the following conclusions.

1. All the esters listed in the table are either mono- or bifunctional alkylating agents. The similar actions of esters activated by either a dihydropyrrolizine, pyrrole, indole or dimethylaminobenzyl structure show that pneumotoxicity is a

TABLE 8.5 Toxicity of Pyrrolic and Related Esters Given i.v. to Rats[a] via a Tail Vein[b]

Compound	Number of alkylating groups[c]	Relative alkylating reactivity[d]	Ability to cause: Acute lung oedema	Ability to cause: Chronic lung damage	Acute LD_{50} (up to 3 days)[e] (mg/kg, approx.)
(276)	2	5.3	+	+	10
(397)	2	120	+	+	20
(398)	2	160	+	+	7.5
(399)	2	6.9	+	+	23
(280)	2	14	+	+	12
(361)	2	8.6	+	+	10
(400)	2	53	+	+	10
(401)	1	53	+	0	38
(354)	1	28	+	0	26
(402)	1	1.2	+	0	33
(403)	1	1.0	+	0	29
(404)	1	80	0	0	63[f]
(405)	1	5.5	+	0	21
(406)	1	20	+	0	2.2
(407)	1	8.6	+	0	23

[a] Results are for male rats except for compound (399) (females) and (354) (both sexes).
[b] Data from Mattocks and Driver (1983) and Mattocks (1978b).
[c] Refers to reactive ester groups except for compound (399), in which one is a reactive hydroxyl.
[d] Rate of reaction with 4-*p*-nitrobenzylpyridine relative to (403), which is set as 1.0.
[e] Deaths usually associated with lung damage but in some cases with peracute pharmacological actions.
[f] No deaths associated with lung damage.

property of esters capable of alkylating tissue constituents, and depends on their chemical reactivity rather than their molecular structure.

2. Characteristic chronic lung toxicity is only caused by the compounds in this series which are bifunctional alkylating agents. This might imply that delayed pneumotoxicity is necessarily associated with some kind of cross-linkage reaction: however, it may simply be that the presence of two reactive groups enables the toxin to bind more strongly than would a monofunctional alkylating agent to tissue constituents.

3. Relative alkylating reactivities are shown in Table 8.5. These figures are not directly comparable for the bifunctional compounds, because in these one alkylating group may be much more reactive than the other. However, the

monofunctional alkylating agents with highest reactivity have low toxicity: the most reactive compound (404) is the only one which does not produce lung oedema, perhaps because much of the injected dose has decomposed before it reaches the lungs.

4. The exceptionally high pneumotoxicity of the indole (406) may be due to its high lipid solubility [lipid–water partition coefficient over three times that of (403)], giving it enhanced affinity for the tissue.

IV. BIOCHEMICAL EFFECTS OF PAs

Biochemical changes in the lungs of animals treated with PAs were dealt with in Section III. Effects occuring principally in the liver are discussed here; experimental studies are summarized in Table 8.6.

TABLE 8.6 Studies of Biochemical Changes in Rats
Following Administration of PAs

Alkaloid or material administered	System investigated	References
Heliotrine	pyridine nucleotide levels	Christie and LePage (1962a,b)
Heliotrine; lasiocarpine	mitochondrial metabolism	Gallagher and Judah (1967); Gallagher (1968)
Lasiocarpine	hepatic RNA synthesis; tryptophan pyrrolase	Reddy et al. (1968)
Lasiocarpine	hepatic protein synthesis	Demarle and Moulé (1971)
Retrorsine	hepatic protein synthesis	Villa-Trevino and Leaver (1968)
Retrorsine	hepatic microsomal enzyme activity	Mattocks and White (1971a)
Senecio jacobaea alkaloids	hepatic microsomal enzyme activity	Shull et al. (1976a)
Jacobine; monocrotaline	hepatic microsomal enzyme activity	Miranda et al. (1980a)
S. jacobaea (in diet)	hepatic microsomal enzyme activity	Miranda et al. (1980b); Garret et al. (1982)
Milk from *Senecio*-fed goats	hepatic microsomal enzyme activity	Miranda et al. (1981b)
S. jacobaea (in diet)	iron, zinc and copper levels and metabolism	Swick et al. (1982c,d)

A. In Mitochondria

In the liver of rats, levels of the pyridine nucleotides NAD and NADP are depleted, resulting in impairment of mitochondrial respiratory activity, from 17 h after a large dose (LD_{70}) of heliotrine (Christie and Le Page, 1962a). The effect is regarded as contributing to the toxic process by disrupting energy generation: this dose of alkaloid would kill many of the animals after 32–36 h. The reduction in pyridine nucleotides is attributed both to the loss of NAD from the cells and to their impaired ability to synthesize NAD (Christie and Le Page, 1962b). The suggestion by Schoental (1975b) that pyridine nucleotide depletion might be due to alkylation of nicotinamide by PA metabolites like dehydroheliotridine appears unlikely, since such reactions are probably reversible (Mattocks and Bird, 1983b) and would require very large doses of alkaloids to produce a significant reduction. Furthermore, the loss is not prevented by giving nicotinamide to rats along with, or after the alkaloid: thus nicotinamide is not protective against heliotrine poisoning (Christie and Le Page, 1962b).

Gallagher and Judah (1967), using isolated rat-liver mitochondria, concluded that PAs (heliotrine and lasiocarpine) in protonated form can compete with other positively charged compounds for sites in mitochondria, thus displacing NAD from these sites; the N-oxides of these alkaloids, which are uncharged, have no effect on mitochondrial oxidations *in vitro* (Gallagher, 1968).

B. On Hepatic Protein and RNA Synthesis

Retrorsine (49 mg/kg), given to rats intragastrically, inhibits the incorporation of radioactive leucine into liver and plasma proteins within 1 h (Villa-Trevino and Leaver, 1968). This inhibition appears to originate in the ribosomes, and could be a result of alkylation of messenger RNA by alkaloid metabolites. Lasiocarpine inhibits protein synthesis in rat liver with a maximum effect 1–3 h after injection (Demarle and Moule, 1971); it also inhibits RNA polymerase activity and nuclear RNA synthesis in the liver (Reddy *et al.*, 1968).

Evidence that protein synthesis is inhibited in rat liver by chronic PA intoxication has been given by Broderick *et al.* (1981): the ability, dependent on protein synthesis, of dexamethazone treatment to increase hepatic ornithine decarboxylase activity, is reduced fivefold in rats given 240 ppm of PAs in the diet for 30 days, compared with normal rats.

C. On Hepatic Microsomal Enzymes

Mattocks and White (1971a) found that pretreatment of rats with slightly less than an acute LD_{50} dose of retrorsine markedly reduces the ability of their liver microsomal enzymes to metabolize retrorsine to pyrrolic metabolites *in vitro*, 20 h later. Similarly, PAs from *Senecio jacobaea* given i.p. to rats (0.5 of LD_{50}) lower the microsomal aminopyrine-*N*-demethylase activity and cytochrome P-450 and microsomal protein levels, after 24 h (Shull *et al.*, 1976a). Pyrrole production from PAs is lowered 1 h after such pretreatment. Hepatic microsomal activity is also inhibited in rats fed *Senecio jacobaea* in their diet, or given milk from goats fed with this plant (Miranda *et al.*, 1980b, 1981b). The PAs jacobine and monocrotaline both lower hepatic microsomal activity in rats, but have almost no immediate effect on the activity of microsomal preparations *in vitro*, suggesting that the inhibitory effect is that of PA metabolites, not the PAs themselves (Miranda *et al.*, 1980a). Aminopyrine-*N*-demethylase activity in rat liver 10,000 × g supernatant is reduced after the preparations have been preincubated with senecionine, seneciphylline or retrorsine (Eastman and Segall, 1981).

These results indicate that metabolites formed from hepatotoxic PAs by microsomal enzymes in the liver, are subsequently able to inhibit the action of these enzymes.

D. On Other Enzymes

Jacobine (which is an epoxide) is a potent inducer of hepatic epoxide hydrolase when given to rats: three daily doses of 80 mg/kg lead to a fivefold increase in this enzyme, but a similar course of treatment of monocrotaline has no effect (Miranda *et al.*, 1980a). Pretreatment of rats with jacobine lowers the ability of their liver microsomes subsequently to metabolize benzo[a]pyrene, as well as the binding of metabolites from the latter to DNA (Williams *et al.*, 1983). Jacobine also induces hepatic glutathione-*S*-transferase, whereas monocrotaline causes a reduction in this enzyme. Increases in both these enzymes are also produced by feeding rats with a diet containing *Senecio jacobaea* (which contains jacobine) (Miranda *et al.*, 1980b).

The activity of steroid-induced tryptophan pyrrolase is inhibited in livers of rats 6 h after an injection of lasiocarpine (80 mg/kg) (Reddy *et al.*, 1968).

E. On Mineral Metabolism

Swick *et al.* (1982b) studied mineral alterations in tissues of rabbits caused by feeding them with a diet containing *Senecio jacobaea* along with supplementary copper and zinc: there is an increase in liver copper, decreased liver zinc and increased plasma iron. Rats fed this plant show increased copper in liver and spleen (Swick *et al.*, 1982d). Iron metabolism is disrupted, with either impaired haemopoesis or increased destruction of erythrocytes. An increased turnover of red blood cells was confirmed by Swick *et al.* (1982c), who studied the subcellular distribution of hepatic Cu, Fe and Zn in rats fed *S. jacobaea* and supplementary copper. In the treated rats the proportion of Cu was increased in nuclei and debris fractions, and lowered in lysosomes and cytosol. Zinc was increased but its distribution was unchanged; Fe was only increased in the cytosol.

An *S. jacobaea* diet did not increase liver copper accumulation in sheep (White *et al.*, 1984a).

V. PHARMACOLOGY OF PAs

A detailed description of the pharmacology of PAs is beyond the scope of this book; the subject has been surveyed briefly by Bull *et al.* (1968) and by McLean (1970). Investigations on PAs and necines are listed in Table 8.7. Pharmacological actions of various other pyrrolizidine derivatives, including semisynthetic esters and quaternary salts, have recently received attention, particularly from Indian workers (see Table 8.8); this subject has been reviewed by Atal (1978).

These alkaloids and derivatives do not generally have exceptional pharmacological effects, although a variety of actions have been recorded. The only comment to be made here is that these effects are produced by a wide variety of structures, and are not directly associated with the cytotoxicity shown by many PAs, although pharmacological actions of PA metabolites might play a part in some aspects of PA-induced lung and heart damage.

VI. OTHER BIOLOGICAL ACTIONS

A. Embryotoxicity

Sundareson (1942) showed that senecionine can cross the placenta. When this PA was injected into pregnant rats twice weekly beginning at day 12 of gestation or later, some litters were born prematurely, and many offspring were born dead or died shortly afterwards. Injections of senecionine into foetuses *in utero* led to necrosis in the mother rat. When four foetuses were each given 1.25 mg (representing about 200–400 mg/kg, which is much higher than an LD_{50} for an adult rat), three of them were still alive after 2 days. The same dose killed a newborn rat in 1 day. Bhattacharyya (1965) found that the livers of embryo rats whose mothers had been injected with PAs (heliotrine, lasiocarpine, retrorsine or monocrotaline) during pregnancy (from day 10 of gestation) were only mildly damaged. Green and Christie (1961) found no liver damage in foetuses from pregnant rats given teratogenic doses of heliotrine. In contrast, when hepatotoxic PAs are given to mothers of neonatal rats, liver damage is more severe in the latter (from PAs transmitted in the milk) than in the mothers (Schoental, 1959; Bhat-

TABLE 8.7 Investigations of Pharmacological Activity of PAs
and PA-containing Plant Materials

Alkaloid or plant tested	Test system, or action investigated	Reference
Heliotrine	smooth muscle *in vitro*; respiration, heart *in vivo*	Harris *et al.* (1957)
Europine, heleurine, heliotrine, jacobine, jaconine, lasiocarpine, monocrotaline, platyphylline, riddelliine, senecionine, seneciphylline, supinine; *N*-oxides of europine, heliotrine and lasiocarpine	rat ileum: smooth muscle inhibition; anticholinergic activity	McKenzie (1958)
Heliotrine, lasiocarpine	neuromuscular block; respiratory failure	Gallagher and Koch (1959)
Lasiocarpine, supinine	rats, mice, cats *in vivo*; rabbit ileum	Rose *et al.* (1959)
Crotalaria agatiflora (alcohol extract)	hypotensive action in dogs	Sharma *et al.* (1967)
C. paniculata seed (fulvine)	guinea pig ileum (papaverine-like activity)	Subramanian *et al.* (1968)
Anacrotine (crotalaburine)	anticholinergic activity	Snehalata and Ghosh (1968)
Anacrotine (crotalaburine)	action on raw paw oedema	Ghosh and Singh (1974)
Usaramine	hypotensive and relaxant action (dogs)	Singh *et al.* (1969)
7-Angelylheliotridine, cynaustraline, heliotrine, heleurine, lasiocarpine, monocrotaline, platyphylline, sarracine, senecionine, spectabiline, supinine	guinea pig ileum: anticholinergic activity	Pomeroy and Raper (1971a)
Necines: desoxyretronecine, hastanecine, heliotridine hydroxyheliotridane, isoretronecanol, platynecine, retronecanol, retronecine, supinidine	guinea pig ileum: cholinomimetic activity	Pomeroy and Raper (1971b)
Heliotrine	guinea pig ileum (ganglion blocking activity); rabbit intestine; frog muscle	Pandey *et al.* (1982)

TABLE 8.8 Pharmacological Actions
of Some Semisynthetic Pyrrolizidine Derivatives

Compound tested	Action investigated	Reference
Pyrrolizidine benzilate ester	psychotomimetic action	Sternbach *et al.* (1974)
Platynecine and heliotridane esters	local anaesthetic	Suri *et al.* (1975a)
Platynecine esters	local anaesthetic; hypotensive, cardiac depressant and spasmolytic action	Gupta *et al.* (1976a)
Pyrrolizidine amides	hypotensive action	Gupta *et al.* (1976b)
Quaternary derivatives	neuromuscular blocking	Suri *et al.* (1976b); Gupta *et al.* (1977b); Siddiqi *et al.* (1979a)
Quarternary derivatives	ganglion blocking	Gupta *et al.* (1977a)
Quaternary derivatives	spasmolytic action	Gupta *et al.* (1979)

tacharyya, 1965). Evidently the rat is much less susceptible to PA hepatotoxicity when *in utero* than it is after birth. This might be because the embryo has a low capacity for metabolic activation of PAs and is consistent with the finding of Mattocks and White (1973) that the ability of liver enzymes to convert retrorsine to toxic metabolites is very low in rats immediately after birth, although it increases rapidly thereafter.

Persaud and Hoyte (1974) tested fulvine in pregnant rats and found it to cause a dose-related incidence of foetal resorptions when given between days 9 and 12 of gestation; however, again, hepatic lesions were not seen in the foetuses. On the other hand, lasiocarpine is able to damage both the maternal and the foetal liver (Newberne, 1968): acute liver necrosis and haemorrhage were often seen in foetuses whose mothers had received this alkaloid (100 mg/kg) on day 13 of gestation; the embryos were often smaller than normal. When mothers were given two doses of 35 mg/kg, on days 13 and 17 of pregnancy, they did not have liver necrosis but the foetal livers were damaged. The reason why lasiocarpine differs from other PAs in having a greater effect on the foetal liver is not known. One possibility is that foetotoxicity is caused chiefly by toxic metabolites formed in the maternal liver, and that the proportion of such metabolites reaching the foetus from lasiocarpine is greater than that from other PAs.

B. Teratogenicity

Green and Christie (1961) showed that single i.p. injections of heliotrine into rats during their second week of pregnancy cause a variety of dose-related foetal abnormalities; these include retardation of development, musculo-skeletal defects, especially in the ribs, and (after high doses) hyperplasia of the jaw, and cleft palate. Doses above 200 mg/kg kill many foetuses. Fulvine also is teratogenic in rats within the dose range 20–80 mg/kg (Persaud and Hoyte, 1974): all foetuses are abnormal after maternal rats are given 80 mg/kg; larger amounts are often lethal to the embryos.

Peterson and Jago (1980) compared the effects of heliotrine with those of its pyrrolic metabolite dehydroheliotridine (DHH), given i.p. to rats on day 14 of pregnancy. A dose of 40 mg/kg DHH had effects on the embryos equivalent to those of 200 mg/kg heliotrine, i.e. DHH is about 2.5 times as effective on a molar basis. Mother rats given up to 90 mg/kg of DHH had no pathological abnormalities. Some embryos died within 24 h after the DHH injection, and a few at later times. DHH retards embryo growth; its main teratogenic effects are skeletal, causing retarded ossification, distorted bones, cleft palate and defective feet. The characteristic antimitotic effects of DHH were not marked in the embryonic livers.

It may be concluded that at least some hepatotoxic PAs are powerful teratogens in rats, and the potent activity of DHH is consistent with the idea that it is this or a similar pyrrolic metabolite, formed from the alkaloid in the maternal liver, which is chiefly responsible for developmental defects in the embryos. Foetal liver differs markedly from adult rat liver, either in being less able to convert PAs to cytotoxic metabolites, or in its response to such metabolites, or both.

Brink (1982) showed that heliotrine is also teratogenic in *Drosophila*, causing morphological abnormalities in the adult abdomen when larvae are fed continually on low levels of the alkaloid.

9

Effects of Pyrrolizidine Alkaloids in Humans

I. SOURCES OF HUMAN EXPOSURE TO PYRROLIZIDINE ALKALOIDS

Plants containing PAs (see Chapter 1) are found in most parts of the world, although those growing in warmer climates often have higher alkaloid levels. A very wide range of plant material is employed by people either medicinally or for food, and it is not surprising that occasionally plants containing PAs are used, either deliberately or in mistake for other species. Cereal crops and grazing land are sometimes contaminated with pyrrolizidine-bearing weeds, and the alkaloids may find their way into bread and other foods, and into milk from cows feeding on these plants.

Table 9.1 lists plants likely to contain PAs which have been used for food or medicine—chiefly the latter. The list is not exhaustive, and not all the plants are known to have harmed people. A list of plants which have caused poisoning is given in Table 9.2.

A further minor source of human exposure is the deliberate use of PAs as anticancer drugs, under controlled hospital conditions.

TABLE 9.1 Some Plants Containing (or Suspected of Containing) PAs
Which Have Been Used by People as Either Herbal Medicines (M) or Foods (F)

Plant	Number (see Chap. 1 Table 1)	Reason for ingestion	Country or region	References
Apocynaceae				
Holarrhena antidysenterica		M	Sri Lanka	Arseculeratne et al. (1981)
Boraginaceae				
Anchusa officinalis	(9)	M	Europe	Broch-Due and Aasen (1980)
Cynoglossum geometricum		M	East Africa	Schoental and Coady (1968)
Cynoglossum officinale	(18)	M	Iran	Coady (1973)
Heliotropium eichwaldii	(32)	M	India	Datta et al. (1978a,b); Gandhi et al. (1966a)
H. europaeum	(33)	M	India; Greece	International Agency for Research in Cancer (IARC) (1976)
H. indicum	(34)	M	India, Africa, South America and elsewhere	Schoental (1968a); Hoque et al. (1976)
H. ramossissimum (ramram)	(39)	M	Arabia	Macksad et al. (1970); Coady (1973)
H. supinum	(42)	M	Tanzania	Schoental and Coady (1968)
Symphytum officinale (comfrey)	(66)	F, M	Japan (and elsewhere)	Hirono et al. (1978)
S. × uplandicum	(68)	F, M	general	Hills (1976); Culvenor et al. (1980a,b)
Compositae				
Cacalia decomposita (matarique)		M	United States	Sullivan (1981)
C. yatabei	(78)	F	Japan	Hikichi et al. (1978)
Farfugium japonicum	(90)	M	Japan	Furuya et al. (1971)
Ligularia dentata	(93)	F	Japan	Asada and Furuya (1984)
Petasites japonicus	(96)	F, M	Japan	Hirono et al. (1973)
Senecio abyssinicus		M	Nigeria	Williams and Schoental (1970)

(*continued*)

TABLE 9.1 (*Continued*)

Plant	Number (see Chap. 1 Table 1)	Reason for ingestion	Country or region	References
S. aureus	(104)	M	United States	Wade (1977)
S. bupleuroides	(109)	M	Africa	Watt and Breyer-Brandwijk (1962)
S. burchelli		F, M	South Africa	Rose (1972)
S. coronatus		M	South Africa	Rose (1972)
S. discolor	(117)	M	Jamaica	Asprey and Thornton (1955)
S. doronicum	(119)	M	Germany	Roeder *et al.* (1980a)
S. inaequidens		F	South Africa	Rose (1972)
S. jacobaea (ragwort)	(139)	M	Europe	Schoental and Pullinger (1972); Wade (1977)
S. longilobus (*S. douglassi*)	(120)	M (mistaken?)	United States	Stillman *et al.* (1977); Huxtable (1979a)
S. monoensis		M	United States	Huxtable (1980a)
S. nemorensis ssp. *fuchsii*	(129)	M	Germany	Habs *et al.* (1982)
S. pierotii	(164)	F	Japan	Asada and Furuya (1982)
S. retrorsus (*S. latifolius*)	(175)	M	South Africa	Rose (1972)
S. vulgaris (common groundsel)	(195)	M	Netherlands; Europe; Iran	Wade (1977); Watt and Breyer-Brandwijk (1962); Coady (1973)
Syneilesis palmata	(197)	F	Japan	Hikichi and Furuya (1976)
Trichodesma africana	(70)	M	Asia	Omar *et al.* (1983)
Tussilago farfara (coltsfoot)	(198)	M	Japan; China?	Culvenor *et al.* (1976)
		M	Norway	Borka and Onshuus (1979)
Leguminosae				
Crotalaria brevidens		F	East Africa	Coady (1973)
C. fulva	(210)	M	Jamaica	Barnes *et al.* (1964); McLean (1970, 1974)
C. incana	(214)	M	East Africa	Schoental and Coady (1968)

TABLE 9.1 (*Continued*)

Plant	Number (see Chap. 1 Table 1)	Reason for ingestion	Country or region	References
C. juncea	(216)	M, F	India	Chopra (1933); Watt and Breyer-Brandwijk (1962)
C. laburnifolia	(217)	M	Tanzania	Schoental and Coady (1968)
		F	Asia	Coady (1973)
C. mucronata	(223)	M	Tanzania	Coady (1973)
C. recta	(229)	M, F	Tanzania	Schoental and Coady (1968); Coady (1973)
C. retusa	(230)	M, F	Africa; India	IARC (1976); Watt and Breyer-Brandwijk (1962)
C. verrucosa	(240)	M	Sri Lanka	Arseculeratne et al. (1981)
Cassia auriculata		M, F	Sri Lanka; India	Arseculeratne et al. (1981)

A. Pyrrolizidine Alkaloids in Food

1. Plants Eaten Deliberately

About one-third of the plants in Table 9.1 have been used as food. There do not appear to be any records of poisoning following such usage, but this is not to say that no harm has been done. Rose (1972) has listed 35 plants of the genus *Senecio,* of which 15, including *S. burchellii* and *S. inaequidens,* are said to be used as 'spinach' in South Africa, although they are 'not popular'. The alkaloid content of *S. burchellii* is not known, but it causes liver damage in rats and has been responsible for bread poisoning (Willmot and Robertson, 1920) (see below). According to Watt and Breyer-Brandwijk (1962), *Crotalaria juncea* has been used as a green vegetable; *C. retusa,* known to contain the toxic PA monocrotaline, has also been used as a vegetable in India and possibly in East Africa; and the seed of *C. mucronata,* roasted but still containing alkaloid, has been used in Indonesia as a coffee substitute. In Japan, young flower stalks of *Petasites japonicus,* a kind of coltsfoot, which is carcinogenic to rats (Hirono *et al.,* 1983), are used as food; young leaves of *Syneilesis palmata,* various *Cacalia* species, and young *Senecio pierotti* plants are sometimes eaten (Hikichi and

TABLE 9.2 Cases of Human Poisoning Attributed to Ingestion of Plant Material Containing PAs

Plant	Principal alkaloid(s)	Reason for ingestion[a]	Country or region	Number of cases	References
Senecio ilicifolius; *S. burchelli*	senecionine?	cont.	South Africa	80	Willmott and Robertson (1920)
Senecio spp.	heliotrine; lasiocarpine	cont.	South Africa	12	Selzer and Parker (1951)
Heliotropium lasiocarpum		cont.	Central Asia	28	(1952): see McLaen (1970)
Heliotropium lasiocarpum		cont.	Central Asia	61	(1965): see McLean (1970)
Crotalaria fulva	fulvine	med.	West Indies	(many)	Bras *et al.* (1954, 1957, 1961); Bras and Hill (1956)
Senecio coronatus		med.	South Africa	2	Rose (1972)
Crotalaria laburnoides (suspected)	unknown	med.	Tanzania	1	Heath *et al.* (1975)
Crotaloria juncea		med.	Equador	1	Lyford *et al.* (1976)
Heliotropium popovii	heliotrine	cont.	Afghanistan	~1600	Mohabbat *et al.* (1976)
Crotalaria nana	crotananine; cronaburmine	cont.	India	67	B. N. Tandon *et al.* (1976); Siddiqi *et al.* (1978b,c)
Senecio longilobus	riddelliine; retrorsine N-oxide (with others)	med.	United States	2	Huxtable *et al.* (1977); Stillman *et al.* (1977); Fox *et al.* (1978)
Heliotropium eichwaldii	heliotrine N-oxide	med.	India	3	Datta *et al.* (1978a)

[a] Cont., Contamination of cereal crop by seeds of plant; med., plant extract taken as a herbal medicine or 'bush tea'.

Furuya, 1976; Hikichi *et al.*, 1978; Asada and Furuya, 1982). Bush teas (see Section I,B) have been a regular part of the diet in some islands of the West Indies, but those containing PAs have a disagreeable taste and are probably only used medicinally (McLean, 1970). Herbal teas, made from a variety of plants, are also popular in Europe (Anonymous, 1979a). Medical practitioners in the United States have been warned of the possible harmful effects of plant products which may be sold in 'health food' stores (Anonymous 1979b). Many people in industrialised countries are turning to 'natural' foods in the mistaken belief that manufactured 'chemical' additives are necessarily harmful whereas what is 'natural' must be good. Comfrey (*Symphytum* spp.) is a popular herb in many countries with a variety of food uses, as a green vegetable and as a tea (Hills, 1976; Hirono *et al.*, 1978; Culvenor *et al.*, 1980b).

2. Foods Contaminated with PAs

a. Cereal Crops. On many occasions when plants containing PAs have grown among wheat, millet or similar crops, their seeds have been harvested and milled along with the cereal, and the alkaloids have been incorporated into bread. This is now unlikely to occur in industrialized countries, where unwanted weeds are effectively controlled by chemical sprays (Huxtable, 1980a), but it may continue to be a problem in undeveloped regions for some time to come. Various species of *Heliotropium*, *Senecio* and *Crotalaria* plants have been involved, and often their presence as a toxic hazard has unfortunately only been recognised after people have been poisoned (see Section II,A).

b. Milk. PAs have been found in the milk of cows and goats fed or dosed with ragwort (*Senecio jacobaea;* known as tansy ragwort in the United States). After four cows were given this dried plant material via rumen cannula at up to 10 g/kg body weight/day, their milk had PA levels of about 0.5–0.8 ppm (Dickinson *et al.*, 1976). Only one (jacoline) of the five alkaloids in the plant was transferred. The cows lost weight and died with liver damage; however, calves feeding on the milk were unharmed. Suckling calves of cows fed chronic lethal doses of dried ragwort sustained no hepatic lesions attributable to PAs, although minor biochemical changes were detected, indicative of mild liver damage (Johnson, 1976). Rats given milk from these cows were also unaffected. The milk from goats fed 1% of their body weight per day of ragwort contained 0.33–0.81 ppm of PAs (Deinzer *et al.*, 1982). In another experiment, milk containing about 7.5 ng PAs/g (dry weight), from goats fed ragwort, was given to two calves: these were not affected, but rats given the dried milk at a level of 80% in their diet developed some chronic liver lesions suggestive of PA toxicity (Goeger *et al.*, 1982b). Rats thus fed for one week showed minor changes in their hepatic drug-metabolizing enzymes (Miranda *et al.*, 1981b).

Ragwort is well known as a poisonous weed contaminating pastures in Great Britain (Forsyth, 1968), and it has become a problem also in the western United States (Dickinson *et al.*, 1976); alkaloids from *Senecio alpinus* have been found in animal feedstuffs in Switzerland (Luthy *et al.*, 1981); many other PA-containing plants cause poisoning of livestock in other parts of the world. Hence it is likely that from time to time PAs are present in milk supplies used for human consumption. Young animals are especially susceptible to PA poisoning (Schoental, 1959), and thus children might occasionally be at risk from PAs in milk. The risk is probably negligible when bulk distribution of supplies leads to high dilution of foreign constituents. However, another possibility is that suckling children might be affected by PAs in the milk of mothers who themselves have ingested foods or herbal medicines containing these alkaloids (cf. Schoental, 1959).

Nevertheless, there is no evidence available to indicate that people have been harmed by PAs in milk.

c. Honey. Specimens of honey produced by bees from the nectar of *Senecio jacobaea* (ragwort) in the western United States have been found to contain all the PAs present in the plant (Deinzer *et al.*, 1977). The total alkaloid levels were very low: 0.3–3.9 ppm. A person would not suffer acute toxic effects from PAs at this level, and ragwort honey is of poor quality and bitter taste. An 'average annual human intake' of honey (600 g), with the highest alkaloid level quoted, would contain well under 3 mg of PAs. Honey from *Echium plantagineum* (Patterson's curse; salvation Jane), a widespread weed in Southern Australia, also contains a number of PAs from the plant, chiefly echimidine (Culvenor *et al.*, 1981).

d. Meat. There is no record of PAs being detected in meat products from livestock which might have ingested PAs. Experience with laboratory animals suggests that levels of PAs in the tissues fall very rapidly after intake. Metabolism by liver enzymes is fast, at least in the rat, and toxic metabolites are probably quickly deactivated. During the chronic ingestion of PAs, the *effects* of the alkaloids are cumulative in the liver, but the alkaloids themselves are not. Thus we anticipate that concentrations of PAs in tissues would only be high if the animals were killed very soon after a massive intake.

B. In Medicines

1. Herbal Medicines

Plants have traditionally been employed in medicine throughout the world, and they are still used widely today, especially in countries where modern medical facilities are sparse. The sale of amorphous plant materials in markets by stallholders with no knowledge of their botanical origins or medicinal effects can lead to populations in rural Africa and elsewhere being exposed to a variety of unknown substances (cf. Maclean, 1965). Information about these materials is not easy to acquire owing to secrecy imposed by traditional medicine men (Coady, 1973), and because of the sheer difficulty of identifying plant materials which are known only by native names. Sometimes different names are used in different localities, or different plants may be given the same name. For instance, in Mexico a number of plants used for preparing herbal tea are known as 'gordoloba', including (perhaps because of misidentification) *Senecio longilobus*, which contains PAs (Huxtable, 1979a). The absence of both fruit and flowers from specimens collected at markets can make botanical identification difficult (R. F. Sturrock, unpublished).

Examples of PA-containing plants which have been used medicinally are given in Table 9.1. Schoental (1972) found evidence that species of *Senecio, Crotalaria, Cynoglossum* and others are used in East Africa. Schoental and Pullinger (1972) considered that at least some PAs found in such plants (e.g. retrorsine, isatidine, monocrotaline) were without medicinal value. *Senecio coronatus* (alkaloid content not known) is given to babies in Southern Africa (Rose, 1972). Schoental (1968a) quotes over a dozen vernacular names for *Heliotropium indicum*, a widely used medicinal herb with applications externally for sores, snake bite, etc. and internally for many purposes including prevention of abortion *and* as an abortifacient! According to Schoental, this plant has greater toxicity (to animals) than does its major alkaloid (indicine); the dried plant is said to be less toxic after storage.

Bush teas were formerly very popular with the poorer populations in parts of the West Indies. Their use in Jamaica has been described by McLean (1970, 1974). They are hot drinks made by water infusion of the leaves of bushes growing in the garden or wild scrub. Not all are poisonous. The majority used in the regular diet are harmless. Those containing PAs (e.g. *Crotalaria fulva*) are bitter, and have been used chiefly as medicines when they are taken as a single large dose by already sick people—often children. These have often caused serious liver damage, and death (see Section II, below). R. F. Sturrock (unpublished observation) reported in 1969 that bush teas were widely used on the island of St. Lucia. Plants growing there include the PA-containing species *Crotalaria mucronata, C. retusa, C. spectabilis, C. verrucosa, Erechtites hieracifolia, H. curassavicum* and *H. indicum;* some of these might have found their way into bush teas, although they were not identified as such. Bush teas have been used for treating fevers, coughs, colds, etc. and in pregnancy; however, in recent years education campaigns have greatly diminished their use in the West Indies.

The use of herbal remedies is not confined to primitive cultures. Indeed, interest in them is growing in industrialised countries, as people endeavour to 'return to nature' (Bouissou, 1973; Anonymous, 1979a). Teas and remedies containing PAs are still sometimes obtainable from herbalists and 'health food' shops in Western countries. PA-containing medicinal plants, many of them used in Europe, are the subject of a recent comprehensive review (Danninger *et al.*, 1983). *Senecio jacobaea* (ragwort) has been obtainable from herbalists in England (Schoental, 1968a; Burns, 1972). Comfrey is still widely available, although there have been warnings about its use (Anonymous, 1979c).

Thus some of the plants used medicinally both in undeveloped and in industrialised societies may contain toxic PAs. Rose (1972) observed that African inhabitants seem 'unaware of the toxicity of such plants', and this probably applies to users worldwide, because the effects of PAs are subacute or chronic, making it difficult to associate the eventual illness with the plant taken.

2. Drugs

The cytotoxic and antimitotic activity of many PAs has prompted investigations into their possible uses as anticancer agents. So far only one, indicine *N*-oxide, has reached clinical trial in cancer patients (see Section II, and Chapter 11). It is probable that other PAs or synthetic analogues will be tested for this purpose in the future.

II. HUMAN POISONING BY PAs

There have been relatively few reports of human intoxication by PAs, although some of the incidents described have involved large numbers of people (Table 9.2). It is possible, as Stillman *et al.* (1977) have suggested, that the recorded cases might represent the tip of an 'iceberg' of people affected by these alkaloids. Medical practitioners are unfamiliar with such poisoning, and when occasional cases are seen the condition might be attributed to other causes.

A. Cases of Human Poisoning

1. Early Reports of 'Bread Poisoning'

Willmot and Robertson (1920) recognised an illness among the poor white population of the Cape Province, South Africa, which they called 'Senecio disease'. Seeds of *Senecio ilicifolius* and *S. burchellii* plants growing in the wheat fields were harvested with the grain and, if winnowing was not efficient, were ground with the meal and incorporated into bread. Some 80 cases occurred in 10 years, and many died. Many afflicted people were children. Deaths occurred from 14 days to over 2 years after the onset of symptoms, which predominantly consisted of abdominal pain with vomiting and the development of ascites. Selzer and Parker (1951) gave details of 12 similar cases seen in Cape Town, half of them children. All had eaten bread made from contaminated wheat, and said that it tasted 'musty' or bitter. The illness was severe, and was described as clinically similar to Chiari's syndrome (rapid accumulation of ascites: abdominal pain; nausea; hepatomegaly). Six patients died, and necropsy showed severe liver damage which appeared to the authors to originate in the vessels rather than the hepatic parenchymal cells. [Similar liver damage was produced by Willmot and Robertson (1920) in laboratory animals by feeding them with seeds of *Senecio ilicifolius* or *S. burchellii*.] Some patients apparently recovered and were discharged from hospital, but there was no long-term follow-up. It was suggested that the effects of the alkaloids might have been enhanced by a protein-deficient diet, the reported cases being from poor families. However, it is also possible that these households used poorer quality wheat than people who were better off, and that bread formed a larger part of the diet.

Outbreaks of heliotrope poisoning in Central Asia have been described by Russian authors; for a discussion, see McLean (1970).

2. Veno-occlusive Disease in the West Indies

An acute disease of the liver, predominantly in children, was described by workers in Jamaica in the 1950s, and named veno-occlusive disease (VOD) (Bras *et al.*, 1954; Bras and Hill, 1956). It was shown to result from the taking of bush teas prepared from leaves of bushes, especially *Crotalaria fulva*, containing pyrrolizidine alkaloids (Bras *et al.*, 1957). In its early stages, VOD is characterised by centrilobular necrosis of the liver, leading to occlusion of central and sublobular hepatic veins, congestion of the liver, and development of ascites. Many patients with VOD died; among survivors, cirrhosis of the liver developed; in 1961, VOD accounted for 30% of a group of 77 cases of liver cirrhosis in Jamaica (Bras *et al.*, 1961). For further descriptions of VOD, see McLean (1974) and McLean and Mattocks (1980).

3. Veno-occlusive Disease in India and Afghanistan

Two cases of VOD in India were described by Gupta *et al.* (1963). Both were men who had taken herbal medicine of unknown constitution: with treatment, both recovered almost completely within 3 months. Although it was not proven that PAs had been taken by these patients, the resemblance of the clinical features to those of VOD in Jamaica and elsewhere was pointed out.

Outbreaks of VOD occurred in villages in central India in 1973, and again in 1975 (B. N. Tandon *et al.*, 1976; R. K. Tandon *et al.*, 1976). Most of the victims were agricultural workers; the area was subject to drought and the people were undernourished. The staple diet, a form of millet known as 'gondli', was contaminated with seeds of a wild plant, later identified as *Crotalaria nana* (Siddiqi *et al.*, 1978b,c). In all, 67 cases were seen, of whom 28 died. People affected were of all ages and both sexes: children did not predominate. The clinical picture was of liver damage, typical of VOD as described above. Examination of a sample of millet from this incident showed that it contained a number of small but conspicuous, black *Crotalaria* seed pods (Fig. 9.1). Some had split and were empty. Other contained up to 11 yellow seeds, much smaller than millet grains, and some of these seeds were also distributed among the millet. These seeds contained toxic PAs similar to monocrotaline, but we found no PAs in the empty pods. The PAs crotananine and cronaburmine were later identified in this species by Siddiqi *et al.* (1978b,c). It is possible that people using this millet might avoid the obvious black pods, but leave the inconspicuous toxic seeds mixed with the grain. If a container full of the grain was shaken, there would be a tendency for the pods to rise to the top, whereas the seeds would accumulate near the bottom. This could lead to meal from the bottom being more toxic than meal from the middle of the container, and people using the same batch of grain being differently affected.

The largest known outbreak of pyrrolizidine poisoning occurred in Afghanistan in 1974 (Mohabbat *et al.*, 1976; H. D. Tandon *et al.*, 1978). It affected a population of some 35,000 in 98 villages in a remote area of northwest Afghanistan. Food was scarce following a prolonged drought, there was little meat available, and the diet consisted largely of bread made from locally grown wheat. The wheatfields were heavily contaminated with a *Heliotropium* plant (*H. popovii* subsp. *gillianum* H. Riedl), and the seeds of this, containing up to 1.49% of toxic PAs (mainly heliotrine N-oxide), were subsequently found in samples of wheat grain. Many people developed characteristic symptoms of VOD, with enormous abdominal distention and emaciation. Among 7200 inhabitants examined, over 1600 were affected, and many died. Over 60% of the patients were males and 46% were children aged under 14. The fatal cases died 3–9 months after the onset of abdominal distention.

Datta *et al.* (1978a) described six cases of VOD in India, of which three had taken *Heliotropium eichwaldii* medicinally. The herb, known as 'hathisunda' ('elephant's trunk'), contained between 1 and 2% by weight of heliotrine N-oxide. Two of the patients were brothers who had suffered from epilepsy for some years and had been taking medically prescribed phenobarbitone and phenytoin sodium prior to taking the herb (Datta *et al.*, 1978b). Phenobarbitone accelerates the conversion of PAs to toxic metabolites in experimental animals and can enhance the hepatotoxicity of some PAs (Mattocks and White, 1971a; Mat-

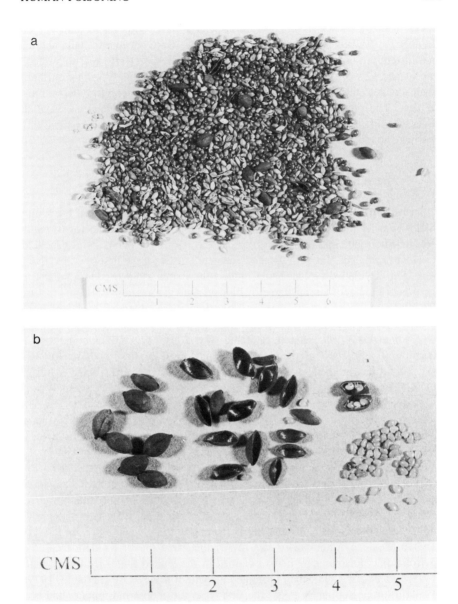

Fig. 9.1. (a) Indian millet sample containing black pods of *Crotalaria nana*. A few of the small *C. nana* seeds are seen at upper left. (b) *Crotalaria nana* pods (left), from millet sample; also opened pods (centre) and the small, poisonous seeds which they contained (right).

tocks, 1972b). A water infusion of the dried leaves was taken daily by one brother for 20 days and by the other for 50 days. Both became ill, with abdominal distention and jaundice, after 45 to 90 days respectively, and they died 2 weeks and 12 weeks, respectively, after admission to hospital. Neither attributed their illness to the herb, even though it had caused them discomfort and vomiting at first. The post mortem showed centrilobular haemorrhagic necrosis of the liver. A third patient, who had taken *H. eichwaldii* for vitilago, remained alive but was not followed up.

4. Other Cases of Veno-occlusive Disease

Lyford *et al.* (1976) described the first case to be diagnosed in the United States. Earlier the female patient, while in Quito, Equador, had for 6 months consumed a medicinal tea made from the leaves of local plants, which included *Crotalaria juncea*. Her daily intake had been 2 litres for the first 6 weeks, then 1 litre. She developed typical VOD, with voluminous ascites, centrilobular congestion of the liver (demonstrated by biopsy), and increased portal vein pressure. An interesting feature of this case was the complete recovery of the patient within a year after ceasing to take the tea.

A somewhat similar case in Scotland with a less fortunate outcome was intriguing because although the patient, a female aged 26, developed severe VOD and subsequently died, the source of poisoning was never fully identified (McGee *et al.*, 1976). The woman had for about 2 years taken extremely large amounts of a herbal tea, known as *maté*, obtained from 'health food' stores. This tea is made from the leaves of *Ilex paraguayensis,* a plant from South America which does not contain PAs. Trace amounts of PAs were detected in samples of teas from the patient's supply. Other samples of *maté* purchased in Glasgow at the time of her illness contained (as expected) about 1% caffeine, but no PAs (our unpublished results). It is possible that the woman had purchased a batch of *maté* tea heavily contaminated with leaves from another plant, containing PAs, but it seems more probable that she had access to other material, of unknown origin, which was the source of her poisoning.

Two Mexican American infants in Arizona were the first reported cases of pyrrolizidine poisoning by herbs obtained in the United States (Stillman *et al.,* 1977; Fox *et al.,* 1978); they prompted a spate of reviews by Huxtable (1979a, 1980a,b) on poisoning by PAs. The first case, a 6-month-old female, had been given large amounts of a tea made from a herb marketed locally as 'gordoloba yerba'. A reputedly harmless plant, *Gnaphalium macounii,* is used under the name ''gordoloba' as a cough medicine by Mexican Americans. On this occasion, the plant used was identified as *Senecio longilobus* (*S. douglasii*). and it contained a high level (1.3%) of hepatotoxic PAs, including riddelliine (base) and retrorsine *N*-oxide (Huxtable, 1980a). Evidently this plant resembles *Gnaphalium,* and is sometimes marketed in mistake for it (Fox *et al.,* 1978). The child had acute hepatocellular disease, ascites, portal hypertension, and a right pleural effusion (Stillman *et al.,* 1977). Her condition improved after treatment with furosemide and a high-sodium diet. However a needle biopsy after 2 months revealed extensive hepatic fibrosis, progressing to cirrhosis over 8 months (Fox *et al.,* 1978).

The other infant, a boy aged 2 months, had also been given *S. longlilobus* over a period of 4 days, in mistake for gordoloba. The plant contained 1.5% of PAs (Fox *et al.,* 1978). The child was ill for 2 weeks before being admitted to hospital, and died 6 days later. His condition was first mistaken for Reye's syndrome (encephalohepatitis); this diagnosis was revised when jaundice, ascites and liver necrosis were observed. An unusual feature in this patient was the presence of free circulating bilirubin.

Following these incidents, Huxtable (1980a) suggested that other, undetected cases of poisoning by PAs in the United States might have occurred, and he cited five examples of Mexican Americans (four of them children) who had sustained serious liver damage after consuming 'gordoloba' or other herbal teas. Three of these patients died with liver failure or complications arising from it.

Kumana *et al.* (1983) in Hong Kong reported four cases of hepatic VOD in Chinese women taking a herbal medicine: unsaturated PAs and *N*-oxides were detected and estimated but not identified. Total doses of alkaloids causing hepatotoxicity were 12–26 mg/kg, taken over periods of up to 46 days: one woman died.

The attention of Heath *et al.* (1975) was drawn to an African youth from Tanzania who died from primary pulmonary hypertension after being suspected of having ingested a herbal remedy prepared from *Crotalaria laburnoides* seeds. Rats which were subsequently fed some of the seeds developed lung damage and right ventricular hypertrophy, which are typical effects of some *Crotalaria*-derived PAs, such as monocrotaline, in laboratory animals (see Chapter 8). This case is particularly interesting because it is possibly the only example on record of a human subject whose death might have been due to the pneumotoxic (rather than hepatotoxic) action of PAs.

B. Clinical Features and Pathological Effects in Humans

The major acute effects of PAs in humans are manifested in the liver as veno-occlusive disease (VOD). The victims have often been children, partly because they have received large doses of medicinal teas (as in Jamaica; Bras and Hill, 1956), and probably also because the young are especially susceptible to the effects of PAs (Schoental, 1959). In some outbreaks of VOD (e.g., in Afghanistan; Mohabbat *et al.*, 1976), but not in all, more men than women have been affected. This might reflect a greater sensitivity on the part of males, but other explanations are possible, e.g. the men may have eaten a larger share of the affected food than women.

1. Clinical Course of Illness

Evidence of toxicity may not become apparent until some days after the ingestion of PAs. The acute illness has been likened to the Budd–Chiari syndrome (thrombosis of hepatic veins leading to liver enlargement, portal hypertension, and ascites) (Selzer and Parker, 1951; McLean, 1970). Early clinical features include nausea and acute epigastric pain; acute abdominal distension (due to ascites) with prominant dilated veins on the abdominal wall (for illustrations, see Stillman *et al.*, 1977; McLean and Mattocks, 1980); sometimes fever and vomiting; and hepatomegaly without splenomegaly (although the latter may be seen in later stages: Bras and Hill, 1956). Jaundice has been seen occasionally (Fox *et al.*, 1978; Datta *et al.*, 1978a). Severe ascites may be accompanied by oedema of the lower limbs in the later stages. There is biochemical evidence of liver dysfunction.

Death may ensue from 2 weeks to over 2 years following poisoning (Willmot and Robertson, 1920), but it is possible for patients to recover almost completely from acute VOD if the alkaloid intake is discontinued (Bras and Hill, 1956; Lyford *et al.*, 1976).

The lungs are sometimes affected: pulmonary oedema was seen by Fox *et al.* (1978) in a child, and pleural effusions, common in experimental animals given some PAs (Barnes *et al.*, 1964), have been observed in patients by McGee *et al.* (1976) and Stillman *et al.* (1977). Lung damage appears to have been prominant, and fatal, in at least one case (Heath *et al.*, 1975).

Chronic illness, which may develop in survivors of acute VOD or in people ingesting small amounts of PAs over a long period, proceeds through fibrosis of the liver to cirrhosis which is 'clinically indistinguishable from any other type of cirrhosis' (Bras and Hill, 1956).

Diagnosis of disease as being due to PAs is often difficult, because of the time interval before its onset. The chronic illness can be especially difficult to associate with its cause. VOD has been confused with viral hepatitis (Datta *et al.*, 1978a) and, because of neurological symptoms, with Reye's syndrome (Fox *et al.*, 1978). Clues to the cause of the illness may be provided by the association of its onset with the taking of herbal medicine, the identification of known PA-containing plant materials in food or medicine, and the analytical detection of PAs in these commodities. The contamination of local pastures with weeds containing PAs, and especially the occurrence of similar disease in grazing livestock, are important evidence (H. D. Tandon *et al.*, 1978). Liver biopsy may be necessary to confirm the nature of the condition in patients.

2. Pathology

Occlusion of the hepatic veins was at first thought to be the primary event in VOD (Bras *et al.*, 1954; Bras and Hill, 1956). It is now recognised that, as in animals, the initial damage is to the hepatic parenchymal cells (McLean, 1970, 1974; H. D. Tandon *et al.*, 1978). Toxic metabolites formed in the liver cells cause centrilobular necrosis, escape into the bloodstream to injure the vascular endothelium downstream, and may also reach the lung to damage vessels there. In the liver, the lumen of sublobular veins, especially the smallest, is blocked by fibrin from breakdown of cells and by thickening of vessel walls, obstructing the blood flow, increasing portal pressure, and leading to massive centrilobular congestion. Examples, with sections of liver biopsy samples, are given by H. D. Tandon *et al.* (1978).

In later (subacute) stages of the disease, there is persistent fibrosis in centrilobular areas, developing into chronic cirrhosis (Bras and Hill, 1956). H. D. Tandon *et al.* (1978) have described a micronodular nonportal cirrhosis in the advanced disease. There was no bile duct proliferation, and regeneration nodules were uncommon.

A curious feature of the chronic disease is that megalocytosis—the presence of giant, abnormal hepatic parenchymal cells—which is a striking feature of PA-poisoning in experimental animals (Schoental and Magee, 1959) and a manifestation of the powerful antimitotic effect of these alkaloids, has never been observed in human subjects (H. D. Tandon *et al.*, 1978; McLean and Mattocks, 1980). The mitotic rate in blood cells is lowered in rats given fulvine, but not in people poisoned by this alkaloid (Martin *et al.*, 1972).

There is no evidence as to whether cancer develops in people exposed to PAs. Some PAs have mutagenic activity (see Chapter 10), and chromosome abnormalities have been found in lymphocytes from children in acute or recovery phases of VOD, up to 14 weeks after exposure to fulvine, as well as in rats given this alkaloid (Martin *et al.*, 1972).

3. Effects of Indicine N-Oxide (INO)

This is the only PA which has been deliberately given to people under clinically controlled conditions, as a potential anti-cancer drug. Its effects are not typical of hepatotoxic PAs. It is highly water soluble, has relatively low toxicity, and 40% of it is excreted unchanged in the urine (Kovach *et al.*, 1979). The chief toxic effect of INO is myelosuppression, which is probably cumulative with successive doses and is exacerbated in patients previously treated with nitro-soureas (Kovach *et al.*, 1979). This effect is prbably not due to pyrrolic metabolites which are responsible for the hepatotoxicity of other PAs. However, two patients who were given INO and later died did develop acute liver damage, with centrolobular necrosis and congestion (Letendre *et al.*, 1981).

C. Relationship between Dose Level and Toxic Effects

In only a few instances have estimates been made of the amounts of PAs ingested by people (Table 9.3). The alkaloids involved were predominantly either heliotrine *N*-oxide from *Heliotropium* plants or mixtures of retronecine-based PAs from *Senecio* plants. From the limited toxicity data available for animals (see Chapter 7), it is possible to estimate that these materials should have oral LD_{50} values in male rats of around 400 and 50 mg/kg, respectively, making the *Senecio* alkaloids about eight times as toxic as the *Heliotropium* alkaloids. A comparison of incidents (i) and (ii) with (iii) (Table 9.3) suggests that a some-what similar toxicity ratio applies in the case of humans. A complication is that the victims in incidents (i) and (ii) were children, probably more susceptible to PAs than adults, but the susceptibility of the patients of Datta *et al.* also may have been raised by their taking phenobarbitone. It might be concluded from these limited data that people are more susceptible to the acute effects of PAs than male rats, and that doses above 12 mg/kg of *Senecio* alkaloids like retrorsine, or 60 mg/kg of heliotrine *N*-oxide, may cause illness, especially in children.

An inconsistency exists in the report of Mohabbat *et al.* (1976) on the Afghan incident (iv). Samples of contaminated wheat were stated to contain an average of 40 heliotrope seeds (300 mg) per kilogram. This would represent up to 4.5 mg of PAs/kg of wheat ($4.5 \times 10^{-4}\%$), and this is in agreement with the authors' calculation that an adult consuming 700 g of wheat daily might ingest 1.46 g of PAs in 2 years. But it is also stated that samples of wheat flour contained (up to) 0.186% of alkaloids. This exceeds the earlier estimate by a factor of over 400 and, as pointed out by Anderson (1981) it represents a *daily* intake of 1.3 g alkaloids per person! There might be an error here; yet the higher estimate is more consistent with the results of Datta *et al.* for the production of acute VOD. It has been pointed out (Section II,A,3) that an uneven distribution of con-taminating seed may occur in a batch of grain, and it seems likely that some of the victims who succumbed to VOD in Afghanistan had indeed taken PAs at a very much higher level than the estimate of 1.46 g in 2 years might suggest.

III. RISKS TO PEOPLE FROM EXPOSURE TO PAs

A variety of circumstances can result in ingestion of PAs by people. A large intake of PAs during a relatively short time can lead to acute VOD. Chronic

TABLE 9.3 Some Estimates of PA Intake in Acute Cases of VOD

Incident	References	Plant	Probable alkaloids (m, minor component)	Number of victims	Age and sex	Estimated alkaloid intake			Outcome
						Total dose (mg)	Time range of intake	Dose (mg/kg) overall	
(i)	Stillman et al. (1977); Huxtable et al. (1977)	Senecio longilobus (S. douglasii)	riddelliine retrorsine N-oxide (m) senecionine NO (m) seneciphylline NO	1	6 months F (6 kg)	70–147	2 weeks	12–25	recovered
(ii)	Fox et al. (1978)	as above	as above	1	2 months M	66	4 days	13–17[a]	died
(iii)	Datta et al. (1978a,b)	Heliotropium eichwaldii	heliotrine NO (m) similar to lasiocarpine	1	20 years M	4000	20 days	67[b]	died
(iv)	Mohabbat et al. (1976)	Heliotropium popovii subsp. gillianum	heliotrine NO	1 very many	23 years M various	10,000 at least 1460	50 days up to 2 years	167[b] at least 24[b,c]	died some died
(v)	Kumana et al. (1983)	unknown	unknown	4	23–28 years	630–1380	19–45 days	12–36	1 died

[a] Assuming body weight 4–5 kg.
[b] Assuming body weight 60 kg.
[c] Might be much higher: see text.

intake of PAs at a low level may cause no early signs of toxicity but ultimately produce cirrhosis of the liver. Some PAs are carcinogenic (see Chapter 11), and there has been concern that chronic PA intake may lead to human cancer. What follows is an attempt to assess the risks and what may be done to reduce them.

A. Risks of Acute Poisoning (VOD)

Recognised cases of VOD have resulted from the consumption of herbal teas, medicines, or cereals containing PAs. From the information available (Section II,C), a conservative estimate is that a cumulative dose (taken within a few weeks) of a retronecine based macrocyclic PA (such as might be found in *Senecio* or *Crotalaria* species) exceeding 10 mg/kg, or of a *Heliotropium* alkaloid such as heliotrine exceeding 50 mg/kg, might cause acute liver damage in people, children being more susceptible than adults.

An illustration of the amounts of plant material or of bread contaminated by it which might contain these doses, is given in Table 9.4. In the past, such high levels have been reached by the consumption of herbal medicines or heavily contaminated grain. The use of herbal medicines is probably declining in 'developing' countries as modern medicine becomes established, although paradoxically it may be on the increase in some industralised countries. Modern farming methods are eliminating weeds among crops, and bulk marketing of grain leads to the dilution of toxic contaminants. Thus, it is likely that the worldwide incidence of acute PA intoxication will continue to decline.

TABLE 9.4 Hypothetical Alkaloid Intake
Which Would Result from Consumption of a Plant
Containing 1.0% PAs, Assuming All the Alkaloids Ingested

| | Dose | | | Weight of material to contain the stated dose |
Person	mg/kg	mg	Plant (g)	Bread (kg) contaminated with 1% plant
10 kg child	10	100	10	1.0
	50	500	50	5.0
60 kg adult	10	600	60	6.0
	50	3000	300	30.0

B. Risks of Chronic Intoxication

Some sources of PAs which may be ingested regularly by people contain such low levels of alkaloids that these are unlikely to cause acute intoxication. The main cause for concern is that these alkaloids might have chronic toxic effects. Some examples will illustrate the dose levels which might be encountered. Dickinson et al. (1976) found up to 0.84 ppm of PAs in milk from cows given ragwort. At this level, a 10-kg child would have to drink 12 litres, and a 60-kg adult 71 litres of milk to receive 1 mg of PAs/kg body weight. Samples of honey have been found to contain up to 3.9 ppm of PAs (Deinzer et al., 1977). A 60-kg person would have to consume 15 kg of this honey to receive a dose of 1 mg/kg.

Comfrey (*Symphytum* spp.) is a popular herb (Hills, 1976; Jones and Gillie, 1981) which has received attention as a potential hazard (Anonymous, 1979c). Table 9.5 lists levels of PAs found in some samples of comfrey. The largest amounts are in the root, which is commercially available as a tea, although it is less widely used than comfrey leaf. Roitman (1981) found 8.5 mg of PAs in a cup of comfrey root tea, representing about one third of the alkaloids in the root used. The acute LD_{50} of the mixture of PAs in comfrey is about 550 mg/kg for rats (Culvenor et al., 1980a), twice that of heliotrine, so it could well be over 100 mg/kg for humans, and to receive this a person weighing 60 kg would have to drink over 700 cups of Roitman's tea. Moreover, PA levels are much lower in mature comfrey leaves (Mattocks, 1980) and lower still in commercial comfrey leaf tea: thus it is improbable that anyone could be *acutely* poisoned by comfrey, and in fact there is no evidence that any of the thousands of comfrey users have sustained liver poisoning as a result (Anonymous, 1979c).* Alkaloids in comfrey preparations applied to the body externally are absorbed through the skin, but at under 5% of the level ingested orally (Brauchli et al., 1982). Moreover, these remain largely in the form of *N*-oxides, which are much less toxic than the corresponding basic PAs; thus dermal absorption is unlikely to be harmful. Other plants containing PAs which are occasionally eaten or used medicinally include varieties of coltsfoot (Borka and Onshuus, 1979; Anonymous, 1981). As with comfrey, their alkaloid content is low: *Tussilago farfara* contains up to 0.015% of senkirkine (Culvenor et al., 1976a; Borka and Onshuus, 1979), and *Petasites japonicus* contains about 0.01% of petasitenine (Hirono et al., 1977). Thus, levels of PAs ingested by people using these plants are likely to be similar to those from comfrey.

What, then, might be the risks of chronic illness arising from the ingestion of such low levels of PAs? Most of the alkaloids and plants mentioned above have

---

*Since this was written, Ridker et al. (1985) have observed hepatic VOD in a woman whose minimum daily intake of PAs, including comfrey alkaloids, was estimated to be about 15 μg/kg (for at least 4 months); however, she may have had other sources of exposure.

TABLE 9.5 Measurements of PAs in Samples of Comfrey (*Symphytum* spp.)

| Reference | Comfrey sample | % (Dry wt.) of alkaloids + N-oxides | | |
		Fresh leaves	Dried leaves	Roots
Pedersen (1975a)[a]	*S. asperum*	0.009	0.059	—
	S. officinale	0.006	0.062	—
	S. × uplandicum	0.009	0.09[b]	—
Long (in Hills, 1976)	various	—	0.013–0.062	—
	tea	—	0.009–0.03	—
Mattocks (1980)[c]	*S. × uplandicum*	—	0.22 (y)	—
		—	0.05 (m)	—
Culvenor *et al.* (1980b)[c]	*S. × uplandicum*	0.15 (y)	—	—
		0.05 (int)	—	—
		0.01 (m)	—	—
Roitman (1981)	(commercial)	—	undetectable (<0.005)	0.14–0.37
	S. asperum	0.01	—	—
Brauchli *et al.* (1982)	*S. officinale*	—	—	0.07

[a] Results estimated by titration. Results by weighing were about twice these.

[b] Pederson gave his results in parts per thousand (0/00), and this figure was misinterpreted by Culvenor *et al.* (1980a) as 0.9%.

[c] y, Young leaves; int, intermediate; m, mature leaves.

been shown to be carcinogenic (see Chapter 11). However, the amounts given to animals to produce tumours are often very large. For example, four out of 20 rats given symphytine (present in comfrey) developed liver tumours (Hirono *et al.*, 1979): these animals received a total of six times the LD_{50} dose during 1 year. An equivalent dose for a 60-kg human (assuming LD_{50} 200 mg/kg) would be 72 g of alkaloids, or over 23 daily cups of tea, each containing 8.5 mg of alkaloids, for 1 year! Czygan (1983) has concluded that the senkirkine content of coltsfoot cough tea is too low to be a health hazard to humans. However, an alterantive point of view favours the most extreme caution: Wolf (1983) calculated from results of Hirono *et al.* (1976) that a 'no effect' level of senkirkine for humans would be 2 mg/year (the content of 100 cups of coltsfoot tea): he then applied a 10^{-4} safety factor, implying that the annual intake should be as low as 2×10^{-7} g! Liver cancer is rare in countries where comfrey is widely used, and from the foregoing it appears that the risks from exposure to PAs in comfrey and similar herbs, or in milk, is probably slight, especially when set against other 'accepted' risks such as that from smoking.

In regions of the world, such as East Africa, where primary liver cancer is common, it may be associated with the use of herbal remedies (Williams *et al.*, 1967), but this is not proven. Other, more powerful carcinogens, such as aflatoxins, are probably present in the diet: PAs might act additively with these (Newberne and Rogers, 1973) to increase the total risk. Culvenor (1983) has pointed out that in some outbreaks of PA poisoning, people have received doses comparable with a carcinogenic dose in rats: thus a long-term study of survivors of such outbreaks might show whether these PAs are carcinogenic in humans.

C. How Can the Risks Be Reduced?

Little can be done to prevent toxic damage by PAs once the alkaloids have been ingested. The effects of the alkaloids, though not the alkaloids themselves, are cumulative in the body. The only really effective approach to PA intoxication is to prevent it happening in the first place. In the cases of herbal medicine usage and grain contamination, people can be educated to recognise poisonous plants. Public health campaigns against the use of *Crotalaria fulva* have been very effective in reducing VOD in Jamica. Better quality control may be needed for commercially produced herbal teas. Plants should be properly identified. Sensitive screening tests are available for toxic PAs (see Chapter 2).

Finally, medical practitioners, especially in high-risk areas, should be educated to recognise the relatively uncommon syndrome of acute PA intoxication. If outbreaks of VOD were recognised in their early stages and the plant sources identified, further cases could be prevented.

10

Mutagenic, Antimitotic and Other Cell-damaging Effects of Pyrrolizidine Alkaloids, Pyrrolic Derivatives and Analogues

I. MUTAGENIC AND CLASTOGENIC ACTIVITY

In recent years a number of different short-term tests have been developed in attempts to identify carcinogenic chemicals. Some of these have been applied to various PAs and derivatives: see Tables 10.1 and 10.2. Earlier experiments on mutagenic and chromosome damaging effects of PAs, also listed in the tables, have been discussed by Bull *et al.* (1968), McLean (1970) and Clark (1976).

A. Mutagenicity

A number of PAs have been shown to be powerful, dose-dependent mutagens in *Drosophila melanogaster* (see Table 10.1). All these are hepatotoxic alkaloids, although the degree of mutagenicity is not necessarily proportional to

TABLE 10.1 Mutagenicity and Cell Tests on Pyrrolizidine Alkaloids
and Source Plants

Alkaloid	Type of test[a]	Response[b]	Reference
Clivorine	A	+	Yamanaka *et al.* (1979)
	HPC	+	Williams *et al.* (1980)
Echimidine	D	+	Clark (1960)
Echinatine	D	+	Clark (1960)
Fukinotoxin (see petasitenine)			
Fulvine	D	+	Cook and Holt (1966)
Heliotrine	D	+	Clark (1959, 1960); Brink (1966, 1969)
	P	+	Avanzi (1961)
	F	+	Alderson and Clark (1966)
	CC	+	Bick and Jackson (1968); Kraus *et al.* (1985)
	A	+	Yamanaka *et al.* (1979)
	B	+	Green and Muriel (1975)
	TM	+	Stoyel and Clark (1980)
	CM/CC	+	Takanashi *et al.* (1980)
Integerrimine	D	+	de Paula Ramos and Marques (1978)
Jacobine	D	±	Clark (1960)
	P	+	Avanzi (1961)
Lasiocarpine	D	+	Clark (1960)
	P	+	Avanzi (1961, 1962)
	F	+	Alderson and Clark (1966)
	A	+	Koletsky *et al.* (1978); Yamanaka *et al.* (1979)
	HPC	+	Williams *et al.* (1980)
	CM/CC	+	Takanashi *et al.* (1980)
	TM	+	Stoyel and Clark (1980)
Ligularidine	A	+	Yamanaka *et al.* (1979)
Lindelofine	A	0	Yamanaka *et al.* (1979)
Lycopsamine	A	0	Yamanaka *et al.* (1979)
Monocrotaline	D	+	Clark (1960)
	P	+	Avanzi (1961, 1962)
	CC	+	Umeda and Saito (1971)
	B	+	Green and Muriel (1975)
	A	0	Yamanaka *et al.* (1979)
	HPC	+	Williams *et al.* (1980)
	CT	+	Styles *et al.* (1980)

(*continued*)

TABLE 10.1 (*Continued*)

Alkaloid	Type of test[a]	Response[b]	Reference
Petasitenine (fukinotoxin)	A	+	Yamanaka *et al.* (1979)
	HPC	+	Williams *et al.* (1980)
	CM	+	Takanashi *et al.* (1980)
Platyphylline	D	0	Clark (1960)
Retrorsine	A	+	Wehner *et al.* (1979)
	D	+	Cook and Holt (1966)
	CT	+	Styles *et al.* (1980)
Rosmarinine	CT	0	Styles *et al.* (1980)
Senecionine	D	+	Clark (1960)
	A	0	Yamanaka *et al.* (1979)
Seneciphylline	A	0	Yamanaka *et al.* (1979)
	P	+	Avanzi (1961)
	D	+	Candrian *et al.* (1984a)
Senkirkine	A	+	Yamanaka *et al.* (1979)
	HPC	+	Williams *et al.* (1980)
	CM/CC	+	Takanashi *et al.* (1980)
	D	+	Candrian *et al.* (1984a)
Supinine	D	±	Clark (1960)
	P	+	Avanzi (1961)
Various	HPC		Mori *et al.* (1985)
Mixed alkaloids from *Senecio jacobaea*	A	0	White *et al.* (1984b)
Plant extracts			
Senecio nemorensis ssp. *fuchsii*	CM	+	Habs *et al.* (1982)
	A	0	Habs *et al.* (1982)
	A	+	Pool (1982)
Senecio jacobaea	A	+	White *et al.* (1983, 1984b)
Senecio longilobus	A	0	White *et al.* (1983)
Symphytum officinale (comfrey)	A	0	White *et al.* (1983)

[a] Types of tests:

A	*Salmonella* ('Ames') test
B	Other bacterial tests
CC	Clastogenic activity in cultured cells
CM	Mutagencity in cultured mammalian cells
CT	Cell transformation test
D	Mutagenicity in *Drosophila*
F	Tests in fungus (*Aspergillus nidulans*)
HPC	Hepatocyte primary culture/DNA repair test
P	Chromosomal aberrations in plant cells
TM	Transplacental micronucleus test

[b] Levels of activity: +, active; ±, marginally active; 0, inactive.

the degree of hepatotoxicity in mammals. The N-oxides of heliotrine, lasiocarpine and monocrotaline are also mutagenic, but less so (Clark, 1960): this might be because these less lipophilic compounds are less susceptible to enzymic transformation into active metabolites. According to Clark (1976), the mutagenic effect of feeding *Drosophila* males for 24 h with a medium containing 10^{-3} M monocrotaline is comparable to about 1000 R of X-rays. Heliotrine is also teratogenic in *Drosophila* (Brink, 1982).

Candrian *et al.* (1984a) have found that the milk of lactating rats given seneciphylline is slightly mutagenic to *Drosophila*.

Lasiocarpine and heliotrine are mutagenic to the mould *Aspergillus nidulans* (Alderson and Clark, 1966).

Heliotrine, lasiocarpine, petasitenine and senkirkine are also mutagenic to cultured Chinese hamster cells in presence of liver microsomes (S9); mutations are also induced by exposing the cells to the alkaloids for a long period without S9 (Takanashi *et al.*, 1980). Thus these alkaloids might themselves be weakly mutagenic or the cells may possess weak activating ability.

Mutagenicity of PAs has proved more difficult to demonstrate in bacterial systems. Green and Muriel (1975) obtained negative results with heliotrine in the Ames test using *Salmonella typhimurium;* however, both heliotrine and monocrotaline are able to kill repair-deficient strains of *E. coli* in the presence of liver microsomes. The type of damage is comparable to that produced by mitomycin C, but much larger amounts of the alkaloids are needed: 10 mg/ml of monocrotaline gives similar results to those from 0.25 μg/ml of mitomycin C. One interpretation of this might be that only a very small proportion of the alkaloid is activated by the microsomal system. Chemically, the reactive metabolites of PAs and of mitomycin C are similar, bifunctional alkylating agents, and it is likely that they have similar mechanisms of mutagenic action, by cross-linking DNA strands (Culvenor *et al.*, 1969a; Mattocks, 1969a).

PAs are negative in the Ames test in absence of metabolic activation. Even with liver microsomes present, results have been equivocal. Yamanaka *et al.* (1979) found that in some cases it is necessary to preincubate the bacteria with the alkaloid and an S9 liver homogenate fraction, to give a positive response: the number of revertants is dependent on the amount of S9 and the preincubation time. Even so, some hepatotoxic alkaloids which are mutagenic in other tests were inactive. Thus, PAs with heliotridine or otonecine as base moiety are positive, but retronecine-based PAs including senecionine, seneciphylline and monocrotaline (which is known to be carcinogenic) are unresponsive; so also are retronecine-based PAs from *Senecio jacobaea* (White *et al.*, 1984b). Paradoxically, the retronecine-based alkaloid, retrorsine, is positive in the standard Ames test (Wehner *et al.*, 1979).

TABLE 10.2 Mutagenicity Tests on PA Derivatives, Analogues and Related Compounds

Compound type	Compound	Type of test[a]	Response[b]	Reference
Semisynthetic PA	Retronecine bis-p-chlorobenzoate	P	+	Kak et al. (1973); Kaul and Kak (1974)
Synthetic PA analogue	Synthanecine A bis-N-ethylcarbamate	CT	+	Styles et al. (1980)
Necines	Retronecine	A	0	Yamanaka et al. (1979)
	Heliotridine	D	0	Clark (1959)
Necic acids	Viridofloric acid	A	0	Yamanaka et al. (1979)
		HPC	0	Williams et al. (1980)
	Heliotric (heliotrinic) acid	D	±	Clark (1959)
Dehydro-alkaloid	LX-201	A	+	Yamanaka et al. (1979)
		HPC	+	Williams et al. (1980)
Dehydronecines	Dehydroretronecine	CT	+	Styles et al. (1980)
		A	±	Ord et al. (1985)

276

Pyrrole and derivatives	Dehydroheliotridine	SCE	+	Ord et al. (1985)
		CM	+	Bick and Culvenor (1971)
	Pyrrole	HPC	0	Williams et al. (1980)
	2,3-Bishydroxymethyl-1-methylpyrrole	CT	+	Styles et al. (1980)
		A	±	Ord et al. (1985)
		SCE	+	Ord et al. (1985)
	2-Hydroxymethyl-1-methylpyrrole	A	0	Ord et al. (1985)
		SCE	±	Ord et al. (1985)
	3-Hydroxymethyl-1-methylpyrrole	A	+	Ord et al. (1985)
		SCE	±	Ord et al. (1985)

[a] Types of tests:

A	*Salmonella* ('Ames') test
CM	Mutagencity in cultured mammalian cells
CT	Cell transformation test
D	Mutagenicity in *Drosophila*
HPC	Hepatocyte primary culture/DNA repair test
P	Chromosomal aberrations in plant cells
SCE	Sister chromatid exchange

[b] Levels of activity: +, active; ±, marginally active; 0, inactive.

B. Chromosome Damage

Some PAs are clastogenic in both plant and animal cells. Heliotrine, lasiocarpine and monocrotaline cause chromosome damage in root tip cells of onion and pea plants (*Allium cepa* and *Vicia faba*) (Avanzi, 1961, 1962). The semi-synthetic ester retronecine bis-*p*-chlorabenzoate induces chromosome damage, decreases seed fertility, and induces some mutations in barley (Kak *et al.*, 1973), and chromosome aberrations in *Allium cepa* root tip cells (Kaul and Kak, 1974).

With mammalian cells, Bick and Jackson (1968) found that heliotrine at $2 \times 10^{-4}\,M$ suppresses cell division in leucocyte cultures from the marsupial *Potorus tridactylus,* and at $5 \times 10^{-5}\,M$ it produces chromatid breaks equivalent to the effects of 300 R of X-rays given at 25 R/min (Bick, 1970). There is evidence that heliotrine is able to break chromosomes in the G_1 phase of the cell cycle, before DNA synthesis. Bick and Culvenor (1971) found dehydroheliotridine, a metabolite of heliotrine, to be 10 times more active than the latter alkaloid in *Potorus* leucocytes, the action being in the S or G_2 periods of the cell cycle.

PAs applied directly without addition of a metabolic activating system can damage chromosomes in a cultured Chinese hamster lung cell line; the nucleus and cytoplasm are also enlarged (Takanashi *et al.,* 1980). Heliotrine ($10^{-3}\,M$) and petasitenine ($10^{-2}\,M$) induce interchromosomal exchanges, chromatid gaps, fragmentations and other aberrations. Lasiocarpine ($5 \times 10^{-4}\,M$) produces mainly chromatid gaps, and senkirkine ($10^{-2}\,M$) has similar but weaker action. The effects are greater after 48 h than 24 h; however, similar results are produced following only 1 h treatment with heliotrine ($5 \times 10^{-4}\,M$) in the presence of rat liver microsomes (S9).

The micronucleus test (Schmid, 1975) is a convenient screening method for chromosome-damaging chemicals. In a transplacental micronucleus test, Stoyel and Clark (1980) gave PAs i.p. to pregnant female mice. The test depends either on the alkaloid reaching the foetus and being metabolically activated by the foetal liver, or on active metabolites formed in the parent liver being transported to the foetus. [Peterson and Jago (1980) have shown that heliotrine and its metabolite, dehydroheliotridine, can cross the rat placenta into 14-day foetuses.] In this test, heliotrine (225 mg/kg) and lasiocarpine (86 mg/kg) both produce high levels of micronucleated erythrocytes in foetal liver 20 h after the injection. In a direct test, when these alkaloids are given to male mice, heliotrine gives a large increase in micronucleated erythrocytes in the bone marrow (comparable to that induced by ethyl methanesulphonate), but lasiocarpine does not; the transplacental test is thus regarded as a more relaible test for chromosome damage brought about by possibly short-lived reactive metabolites.

Martin *et al.* (1972) found chromosome damage in blood cells from children in Jamaica who had veno-occlusive disease, probably caused by the PA fulvine. Similar chromosome damage is also seen in rats which have been given fulvine.

Dehydroretronecine, a pyrrolic alkylating agent formed by metabolism of retronecine-based PAs, is able to induce sister chromatid exchange in human lymphocytes (Ord *et al.*, 1985).

Kraus *et al.* (1985) found that heliotrine, but not senkirkine or tussilagine, could induce chromosome damage in human lymphocytes.

C. Indicators of DNA Damage

Electron-microscopic studies have shown nuclear abnormalities in liver cells of rats within 30 min after being given lasiocarpine (Svoboda and Soga, 1966); this alkaloid inhibits protein synthesis and RNA synthesis in liver when given to rats 2–4 h previously (Reddy *et al.*, 1968). Frayssinet and Moule (1969) investigated the effects of lasiocarpine on transcription in liver cells. At 3 h after an LD_{50} dose of the alkaloid is given to male rats, DNA-dependent RNA polymerase activity in their livers is inhibited. However, in an *in vitro* system this alkaloid fails to interact with DNA, indicating that a metabolite, not lasiocarpine itself, is active *in vivo*.

Curtain (1975) showed that there is a significant lowering (3.5%) of heavy satellite DNA in livers of sheep poisoned by PAs. This is consistent with its being brought about by a metabolite such as dehydroheliotridine (DHH), since the latter is able to depress the replication of satellite DNA in cultured ovine kidney cells. Black and Jago (1970) have shown that DHH can alkylate DNA *in vitro*. Curtain and Edgar (1976) found that whereas DHH binds equally to both main and satellite band DNA from cultured sheep lymphocytes *in vitro*, radiolabelled DHH preferentially binds to the satellite DNA in synchronised primary cultures of ovine kidney cells, suggesting that the attack of DHH occurs during mitosis.

Petry *et al.* (1984) isolated hepatic DNA from male rats given i.p. injections of monocrotaline. There was no evidence of DNA single-strand breaks, but DNA–protein and persistent DNA–DNA dose-dependent cross links were detected.

Candrian *et al.* (1985) have shown that radioactivity from tritium-labelled senecionine and seneciphylline can become covalently bound to rat liver DNA *in vivo*.

A DNA repair test using a primary culture of rat hepatocytes gives positive results with hepatotoxic PAs, including the carcinogen monocrotaline (Williams *et al.*, 1980; Mori *et al.*, 1985). The cells are exposed to the test compound for 18 h in presence of tritiated thymidine: the extent of DNA-repair synthesis is then measured by autoradiography. The test is considered more sensitive for carcinogenic PAs than the Ames test, which fails with monocrotaline.

Armstrong and Zuckerman (1972) exposed human embryo cell primary cultures to the PAs lasiocarpine and retrorsine. The liver, but not the lung tissue, is able to convert PAs to pyrrolic metabolites (Armstrong and Zuckerman, 1970). Synthesis of DNA and RNA is inhibited by both alkaloids in the liver but not the lung cells, as shown by autoradiography using tritiated thymidine. However, the metabolite dehydroretronecine (retronecine pyrrole) inhibits DNA, RNA and protein synthesis in cultured human liver cells (Armstrong *et al.*, 1972); after 24 h exposure to the alkaloid, there are marked morphological changes in the nucleus and nucleolus suggesting chromosome damage associated with alkylation of DNA.

D. Cell Transformation

A mammalian cell transformation test has been used to test chemicals for carcinogenic potential (Styles, 1977). In presence of hepatic metabolizing enzymes (S9), this test gives positive results with the carcinogenic PAs retrorsine and monocrotaline, and with the hepatotoxic PA analogue, synthanecine A bis-*N*-ethylcarbamate, but not with the non-hepatotoxic PA rosmarinine (Styles *et al.*, 1980). Retrorsine is also positive in the absence of S9, showing that the cells used are themselves capable of activating the alkaloid. The pyrrolic alcohols dehydroretronecine (DHR) and 2,3-bishydroxymethyl-1-methylpyrrole, which are potential reactive metabolites from the toxic alkaloids and the synthanecine carbamate, respectively, are also effective in transforming cells, but they are markedly less toxic to the cells than the parent compounds, suggesting that the high cytotoxicity of the latter may be due to their more reactive primary metabolites (pyrrolic esters).

E. DNA-damaging Activity in Relation to Structure

The PAs which have proved to be mutagenic or to damage DNA possess structural features characteristic of hepatotoxic alkaloids (i.e. they are esters of unsaturated necines), and many of them have been shown to be carcinogenic (see Chapter 11). Not all tests of PAs for DNA-damaging activity are equally conclusive. The Ames test, in particular, has proved somewhat unreliable with these compounds, possibly reflecting their generally weak carcinogenicity. Nevertheless, most hepatotoxic PAs are reactive in at least one type of test. Conversely, the non-hepatotoxic PAs rosmarinine, platyphylline, and lindelofine give negative results (Table 10.1), as do necines and necic acids (Table 10.2).

Metabolic activation is necessary for PAs to exert maximum mutagenicity or DNA-damaging activity. In some *in vitro* experiments, results are positive in the absence of an added metabolizing system. Although it cannot be ruled out that the intact alkaloids might themselves possess weak activity, the presence of activating systems in the treated cells can account for these results.

There is evidence that DNA damage is caused by the same type of pyrrolic metabolites as those which are responsible for PA hepatotoxicity (see Chapter 12). For example, the clastogenic activity of the metabolite dehydroheliotrine in cultured cells is 10 times that of its parent PA, heliotrine (Bick and Culvenor, 1971). The ability to damage DNA is not a property of pyrroles as such—pyrrole itself is inactive (Williams *et al.*, 1980)—but of alcohols or esters rendered chemically reactive by their attachment to a pyrrolic nucleus. Factors which can influence the activity of these types of metabolites are their reactivity, and whether they are monofunctional or bifunctional alkylating agents. The latter can not only react with DNA (Black and Jago, 1970), but can also cross link DNA strands (White and Mattocks, 1972) in a manner similar to the active form of mitomycin C (Iyer and Szybalski, 1964). Such cross linkage has been demonstrated in the liver DNA of rats given monocrotaline (Petry *et al.*, 1984). Activated PAs cause DNA damage in *E. coli* which resembles that due to bifunctional alkylation by mitomycin C (Green and Muriel, 1975). Like mitomycin C, bifunctional pyrrolic metabolites such as dehydroretronecine are very effective inducers of sister chromatid exchange in human lymphocytes (without the need for metabolic activation), albeit at much higher concentration levels (10^{-4}–10^{-3} M compared with 10^{-7} M for mitomycin C), whereas monofunctional pyrroles are less active (Ord *et al.*, 1985). Such pyrroles also resemble mitomycin C in being only weakly positive in the Ames test; for this neither an activating system nor bifunctionality are essential, but these compounds are only effective against error-prone plasmid strains of *Salmonella,* suggesting that they are non mutagenic under normal repair conditions (Ord *et al.*, 1985). However this activity appears to be related to chemical stability: thus the most chemically reactive molecule tested, 2-hydroxymethyl-1-methylpyrrole, is inactive, perhaps because it polymerizes before it can penetrate the bacterial cell wall. This possibility is illustrated by the fact that when amoebae are treated with dehydromonocrotaline (a reactive pyrrolic metabolite from monocrotaline), their mucopolysaccharide surface quickly becomes coated with pyrrolic polymer, as seen in electron micrographs (M. J. Ord and A. R. Mattocks, unpublished results).

II. ANTIMITOTIC ACTIVITY

The most characteristic chronic effect of hepatotoxic PAs in the liver of animals is to induce the development of greatly enlarged hepatocytes, often called megalocytes (Bull and Dick, 1959); this is a consequence of a powerful antimitotic action exerted by the compounds on liver cells. It is a cumulative effect in animals receiving repeated small doses (e.g. 0.1 LD_{50}) of PAs such as heliotrine or lasiocarpine (Bull and Dick 1959); and it can also be produced by a single sub-

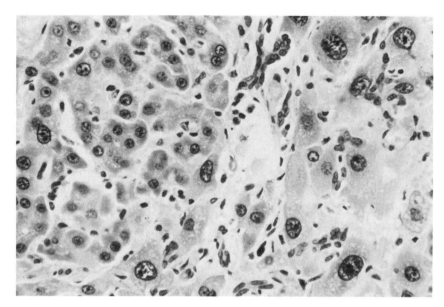

Fig. 10.1   Liver of female rat which died, age 43 days, after being given retrorsine (10 mg/kg, i.p.) at age 2 days. Many giant hepatocytes are present. Haematoxylin and eosin × ca. 225.

lethal dose of alkaloid (Schoental and Magee, 1957, 1959). Very young animals are particularly susceptible (Fig. 10.1). The lesion appears within a few weeks and may persist for the lifetime of the animal—up to 2 years in the rat. Repeated doses of some other hepatotoxins or hepatocarcinogens can cause liver cell enlargement in rats (Schoental and Magee, 1959), but their effects are usually less marked and, with the exception of N-nitrosopyrrolidine (Hendy and Grasso, 1977), less persistent than those produced by PAs.

A. Megalocytosis *in Vivo*

Hepatic megalocytosis has been studied chiefly in rats, but it has also been observed in mice, sheep, horses and pigs poisoned by PAs (McLean, 1972). It has also been observed in trout (Hendricks *et al.*, 1981) but not in cases of human intoxication by PAs (H. D. Tandon *et al.*, 1978; McLean and Mattocks, 1980), although cultured human foetal liver cells become enlarged after exposure to PAs (Sullman and Zuckerman, 1969; Armstrong *et al*, 1972). Megalocytes are occasionally seen in other tissues; Hooper (1974) found enlarged bronchiolar epithelial cells and alveolar cells in the lungs, and tubular epithelial cells in the kidneys, of mice fed *Senecio jacobaea*. Megalocytes have been seen in renal tubules and glomeruli of pigs poisoned by *Crotalaria retusa* seed (containing monocrotaline) (Hooper and Scanlan, 1977) and in renal tubules of monkeys given retrorsine (van der Watt *et al.*, 1972).

The fine structure of the enlarged hepatocytes produced by single doses of PAs has been described by Svoboda and Soga (1966) and by Afzelius and Schoental (1967); for a discussion see also McLean (1970). Allen *et al.* (1970b) described ultra-structure and biochemical changes in hepatic megalocytes of rats fed *Crotalaria spectabilis* (monocrotaline) for 8 months: the cells are enlarged about 2.5 times; liver DNA (but not RNA) is doubled; a postulated sequence of morphological changes in the nuclei of megalocytes is described by these authors.

Hepatic megalocytosis is produced by PAs which (at higher dose levels) are acutely hepatoxic, including semisynthetic derivatives (Schoental and Mattocks, 1960; Culvenor *et al.*, 1976b), and by hepatotoxic synthanecine esters (Mattocks, 1971d), but not by non-hepatotoxic PAs such as platyphylline (Downing and Peterson, 1968; Jago 1970; Culvenor *et al.*, 1976b).

Suppression of acute PA hepatotoxicity by enzyme inhibitors (e.g. chloramphenicol) can improve the survival of rats and so enhance the development of megalocytosis (Allen *et al.*, 1972).

The development of megalocytosis appears to be the result of two distinct events: the persistent inhibition of mitosis induced by the alkaloid, and a subsequent stimulus for the cells to divide (Jago, 1969). The stimulus might be physical injury to the liver, such as partial hepatectomy (Peterson, 1965), or the necrogenic action either of the PA itself or of a different hepatotoxin (Jago, 1969). [Schoental and Magee (1959) showed by serial liver biopsies in rats given single doses of lasiocarpine that large parenchymal cells could develop even though there was no distinct necrosis. Nevertheless, this alkaloid is *capable* of causing necrosis, and it must be presumed that some injury had been done to the cells.]

The sequence of events has been investigated in a series of studies. Peterson (1965) gave lasiocarpine N-oxide to rats as a single i.p. dose of 0.4 LD_{50}, and performed partial hepatectomy 3 months later. The subsequent wave of regeneration resulted in an increase in the size of hepatic parenchymal cells rather than an increase in the number of cells: the few mitotic figures seen were mostly abnormal. The cells had entered into mitosis but were incapable of completing it. Downing and Peterson (1968) have developed a bioassay to assess the antimitotic effect. Rats are given single i.p. doses of PAs, related to their acute LD_{50} (range, 0.01 to 0.4 LD_{50}). Partial hepatectomy is performed after 4 weeks, as a stimulus for cell replication; 48 h later, colchicine is injected to arrest mitoses, and after a further 8 h the rats are killed and the number of mitoses per 1000 hepatic parenchymal cell nuclei (the mitotic index) is counted. In this way the doses of PAs are determined which can produce a 50% inhibition of mitotic index compared with that of control rats. For heliotrine, this is 0.05 LD_{50}; for lasiocarpine N-oxide, 0.01 LD_{50}. The inhibition persists for at least 8 weeks after administration of the alkaloid.

Jago (1969) also showed that megalocytosis can be produced by separate actions of mitotic inhibition and a stimulus for regeneration. When rats are given a low dose of lasiocarpine (0.2 LD_{50}) once every 4 weeks, and the hepatotoxin carbon tetrachloride is given repeatedly (up to 3 doses/week) over the same period, hepatic megalocytosis develops in 8–16 weeks; it is not seen when rats are given either compound alone, although lasiocarpine produces a gradual, slight increase in the size of liver cell nuclei. A single dose of lasiocarpine (0.4 LD_{50}) can alone produce extensive hepatic megalocytosis in 4 weeks when given to very young (2-week-old) rats, in which the liver tissue is rapidly growing, but not in 10-week-old rats. In the young rats, lasiocarpine is able to reduce the mitotic index in liver cells to 0.04 after 1 day, and to 0.21 after 2 days (control value, 1.61). At 4 weeks, there are two populations of parenchymal cells: megalocytes, in which the mitotic index is 1.72 but all the mitoses are abnormal, and nodules of small, regenerating cells with a high mitotic index (3.33). This suggests either that some cells are able to escape the antimitotic action of the alkaloid, or that the damage can be repaired.

A study of chronic PA intoxication of vervet monkeys (van der Watt *et al.*, 1972) also showed evidence of limited resistance to mitotic inhibition. In the livers of three animals which survived intermittent exposure to retrorsine for a year, foci of atypical regenerating hepatocytes developed. This only occurred after 60% of the liver mass had been lost.

Jago (1970), taking advantage of the greater susceptibility of young rats, developed a method for assessing the chronic hepatotoxicity of PAs. Weanling rats aged 2 weeks are each given a single i.p. dose of alkaloid. The mitotic index of liver parenchymal cells is determined 2 days later by histological examination. Seven PAs were tested by this method, and for comparison further rats were given graded single doses of the alkaloids to determine acute toxicity and the development of hepatic megalocytosis up to 4 weeks later. Lasiocarpine and its *N*-oxide, heliotrine, anacrotine, monocrotaline, and senecionine were all found to suppress mitosis and to produce hepatic necrosis and megalocytosis at appropriate dose levels. Platyphylline was neither antimitotic nor hepatotoxic. A further, large group of PAs and derivatives has been screened using this procedure (Culvenor *et al.*, 1976b).

B. Antimitotic Action of Pyrrolic Metabolites

The pyrrolic alcohol dehydroheliotridine is a secondary metabolite probably formed by hydrolysis of very reactive dihydropyrrolizine esters resulting from microsomal oxidation of heliotridine-based PAs (Jago *et al.*, 1970). Dehydroheliotridine is not acutely hepatotoxic, but when it is given to young rats as a single large (up to 93 mg/kg) i.p. dose it produces marked mitotic inhibition (Peterson *et al.*, 1972). The mitotic index of the liver is reduced by a factor of 10–20 compared with control rats, and to less than one-tenth of controls in the epithelium of renal tubules and the dorsal surface of the tongue. Dehydroheliotridine induces moderate megalocytosis in the rat liver, and some in the kidney, but this is greatly enhanced in the liver by concurrent administration of carbon tetrachloride. Thus, dehydroheliotridine is antimitotic, but unlike lasiocarpine it lacks the tissue damaging action which can stimulate regeneration. Dehydroretronecine (201), a secondary metabolite of retronecine-based PAs and enantiomeric with dehydroheliotridine, has similar inhibitory action on mitosis (Hsu *et al.* 1973a; Allen and Hsu, 1974). So also have the analogous synthetic pyrrolic alcohols, 2,3-bishydroxymethyl-1-methyl pyrrole (279) and 2,3-bishydroxymethyl-5-methyl-1-phenylpyrrole (410). If either of the latter are given to rats they produce neither liver necrosis nor megalocytosis: however, if sufficient of the hepatotoxin, dimethylnitrosamine, to cause liver necrosis is given several days later, hepatic megalocytosis develops (Mattocks, 1981b).

(201) R = OH
(345) R = H

(279) R = H
(408) R = MeO

(409) R = Me
(410) R = Ph

(411)

Dehydromonocrotaline (monocrotaline pyrrole), the putative primary metabolite from the PA monocrotaline, is much more reactive and toxic than its hydrolysis product dehydroretronecine. When dehydromonocrotaline (15 mg/kg) is injected into the liver of a rat via a mesenteric vein, it causes severe damage to that organ, and bizarre giant hepatocytes subsequently develop, similar to those caused by much greater doses of monocrotaline (Butler *et al.*, 1970). A smaller amount (2 mg/kg) of dehydromonocrotaline (Hsu *et al.*, 1973a) produces a severe depression of the mitotic index in the liver 4 weeks later, as measured by the method of Downing and Peterson (1968) following partial hepatectomy.

A suggested scheme of antimitotic action of a hepatotoxic PA *in vivo*, consistent with the above results, in shown in Fig. 10.2. The PA is irreversibly converted by hepatic microsomal enzymes into a pyrrolic ester, which can hydrolyse to a pyrrolic alcohol. The latter is able to interact with cells, producing a latent mitotic inhibition. The more reactive primary metabolite may also do this (it could react with tissue constitutents to give products identical with those from the pyrrolic alcohol), but in addition it can cause acute tissue damage. This damage provides a stimulus for regeneration, so that administration of the PA or the pyrrolic ester can induce megalocytosis to a much greater extent than the secondary metabolite alone.

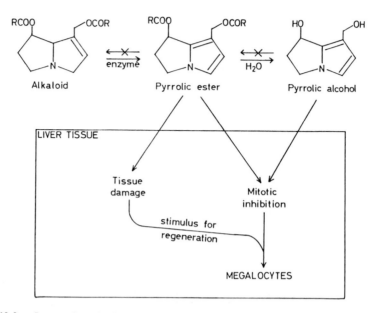

Fig. 10.2.   Suggested mechanism for induction of enlarged hepatocytes (megalocytes) by a pyrrolizidine alkaloid in mammalian liver.

C. Mitotic Inhibition *in Vitro*

PAs and their pyrrolic metabolites can exert their antimitotic activity in cultured mammalian cells. Lasiocarpine and retrorsine produce a long-term inhibition of mitosis in cultures of human embryo lever cells (Armstrong *et al.*, 1972; Armstrong and Zuckerman, 1972). Lung cells, which unlike the liver cells cannot metabolize these PAs to pyrroles, are unaffected. However, the pyrrolic metabolite dehydroretronecine is antimitotic in both the liver and lung cells.

Dehydroretronecine is also highly active in cultures of leucocytes from the marsupial *Potorus tridactylus,* completely suppressing cell division at a concentration of 6×10^{-5} M (Bick and Culvenor, 1971).

Mattocks and Legg (1980) compared the effects of dehydroretronecine and a series of analogous pyrrolic alcohols in a rat liver parenchymal cell line. Cells were exposed to the compounds, then stimulated to divide: colchicine-arrested mitoses were subsequently counted in stained preparations. It was found that the bifunctional alkylating agent DHR (201) and the synthetic pyrroles (279), (408), (409), (410) and (411) are highly effective, suppressing mitosis almost completely at a concentration of 10^{-4} M. Treated cultures kept 3 weeks contain enormously enlarged cells, reminiscent of the giant hepatocytes in animals with chronic PA intoxication, alongside areas of normal sized cells (Fig. 10.3). Simi-

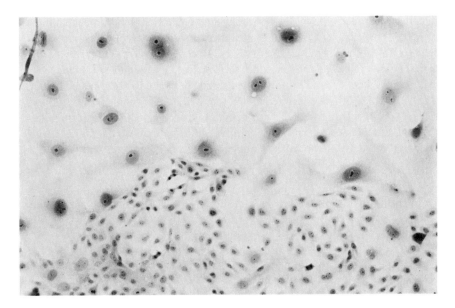

Fig. 10.3.   Giant, alongside normal sized, cells in a cultured rat-liver cell line (BL8L) treated 3 weeks earlier with 2,3-bishydroxymethyl-1-methylpyrrole (10^{-4} M). Haematoxylin and eosin $\times$ ca. 98.

larly, Armstrong and Zuckerman (1972) and Takanashi *et al.* (1980) saw enlarged cells in their treated cultures; these are evidently able to synthesize cell components but are unable to divide. Some monofunctional pyrroles, including dehydrosupinidine (345), a putative metabolite of the PA supinine, also inhibit mitosis, but they are more toxic to the cells than the bifunctional pyrroles. The greater effect on mitotis of the latter, compared with monofunctional pyrroles, reflects the relative abilities of the parent PAs to produce giant hepatocytes in animals: e.g. single doses of many diester PAs can produce megalocytosis (Culvenor *et al.*, 1976b), whereas multiple doses of the monoester supinine are necessary before this occurs (Mattocks, 1978b).

The precise nature of the interaction between pyrrolic alcohols and cells which results in mitotic inhibition is not known, but the fact that the compounds are alkylating agents and the persistence of the antimitotic action support the view that a chemical reaction with some nucleophilic cell constituent is responsible. The effectiveness of some monofunctional pyrroles implies that a cross linking reaction is not necessary, although the stronger binding afforded by alkylation at two adjacent sites on a target molecule might account for the greater activity of the bifunctional pyrroles.

The target is not necessarily DNA. Armstrong and Zuckerman (1972) consider that antimitotic activity is not related directly to inhibition of DNA synthesis, since the latter recovers within a week whereas mitotic inhibition can continue for up to 4 weeks. However, this would not rule out the effect being due to alteration of a small but vital part of the DNA. Autoradiography of tritiated thymidine incorporation indicates that the level of DNA synthesis is reduced in cells that are kept from dividing, but it is not prevented (Mattocks and Legg, 1980), and when cells are counted in relation to their DNA content by cytofluorimetry 24 h after being treated with dehydroretronecine, the proportion with a tetraploid complement of DNA (i.e. equivalent to normal cells in the G_2 and M phases of the division cycle) has trebled to 35.5% compared with 11.5% in control cultures (A. R. Mattocks and R. F. Legg, unpublished).

D. Location of Site of Action in Cell Cycle

Samuel and Jago (1975) used the stimulus to divide, induced in rat liver by a single injection of thioacetamide (Reddy *et al.*, 1969), to investigate the location in the cell cycle of the antimitotic action of lasiocarpine and of its pyrrolic metabolite, dehydroheliotridine. About 52 h after a dose of thioacetamide, a large peak of mitosis occurs, preceded about 14 h earlier by a wave of DNA

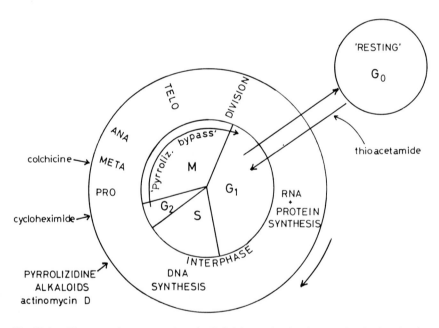

Fig. 10.4.   Diagrammatic representation of cell division cycle, showing postulated point of action of PAs (via pyrrolic metabolites) compared with those of actinomycin D, cycloheximide, colchicine and thioacetamide.

synthesis. Lasiocarpine given as a single small dose, at any time between 2 weeks before and 41 h after the thioacetamide, almost deletes this wave of mitosis (but not the DNA synthesis), but it is ineffective if given later than 8–12 h before mitosis. A dose of dehydroheliotridine (about seven times that of lasiocarpine on a molar basis) has the same effect. It is concluded that the alkaloid acts during the late S or early G_2 phase of the cell cycle (Fig. 10.4). Actinomycin D, which also acts in this region of the cycle, is thought to inhibit the messenger RNA needed for production in late G_2 of proteins essential for division. The long persistence of the antimitotic action of lasiocarpine precludes its being due to a direct effect on RNA polymerase: this alkaloid can inhibit the latter but only for a few days after a large dose (Reddy *et al.*, 1968). Nor does the alkaloid act on the protein synthesis itself—the latter can be inhibited late in G_2 by cycloheximide, preventing mitosis, whereas lasiocarpine has no effect at this time. Samuel and Jago suggest that the active PA metabolite may bind to DNA or chromosomal protein in such a way as to inactivate the section of genome that codes for the proteins of division.

Because the action of the alkaloid is specific and persistent, affected cells can continue to cycle but the M phase is effectively bypassed (McLean, 1970); thus the cells continue to synthesise protein, RNA and DNA, but do not divide. Occasional mitotic figures are seen in the enlarged hepatocytes, but Schoental and Magee (1959) consider that the attempt to divide is usually unsuccessful: such cells often appear to be dying.

11

Carcinogenicity and Antitumour Activity of Pyrrolizidine Alkaloids and Plants Containing Them

I. GENERAL INTRODUCTION

The powerful toxicity of many PAs to animals (Chapter 7), and the striking cytotoxic, mutagenic and antimitotic effects of these alkaloids and their pyrrolic derivatives on cells (Chapter 10) make it desirable that these compounds should be tested both for carcinogenic and anticancer activity.

Surprisingly, little systematic work was done on pyrrolizidine carcinogenicity prior to 1970, although the earlier results, often obtained using inadequate numbers of animals, strongly suggested that some PAs are hepatocarcinogens. In this chapter, all the principal results are considered, although for further discussion of the earlier work, reference should be made to Bull *et al.* (1968) and McLean (1970).

There has recently been increasing interest in the potential antitumour activity of PAs and derivatives. It now appears likely that this activity may be expressed through at least two unrelated mechanisms, involving on the one hand the antimitotic effects of pyrrolic metabolites, and on the other an unknown mechanism associated with some alkaloid *N*-oxides, one of which has recently been brought to clinical trial.

II. CARCINOGENICITY

A. Introduction

No cases of cancer or increased tumour incidence in populations of people or livestock have been associated with outbreaks of pyrrolizidine poisoning or with the ingestion of foods or medicines containing PAs (McLean, 1970; Svoboda and Reddy, 1972; IARC, 1976). Such associations would be in any event be difficult to demonstrate.

Adequate tests for carcinogenicity have been carried out on relatively few of the known cytotoxic PAs (Table 11.1). The early results (up to 1959) were somewhat unsatisfactory. Incidences of approximately 25% of hepatoma and 5% of other tumours were reported in rats and chickens given the alkaloids retrorsine, isatidine, riddelliine, monocrotaline, mixed alkaloids from *Senecio jacobaea,* or the latter plant material. The significance of these results was disputed by Bull *et al.* (1968), who themselves reported finding no neoplasia of hepatic parenchymal cells in rats dosed for long terms with the PAs heliotrine or lasiocarpine. They found difficulty in reconciling a carcinogenic effect with the powerful, persistent antimitotic activity of PAs in the liver.

Since 1970 there has been increasing interest in pyrrolizidine alkaloid carcinogenicity, stimulated by an awareness that they may be present in human food (see Chapter 9). In 1976, a major review (IARC, 1976) concluded that of the PAs tested, the carcinogenicity of four had been established, namely retrorsine, isatidine (retrorsine *N*-oxide), lasiocarpine, and monocrotaline. More recently, tests have been carried out with several other PAs (IARC, 1983), and most of these have yielded liver tumours which were rare in control animals. Thus it is probably correct to class most of the alkaloids tested as hepatocarcinogens, although in some cases confirmatory tests would be desirable. Pathologists differ in their assessment of liver tumours (McLean, 1970). Thus the authors' assessment of tumours in the publications cited may not be universally accepted.

A chronological list of tests which have been carried out on individual PAs and on plants containing PAs is given in Table 11.1.

TABLE 11.1 Summary of Carcinogenicity Tests on PAs or PA-containing Plants

Reference	Material tested	Test animal	Route of administration	Concurrent treatments
Cook et al. (1950)	alkaloids from Senecio jacobaea (Ragwort)	rats	drink[a]	
Schoental et al. (1954)	S. jacobaea alkaloids	rats	drink	
	retrorsine	rats	drink	
	isatidine (retrorsine N-oxide)	rats	drink or i.p. + top[b]	choline
Schoental and Head (1955)	monocrotaline	rats	top or drink or i.p. or i.p. + drink	
Campbell (1956)	S. jacobaea alkaloids (mainly senecyphylline)	chickens	i.v.	
	S. jacobaea: milled plant (less roots)	chickens	diet[c]	
Schoental and Head (1957)	riddelliine	rats	drink + i.p.	betaine; high casein diet
	retrorsine	rats	i.p.	
	isatidine	rats	drink or i.p.	
Dybing and Erichsen (1959)	Senecio aquaticus	rats	diet	
Schoental and Bensted (1963)	retrorsine	rats	i.g.[d] (1 dose)	whole body irradiation; partial hepatectomy
Schoental et al. (1970)	mixed Amsinckia intermedia alkaloids (intermedine, lycopsamine)	rats	i.g.	
	alkaloids from Heliotropium supinum	rats	diet; i.g.	
Harris and Chen (1970)	Senecio longilobus: milled leaves and stem	rats	diet	
Svoboda and Reddy (1972)	lasiocarpine	rats	i.p.	
Schoental and Cavanagh (1972)	Heliotropium ramosissimum: ground plant	rats	diet	

Reference		Animal	Route	
Schoental and Cavanagh (1972)	hydroxysenkirkine	rats	i.p.	
	retronecine	rats	s.c.	
Newberne and Rogers (1973)	monocrotaline	rats	i.g.	aflatoxin B_1; low-lipotrope diet
Hirono et al. (1973)	*Petasites japonicus*: young flower stalks	rats	diet	
Svoboda and Reddy (1974)	lasiocarpine	rats	i.p.	
Schoental (1975a)	heliotrine	rats	i.g.	nicotinamide
Hirono et al. (1976)	*Tussilago farafara* (Coltsfoot): dried	rats	diet	
Shumaker et al. (1976b)	monocrotaline	rats	s.c.	
Hayashi et al. (1977)	monocrotaline	rats	s.c. (1 dose)	
Hirono et al. (1977)	petasitenine	rats	drink	
Rao and Reddy (1978)	lasiocarpine	rats	diet	
Hirono et al. (1978)	*Symphytum officinale* (Russian comfrey): leaves or root, dried and milled	rats	diet	
Hirono et al. (1979)	senkirkine	rats	i.p.	
	symphytine	rats	i.p.	
Kuhara et al. (1980)	clivorine	rats	drink	
Habs (1982); Habs et al. (1982)	alkloidal extract from *Senecio nemorensis*, ssp. *fuchsii* (senecionine; fuchsisenecionine)	rats	i.g.	
Hirono et al. (1983)	*Farfugium japonicum*; *Senecio cannabifolius*	rats	diet	

[a] In drinking water.
[b] Topical application.
[c] Mixed with food.
[d] Via stomach tube.

293

B. Carcinogenicity Tests on Isolated Pyrrolizidine Alkaloids

Experiments up to the present time (1984) which have generally led to the production of liver tumours are summarised below. Alkaloids are listed alphabetically.

1. Clivorine

Kuhara *et al.* (1980) tested this alkaloid in 12 rats of both sexes. The animals were given a solution of 0.005% clivorine as drinking water, continuously for 340 days, and plain water thereafter. Twenty control rats received tap water throughout the experiment. All the treated rats survived over 440 days. Eight developed liver tumours: two of these were hemangioendothelial sarcomas (one with lung metastasis), and the remaining six were described as neoplastic nodules. No liver tumours were found in the control rats.

2. Lasiocarpine

Svoboda and Reddy (1972) gave 0.1 LD_{50} of this alkaloid i.p. to 25 male rats twice weekly for 4 weeks, then weekly for 52 weeks. Of 18 survivors, 16 rats developed tumours 60–76 weeks after the start of the experiment. Eleven had hepatocellular carcinomas, six had squamous cell carcinomas of the skin, and five had pulmonary adenomas. The malignancy of skin and liver tumours was confirmed by transplantation through five generations. Two lung adenomas were the only tumours reported in 25 control rats. The cumulative dose of alkaloid was about 125 mg/rat. It was concluded that lasiocarpine is a complete carcinogen, not only for the liver but also for skin and probably the small intestine of the rat. In another communication apparently concerning the same experiment, the transplantation of skin tumours was described (Svoboda and Reddy, 1974). In a feeding experiment (Rao and Reddy, 1978), 20 male rats were given lasiocarpine (50 ppm) in their diet for 55 weeks. Malignant tumours developed in 17 animals between 48 and 59 weeks. Nine had angiosarcomas of the liver and seven had hepatocellular carcinomas. The average cumulative dose of lasiocarpine was estimated to be 190–200 mg/rat. Hyperplastic nodules develop rapidly in the liver of rats concurrently given lasiocarpine and thioacetamide (Reddy *et al.*, 1976); carbohydrate or total calorie restriction has no effect on the progression of these nodules to carcinoma (Rao *et al.*, 1983).

3. Monocrotaline

Schoental and Head (1955) found liver tumours, one said to be a trabecular hepatoma, in two of six rats which had been given monocrotaline both in the drinking water and by injection. In a well controlled experiment, Newberne and Rogers (1973) gave monocrotaline weekly to 50 weanling male rats by stomach tube. Ten out of 42 rats surviving 55 weeks or more developed hepatocarcinomas. In another group given a low-lipotrope diet, 14 out of 35 survivors had hepatocarcinomas. Some lung metastases were seen. Forty-five surviving control rats had no liver tumours. Shumaker *et al.* (1976b) gave s.c. injections of monocrotaline (5 mg/kg) to 60 male rats on alternate weeks for 1 year. During the succeeding 12 months, 17 of the rats developed tumours at a variety of sites, including 10 pulmonary adenocarcinomas, five hepatocellular carcinomas, four rhabdomyosarcomas and some others.

Hayashi *et al.* (1977) obtained pancreatic tumours in rats given single s.c. injections of monocrotaline (40 mg/kg).

4. Petasitenine

Hirono *et al.* (1977) fed this alkaloid continuously to rats in their drinking water. A solution of 0.01% was given to 11 rats of both sexes, initially 1 month old; 19 control rats were given tap water. Eight out of the 10 rats which survived beyond 160 days developed liver tumours. Five had hemangioendothelial sarcomas of the liver (one with lung metastasis); two of these rats, as well as three others, had liver cell adenomas. Among the 19 control rats, only one subcutaneous fibrosarcoma was observed. Similar liver tumours were not observed in another control group of 266 animals, but were seen in rats fed the source plant of the alkaloid, *Petasites japonicus* (Hirono *et al.*, 1973).

5. Retrorsine

Schoental *et al.* (1954) gave retrorsine to 14 rats in their drinking water (0.03 mg/ml), 3 days weekly for up to 24 months. Hepatomas were found in four out of 10 surviving rats. Out of four female rats, each given six i.p. injections (25–30 mg/kg) of retrorsine, one, which had also received betaine in its drinking water, had a hepatoma 23 months after the last injection (Schoental and Head, 1957). From a group of 95 weanling rats each given a single dose of retrorsine (30 mg/kg) by stomach tube, 29 survived more than 1 year; five of these had hepatomas (Schoental and Bensted, 1963). Members of another group, similarly treated, were later given whole-body X-irradiation: six out of 25 survivors had primary liver tumours, of which one was a hepatocarcinoma with metastases in the lung. Members of a third group were given retrorsine as before, preceded by partial hepatectomy. Hepatomas were found in two out of nine survivors.

A variety of other tumours were found in the foregoing groups of rats. It could be concluded that retrorsine is hepatocarcinogenic, although this has been disputed by Bull *et al.* (1968).

6. Retrorsine N-Oxide (Isatidine)

Twenty-two rats were given isatidine in the drinking water (0.03–0.05 mg/ml) 3 days weekly for about 20 months. The livers of 10 of these were said to contain multiple tumour foci varying up to 10 mm diameter Schoental et al. (1954). Among seven rats similarly treated and also given a diet containing supplementary choline, six had nodular hyperplasia of the liver, and four had whitish nodules, interpreted as trabecular hepatoma. Similar nodules were also seen in the liver of one rat out of five which had received one i.p. injection of isatidine (probably between 15 and 35 mg/kg) when young, followed by three skin applications weekly for 15 months. Isatidine is converted to retrorsine in the rat gut (Mattocks, 1971c), so if the latter is carcinogenic it would not be surprising if the former is, also.

Hyperplastic nodules were observed in livers of another series of rats given isatidine in the drinking water and later monitored by liver biopsy (Schoental and Head, 1957).

7. Riddelliine

This alkaloid has only once been tested (Schoental and Head, 1957), and the results were inconclusive (IARC, 1976). Twenty rats received riddelliine intermittently in the drinking water, and survivors were also given i.p. injections. Nodular livers were seen in some rats, but no hepatomas were diagnosed.

8. Senecionine (with Fuchsisenecionine)

The material tested (Habs et al., 1982; Habs, 1982) was a crude alkaloidal extract from the plant Senecio nemorensis ssp. fuchsii, containing 50% of fuchsisenecionine and 1% of senecionine. Rats of both sexes were given solutions of this extract by stomach tube, five times weekly for 104 weeks, at levels of 8 or 40 mg/kg. A dose-related tumour-inducing effect was observed. Female rats were more sensitive than males. Two liver tumours were found among 20 male rats given doses of 8 mg/kg, and five in 20 male rats given 40 mg/kg. Among 20 females given 8 mg/kg, there were 11 liver tumours, and in 20 given the higher dose there were 29 liver tumours. Of the 40 control rats, one (male) had a neoplastic nodule in the liver. The origins of tumours in the treated animals were classified as: hepatocellular, 19; cholangiogenic, 16; and haemangiogenic, 12.

The nature of the component of the mixture responsible for carcinogenicity was not discussed. Senecionine is known to be hepatotoxic and capable of being converted in rat liver into cytotoxic metabolites (cf. Chapter 6). Fuchsisene-

cionine is a saturated PA of a type not previously found to be cytotoxic. Therefore it is likely that the former alkaloid is the carcinogenic component. However,
this will require confirmation by testing the pure alkaloids individually. The
possibility that one alkaloid might have a synergistic action on the effects of the
other cannot be excluded; moreover, there appear to have been other, unknown
components in the mixture tested.

9. Seneciphylline

A mixture of PAs, extracted from *Senecio jacobaea* and said to consist largely
of seneciphylline, was tested by Campbell (1956), who gave it i.v. to chickens.
After repeated weekly doses of 20–35 mg/kg, liver tumours were reported in six
out of 18 birds.

10. Senkirkine

This alkaloid, which occurs in a number of plants (see Chapter 1), was extracted from flower buds of coltsfoot (*Tussilago farfara*), and given i.p. at a dose
level of 0.1 LD_{50} (22 mg/kg) to each of 20 male rats (Hirono *et al.*, 1979).
Injections were twice weekly for 4 weeks, then weekly for 52 weeks. Twenty
control rats received saline injections. The injection schedule was similar to that
used for lasiocarpine by Svoboda and Reddy (1972). All the treated rats survived
more than 290 days from the start of the experiment; nine of them developed
benign liver cell adenomas; there were no malignant liver tumours. No liver
tumours were observed in the control rats or in another group of 359 rats; thus it
was concluded that the adenomas were caused by senkirkine.

11. Symphytine

This alkaloid, extracted from comfrey (*Symphytum officinale*), was tested
alongside senkirkine (see above) by Hirono *et al.* (1979). An i.p. dose of 0.1
LD_{50} (13 mg/kg) was given to 20 rats, twice weekly for 4 weeks, then once a
week for 52 weeks. All rats survived more than 330 days: one had a liver cell
adenoma and three had hemangioendothelial sarcomas of the liver (two with lung
metastases). A large number of control rats (see senkirkine) had no liver tumours; thus the tumours were attributed to the effects of symphytine.

C. Carcinogenicity Tests on Plant Materials

In addition to experiments with PAs, some plants known to contain cytotoxic
PAs have been tested for carcinogenicity. The plants often contained more than

one alkaloid. Caution should be exercised in attributing the observed carcinogenic or cytotoxic effects to the alkaloids. Other constituents of the plant might be partly or wholly responsible for the observed activity, or might enhance or inhibit the effects of the alkaloid(s). Nevertheless, tests on plants are important, since it is these which sometimes find their way into human or animal food.

Relatively few plants containing PAs have been tested. About half of these have yielded insufficient tumours to warrant their being classed as carcinogenic; however, the results are suggestive, and justify further investigation.

1. Senecio aquaticus

This plant (which contains seneciphylline) was fed to 10 male rats by Dybing and Erichsen (1959). The diet contained 1% of plant material. The animals died or were killed after 216–291 days, and all had severely damaged liver. Whether some of the effects seen should be interpreted as hepatoma is open to question (Bull *et al.*, 1968).

2. Senecio jacobaea

Although alkaloidal extracts from this plant have been used for several experiments (Cook *et al.*, 1950; Schoental *et al.*, 1954; Campbell, 1956), the only experiment in which tumours were reported to have resulted from feeding the plant itself was that of Campbell (1956). The dried, milled plant (except roots) was given to chickens at a level of about 7% in the diet (calculated to represent about 1 mg alkaloids/day/bird); this level was halved after 1 month because of excessive toxicity. Forty-three chickens were fed for 14 weeks. Four birds (three males, one female) developed primary liver tumours.

In view of the widespread occurrence of this plant as a contaminant of farmland, a more thorough carcinogenicity test of this plant would seem to be needed.

3. Senecio nemorensis *ssp.* fuchsii

Preparations of this plant are marketed commercially as a herbal medicine (Habs *et al.*, 1982). The plant itself has not been tested, but a crude alkaloidal extract from it produced liver tumours when given to rats: see senecionine (above).

4. Senecio longilobus

Harris and Chen (1970) gave the dried plant to rats in an extensive series of feeding experiments. The best results were obtained when 100 rats were fed on alternate weeks for 1 year with a diet containing 0.5% of the plant. Among 47 rats surviving more than 200 days, malignant liver tumours were found in 14 males and three females; 16 of these had hepatocarcinomas and one had three angiocarcinomas. Rats fed the plant continuously often died earlier, and damage to the lungs, as well as the liver, was common. The authors pointed out the relative rarity of spontaneous liver tumours in these rats. The results demonstrate that *S. longilobus* is hepatocarcinogenic.

5. Petasites japonicus

Young flower stalks of this plant, of a kind of coltsfoot, which is used in Japan as a food or herbal remedy, were tested by Hirono *et al.* (1973). Twelve male and 15 female rats, initially aged 1 month, were fed a diet containing 4% of the dried plant material for 6 months, then a diet which contained 8% of coltsfoot on alternate weeks and a normal diet during the other weeks. Another group of 11 male and 8 female rats were kept on the 4% coltsfoot diet. Among 25 surviving rats in the first group, 11 developed liver tumours: three of these were hemangioendothelial sarcomas, six liver cell adenomas, and two hepatocellular carcinomas. In the second group, all rats survived more than 220 days, and eight had hemangioendothelial sarcomas, four had liver cell adenoma and one had hepatocellular carcinoma. There was no statistical significance in the difference in tumour incidence between the sexes. No liver tumours were seen in 14 control rats or in another group of 50 controls.

6. Tussilago farfara *(Coltsfoot)*

This plant, widely used as a herbal remedy, was investigated by Hirono *et al.* (1976). Twelve rats of both sexes were fed a diet containing 32% of coltsfoot (dried flower buds) for 4 days, then 16% coltsfoot until the end of the experiment (over 380 days). Five male and three female rats developed hemangioendothelial sarcoma of the liver. One of these also had hepatocellular adenoma and another had hepatocellular carcinoma. In a further 10 rats given 8% coltsfoot diet for 600 days, only one developed hemangioendothelial sarcoma of the liver. Eleven rats fed a 4% coltsfoot diet and 16 control rats given a normal diet developed no liver tumours, nor did 150 control rats in other experiments.

7. Symphytum officinale *(Russian Comfrey)*

This plant, widely used as a food and medicinal herb, was tested by Hirono *et al.* (1978). The plant, dried and milled, was fed to rats of both sexes: leaf material (8, 16 or 33% in the diet) was given for up to 600 days to 88 rats. Root

(0.5–4%) was given for up to 280 days or until death to 87 rats. Among rats given the leaf, 24 were found to have hepatocellular adenoma and one had hemangioendothelial sarcoma. Five papillomas and three carcinomas of the urinary bladder were also reported. Of the rats given comfrey root, 57 had hepatocellular adenoma and two had hemangioendothelial sarcoma. One papilloma and one carcinoma of the urinary bladder were seen. No liver tumours were seen in 129 control rats.

The incidence of hemangioendothelial sarcoma was much lower in rats fed comfrey than in those given *Petasites japonicus* or *Tussilago farfara,* and it was concluded that comfrey is a weaker carcinogen. However, it was pointed out that histological liver changes were frequently seen (but the dietary level of plant was relatively high in these experiments).

8. Senecio cannabifolius

A proportion of rats given 0.2% of this plant (dried leaves and stalks) in their diet developed hemangioendothelial sarcoma of the liver and liver cell adenoma (Hirono *et al.*, 1983). Both the tumour incidence and the hepatotoxicity of the plant were much greater in female than in male rats.

9. Farfugium japonicum

This is a popular edible plant in parts of Japan. Among rats given 20% in their diet, some developed hemangioendothelial sarcoma (some with lung metastasis) and liver cell adenoma (Hirono *et al.*, 1983).

D. Carcinogenicity Tests Leading to Tumours in Tissues Other Than Liver

In most carcinogenicity tests with PAs, interest has been centered mainly on liver carcinogenesis, although tumours in various other organs have been recorded. In a few experiments however, other tissues were mainly affected: these are considered below.

1. Pancreatic Tumours

Schoental *et al.* (1970) gave a mixture of alkaloids extracted from seeds of *Amsinckia intermedia,* and said to be a mixture of intermedine and lycopsamine, to 15 weanling male rats. The animals each received a single dose (500–1500 mg/kg). Three rats subsequently developed tumours of the pancreas: one adenoma and one adenocarcinoma of the islet cells, and one adenoma of the exocrine pancreas. One of six rats given crude alkaloids (300 mg/kg) from an Ethiopian specimen of *Heliotropium supinum* also developed an islet cell adenoma, as did one of two rats fed dried *H. supinum* in their diet for 1 month. The tumour-bearing rats were aged 26–31.5 months. There were no control rats in this experiment, but the authors claimed that such tumours are 'very rare' and were not observed in other control rats of similar ages. Schoental (1975a) reported finding further pancreatic tumours in male rats which had received heliotrine as weanlings. Out of four rats given this alkaloid, one which survived 27 months after two intragastric doses (each 230 mg/kg) had an islet cell tumour. It also had an adenoma of the pituitary, but so did three of eight control rats. Twelve further rats were pretreated with nicotinamide, then given heliotrine (230 mg/kg); six of these were similarly dosed again, 5 days later. Among six rats surviving 22 months or more, three had pancreatic islet cell tumours. The rarity of such tumours was again emphasised. The role, if any, of nicotinamide in the above experiments is not clear, although it was said to prevent liver necrosis.

Hayashi *et al.* (1977) gave single s.c. doses (40 mg/kg) of monocrotaline to 80 young male rats. Thirty control rats received saline. Among the long-term survivors, 16 out of 23 treated rats had insuloma, compared with one of 28 controls.

The foregoing results seem to demonstrate that at least some PAs can cause the development of pancreatic tumours.

2. Tumours of the Central Nervous System

Schoental and Cavanagh (1972) fed *Heliotropium ramosissimum* (containing heliotrine) to a female rat during pregnancy and after parturition, and (6 months later) to five female offspring. One of the latter developed a spinal cord tumour. Ten newborn rats were given s.c. injections of the non-hepatotoxic amino-alcohol retronecine. One of these (given 600 mg/kg) had a spinal cord tumour 6.6 months later. In a third experiment, hydroxysenkirkine was given to five male weanling rats. One developed three similar but apparently separate brain tumours, 14.5 months later.

These results are clearly too limited to establish an association between intake of PAs and the development of central nervous system (CNS) tumours. It should be pointed out that in no other carcinogenicity tests on PAs or plants containing them has an excessive level of CNS tumours been reported.

E. Carcinogenicity of Pyrrolizidine Alkaloid Metabolites and Analogous Synthetic Compounds

Many pyrrolizidine ester alkaloids are converted by hepatic microsomal enzymes into cytotoxic metabolites. Primary metabolites are highly reactive pyrrolic esters. These can hydrolyse to secondary metabolites which are less reactive, more water soluble pyrrolic alcohols, and capable of longer life and being distributed more widely through the body. Some of these compounds and synthetic analogues have been tested for carcinogenicity.

1. Pyrrolic Esters

a. Dehydromonocrotaline (Monocrotaline Pyrrole). This highly reactive ester is prepared chemically from monocrotaline (Mattocks, 1969a). Hooson and Grasso (1976) gave each of 20 rats injections of dehydromonocrotaline (30 μg in tricaprylin) into the right flank, twice weekly for 30 weeks. A second group of rats received 60-μg doses; 20 control rats were given tricaprylin alone. Sarcomas developed at the injection site in three of the rats receiving 30 μg and in four receiving 60 μg doses of the pyrrole. Two sarcomas were also seen at the injection site in control rats. There were no tumours in internal organs. These results were insufficient to demonstrate unequivocally that dehydromonocrotaline was carcinogenic. Initially, the pyrrole caused local necrosis near the injection site, followed later by the appearance of reparative granulation tissue. However, the latter appeared sooner than when known proximate carcinogens had been injected, and was better organised than after the latter. The only respect in which the tissue reaction resembled that due to carcinogens was that some fibroblasts were greatly enlarged. Thus the connective-tissue reaction was intermediate between the response produced by conventional injury and that induced by carcinogens.

Mattocks and Cabral (1979) painted the backs of male BALB/c mice with dehydromonocrotaline at 1- to 2-week intervals; 37 treatments (each 1 μmol) yielded no skin tumours in 16 mice. Fourteen control mice treated with the solvent (acetone) had no tumours. In a later study (Mattocks and Cabral, 1982), 11 female LACA mice were each given 47 applications of 2.5 μmol dehydromonocrotaline; one developed a malignant skin tumour. However, when 10 similarly treated mice were subsequently given 61 twice-weekly applications of the tumour promoter croton oil at the same site, half of them developed malignant skin tumours. This result was significant when compared with a control group (10 mice) in which one skin tumour was seen.

From these results it appears that dehydromonocrotaline may be regarded as an 'incomplete' carcinogen, requiring the effects of a promoter to manifest its carcinogenic potential.

b. Dehydroretrorsine. This pyrrole, a putative metabolite of retrorsine and prepared chemically from it, was applied to the skins of male BALB/c mice at 1- to 2-week intervals; 33 treatments of 0.5 or 1 μmol failed to yield any skin tumours in 15 mice which survived for up to 60 weeks (Mattocks and Cabral, 1979). Thus under these conditions the pyrrole was not found to be carcinogenic.

(412) R = CMe₃
(280) R = Me

(400)

(201)

c. 1-Methyl-2,3-bistrimethylacetoxymethylpyrrole (412). This synthetic ester has similar alkylating reactivity to dehydromonocrotaline, but is somewhat more stable. In skin tests with male BALB/c mice, three out of nine treated mice developed skin tumours (Mattocks and Cabral, 1979). The treatment began with 10 topical doses, each of 1 μmol of the pyrrole; however, this caused severe skin damage with ulceration, scarring and permanent depilation. Subsequently, 25 doses of 0.5 μmol were given. In a later experiment (Mattocks and Cabral, 1982), 22 female LACA mice each received up to 47 topical doses (0.5 μmol) of the compound. The skin was damaged as before, and malignant skin tumours developed in 19 out of 21 surviving mice: a highly significant result.

The two hydrolysis products from this ester, pivalic acid and 1-methyl-2,3-bishydroxymethylpyrrole, when similarly tested, yielded much lower levels of tumours, confirming that the carcinogen was the intact pyrrolic ester.

d. 1-Methyl-2,3-bisacetoxymethylpyrrole (280) and 5-Methyl-1-phenyl-2,3-bisacetoxymethylpyrrole (400). These pyrroles, similar to the previous compound but more chemically unstable, were also tested on the skin of male BALB/c mice: no skin tumours were produced (Mattocks and Cabral, 1979).

2. Pyrrolic Alcohols

a. Dehydroretronecine (Retronecine Pyrrole) (201). Known to be a secondary metabolite formed from monocrotaline (Hsu *et al.,* 1973b), this pyrrole was given s.c., biweekly, to male rats (Allen *et al.,* 1975; Shumaker *et al.,* 1976b). The dosage was 20 mg/kg for 4 months, and then (because the treated rats gained weight at a much lower rate than controls) it was reduced to 10 mg/kg for a further 8 months. Twelve months following the cessation of treatment, 39 out of 60 rats had rhabdomyosarcomas at the injection site; metastases were present in five of these. In another experiment, female 'Swiss' mice were used (Johnson *et al.,* 1978). The dose schedule was once a week for 4 weeks, then twice more after 6 months. Each dose was 20 mg/kg (about 5 μmol/mouse). Out of 16 mice given the pyrrole topically on their shaved backs, six developed skin tumours. The s.c. injection yielded 13 tumour-bearing mice among 21 animals. Out of 55 mice given both topical and s.c. doses, 28 had skin tumours.

The same topical dose schedule yielded only one skin tumour in a group of 34 female BALB/c mice, and no skin tumours among 17 of these mice given 65 weekly topical doses (Mattocks and Cabral, 1982); however, five animals with malignant skin tumours were among 20 female LACA mice given 47 topical doses of dehydroretronecine (each, 5 μmol).

Thus, dehydroretronecine appears to be a direct-acting carcinogen for the skin of rats and mice, although some strains of mouse are less susceptible than others.

Dehydroheliotridine, an enantiomer of dehydroretronecine and a metabolite of heliotridine-based PAs, was given to rats as nine i.p. injections (60–76.5 mg/kg) over 32 weeks by Peterson *et al.* (1983). The animals' lifespan was shortened and a variety of tumours occurred, none of which were seen in the control group. It was concluded that dehydroheliotridine could be responsible for much or all of the carcinogenicity of its parents PAs.

b. 1-Methyl-2,3-bishydroxymethylpyrrole. This synthetic compound is an alkylating agent very similar to dehydroretronecine. Application to the skin of 12 female LACA mice (47 weekly doses of 5 μmol) led to three malignant skin tumours; however, only two skin tumours were seen in 10 similarly treated mice which were subsequently given 61 applications of croton oil (Mattocks and Cabral, 1982). The compound appears to be weakly carcinogenic.

c. 5-Methyl-1-phenyl-2,3-bishydroxymethylpyrrole. A few tumours were seen in the skin of LACA mice painted with this compound (Mattocks and Cabral, 1982). The result was not statistically significant.

F. Discussion

In spite of early doubts, it is now generally accepted that some PAs, and plants containing them, are carcinogenic to experimental animals (chiefly rats), and they are listed among the relatively few carcinogens known to occur in higher plants (Hirono, 1979, 1981). They are not powerful carcinogens. Generally, chronic dosing schedules have been most successful in producing tumours. Regardless of the route of administration, whether given orally or injected by various routes, tumours have most frequently appeared in the liver and not at the site of injection or of topical application of the alkaloid. This strongly suggests that the proximate carcinogens are metabolites formed in the liver. It is possible that the powerful, persistent antimitotic action in the liver, which is the most characteristic feature of chronic poisoning by hepatotoxic PAs, may inhibit tumour formation. McLean (1970) pointed out that interrupted, as opposed to continuous, dosing of alkaloids produced livers containing nodules of regenerating parenchymal cells: presumably these arose from centres which had managed to escape from the mitotic inhibition. Some support for this has been provided by Harris and Chen (1970), who obtained their highest tumour yield by feeding rats a *Senecio* diet on alternate weeks rather than continually; by Hirono *et al.* (1973), who obtained more liver tumours in rats given coltsfoot intermittently than in animals given the same amount continuously; and also by Hirono *et al.* (1978), in feeding experiments with comfrey. On the other hand, Rao and Reddy (1978) found that liver tumours developed in rats during the continual administration of lasiocarpine, in contrast with an earlier claim (Svoboda and Reddy, 1972) that interrupted dosing of this alkaloid was necessary for the expression of carcinogenicity. Also, treatment with lasiocarpine did not prevent the hepatocarcinogenicity of aflatoxin B_1 in rats (Reddy and Svoboda, 1972).

It has been suggested (Schoental and Bensted, 1963) that primary liver tumours might develop from the persistent, enlarged parenchymal cells. However, it is questionable whether such cells are capable of undergoing reproduction, and it seems more likely that tumours arise from cells which have escaped the antimitotic effect of the alkaloid.

A variety of tumour types have been found in the livers of rats given PAs (see Table 11.2). The carcinogenic response appears to depend not only on the nature of the alkaloid but also on its administration schedule (Rao and Reddy, 1978), and on whether it is accompanied by plant material. For example, *Tussilago farfara* produced hemangioendothelial sarcomas whereas its constituent alkaloid, senkirkine, produced benign adenomas. Repeated i.p. administration of lasiocarpine produced hepatocellular tumours, whereas when given in the diet this alkaloid produced vascular as well as hepatocellular tumours. Some treatments, as with *Senecio nemorensis* alkaloids, have yielded a high proportion of cholangiogenic tumours.

It is not clear whether a sex difference really exists in the response of rats to pyrrolizidine hepatocarcinogenesis. Male rats are more susceptible than females to pyrrolizidine hepatotoxicity (Chapter 7). Harris and Chen (1970) found a much higher incidence of hepatic tumours in male rats than in females given *Senecio longilobus;* on the other hand, Habs *et al.* (1982) found female rats significantly more susceptible to *S. nemorensis* alkaloids, which included senecionine, also present in *S. longilobus!*

Tumours of tissues other than liver have been reported in most carcinogenicity tests on PAs. The significance of these compared with levels in control animals does not appear to have been established, except possibly in the case of pancreatic tumours. Thus, whether extrahepatic tissues are susceptible to carcinogenesis by PAs or by PA metabolites formed locally or transported from the liver is in general not known, but cannot be ruled out. It should also be pointed out that the rat is the only mammalian species in which PAs have been tested: the relative susceptibility of other species, including humans, is not known.

G. Molecular Structure and Carcinogenic Activity

The data at present available are insufficient to enable relationships to be established between the types of tumours seen in livers of rats given PAs and the

TABLE 11.2 Liver Tumours Found in Rats Given PAs and Plant Materials (Condensed Results from Various Authors)

| Alkaloid or plant given | Alkaloid type | | Route of administration | Number of rats at autopsy | Number and types of liver tumours found | References |
	Necine	Type of ester[a]				
Clivorine	otonecine	cycl	p.o.	12	2 hemangioendothelial sarcoma 6 neoplastic nodules	Kuhara et al. (1980)
Lasiocarpine	heliotridine	di	i.p.	18	11 hepatocellular carcinoma	Svoboda and Reddy (1972) Rao and Reddy (1978)
			p.o.	20	9 angiosarcoma 7 hepatocellular carcinoma	
Monocrotaline	retronecine	cycl	p.o.	77	24 hepatocellular carcinoma	Newberne and Rogers (1973)
Petasitenine	otonecine	cycl	p.o.	10	5 hemangioendothelial sarcoma 3 adenoma	Hirono et al. (1977)
Retrorsine	retronecine	cycl	various		'hepatoma'	Schoental et al. (1954); Schoental and Head (1957)

					'hepatoma' and 'nodular hyperplasia'	
Isatidine	retronecine	cycl	various		19 hepatocellular tumours 16 cholangiogenic tumours 12 haemangiogenic tumours	Schoental et al. (1954); Schoental and Head (1957)
Senecionine (+fuchsisenecionine)	retronecine (platynecine)	cycl mono	p.o.	60		Habs et al. (1982)
Senkirkine	otonecine	cycl	i.p.	20	9 adenoma	Hirono et al. (1979)
Symphytine	retronecine	di	i.p.	20	1 adenoma 3 hemangioendothelial sarcoma	Hirono et al. (1979)
Senecio longilobus (seneciphylline, retrorsine, etc.)	retronecine	cycl	p.o.	47	16 hepatocellular carcinoma 1 angiosarcoma	Harris and Chen (1970)
Petasites japonicus (petasitenine)	otonecine	cycl	p.o.	46	11 hemangioendothelial sarcoma 10 adenoma	Hirono et al. (1973)
Tussilago farfara (senkirkine)	otonecine	cycl	p.o.	12	3 hepatocellular carcinoma 8 hemangioendothelial sarcoma	Hirono et al. (1976)
Symphytum officinale (various)	retronecine	mono + di	p.o.	175	81 adenoma 3 hemangioendothelial sarcoma	Hirono et al. (1978)

[a] cycl, Macrocyclic diester; di, "open" diester; mono, monoester.

molecular structure of the PAs. Table 11.2 is provided for those who wish to speculate. Carcinogenic activity has been demonstrated in PAs which are macrocyclic or 'open' diesters, and in which the amino alcohol moiety is retronecine, heliotridine or otonecine. A common factor is that these are all esters of unsaturated necines and capable of being metabolized to pyrrolic esters in mammalian liver (see Chapter 6).

It is for this reason that some pyrroles which are putative metabolites of PAs, and also analogous synthetic compounds, have been tested for carcinogenicity.

The results with pyrrolic esters have been equivocal. Dehydromonocrotaline is capable of damaging skin, but appears to be an 'incomplete' carcinogen (Hooson and Grasso, 1976; Mattocks and Cabral, 1979, 1982): its carcinogenicity was only demonstrated after the subsequent multiple application of a tumour promoter (croton oil). However, it is worth considering that the effects of this very reactive compound when applied to skin might not be the same as those following its metabolic formation *within* the liver cell.

The synthetic compound 1-methyl-2,3-bistrimethylacetoxymethylpyrrole (412), which is a chemically similar alkylating agent to dehydromonocrotaline, is clearly carcinogenic to mouse skin (Mattocks and Cabral, 1982). This compound is more cytotoxic than dehydromonocrotaline and can cause severe skin damage: possibly it is the promoting action of this injury which leads to tumour formation. The reason for the enhanced toxicity and carcinogenicity of this compound might lie in its higher chemical stability (Mattocks and Cabral, 1979) and greater lipophilicity (favouring penetration of the cell) compared with the other pyrrolic esters tested.

Dehydroretronecine (DHR) (201), a secondary metabolite of monocrotaline (and probably of other retronecine-based PAs) is carcinogenic to skin; it is regarded as a proximate carcinogenic metabolite since it acts at the site of application, whereas monocrotaline acts at various remote sites (Shumaker *et al.*, 1976b). The effectiveness of DHR as a carcinogen when applied topically to mouse skin appears to be markedly dependent on the strain of animal, as can be seen from Table 11.3. Its relative weakness as a carcinogen is evident when comparison is made with 7,12-dimethylbenzanthracene. It is worth noting that the doses of DHR injected in some carcinogenicity tests have been relatively high in comparison with carcinogenic doses of monocrotaline. For example, it is estimated that the total dose of DHR given to each rat during 1 year by Shumaker *et al.* (1976b) was about 2.1 mmol/kg body weight; the monocrotaline given in a parallel experiment was about 0.4 mmol/kg; the carcinogenic dose of monocrotaline given by Newberne and Rogers (1973) was about 1.2 mmol/kg. The foregoing amounts of DHR could not possibly be generated from these monocrotaline doses, even if conversion was total. It might be inferred that DHR has relatively low activity as a proximate carcinogen, and that the hepatocar-

TABLE 11.3 Skin Tumours in Mice Given Topical Applications of Dehydroretronecine

Mouse strain	Effective number of animals	Maximum number of applications	Maximum total dose of DHR (μmol)	Mice with skin tumours		Reference
				Number	%	
"Swiss"	16	6	31	6	38	Johnson et al. (1978)
BALB/c	34	6	31	1	3	Mattocks and Cabral (1982)
BALB/c	17	65	325	0	0	Mattocks and Cabral (1982)
LACA	20	47	235	5	25	Mattocks and Cabral (1982)
For comparison: using 7,12-dimethylbenz[a]anthracene						
BALB/c	20	24	9	18	90	Mattocks and Cabral (1982)

cinogenicity of monocrotaline is at least partly contributed by other metabolite(s) (dehydromonocrotaline?).

DHR, like its parent alkaloids, has antimitotic activity (Allen and Hsu, 1974; Mattocks and Legg, 1980); thus it is possible that an interrupted dose schedule of DHR might be more effectively carcinogenic than continuous treatments.

III. ANTITUMOUR ACTIVITY OF PYRROLIZIDINE ALKALOIDS, DERIVATIVES AND ANALOGUES

A. Introduction

Because of their cytotoxic and antimitotic properties, pyrrolizidine alkaloids have attracted interest as potential antitumour agents, although their hepatotoxicity would seem to present an obstacle to their therapeutic usefulness. A number of PAs, or plants containing them, have been tested for their effects on tumours in animals *in vivo,* or tumour cells *in vitro.* Another approach has been to test whether PAs administered to animals can diminish the incidence of tumours produced by another carcinogen given at the same time.

Because cytotoxic metabolites of PAs, responsible for liver damage, might be active against tumour cells, some semi-synthetic and synthetic pyrrolic alkylating agents related to these metabolites have also been tested.

B. Effects of PAs on Tumour Induction by Other Carcinogens

Rogers and Newberne (1971) treated rats with sufficient lasiocarpine to significantly inhibit cell division for over 4 months, then fed them a diet containing the carcinogen N-2-fluorenylacetamide (AAF). The alkaloid prevented neither the hyperplasia induced by AAF nor the induction of hepatocarcinoma. Evidently, either the AAF was able to overcome the inhibition of cell division imposed by lasiocarpine, or it was acting at sites not affected by the alkaloid. Likewise, Reddy and Svoboda (1972) found that in rats, lasiocarpine did not prevent the initiation of liver tumours by aflatoxin B_1 given in the diet. The presence of giant liver cells showed that lasiocarpine was exerting antimitotic activity. It was suggested that a proportion of hepatocytes which escaped the influence of the alkaloid were affected by the aflatoxin.

Newberne and Rogers (1973) found that when rats were given monocrotaline and aflatoxin B_1 at the same time, the two carcinogens acted synergistically, giving a higher incidence of liver tumours than after the individual compounds. Bykorez (1969) found that the incidence of malignant liver tumours in rats given the carcinogen p-dimethylaminoazobenzene was about halved when animals were also fed 'heliotrope' seeds. The seeds themselves caused chronic liver damage, but it is not clear whether this or the inhibition of carcinogenesis was due to the presence of a pyrrolizidine alkaloid.

Thus from the evidence available, it is not clear whether PAs are able to inhibit the induction of liver tumours by other carcinogens.

C. Antitumour Effects

1. Various Alkaloids

Tests on the tumour-inhibitory activity of 18 PAs and some derivatives have been reported by Culvenor (1968) and further discussed by Bull et al. (1968). Nine alkaloids which displayed significant activity against one or more of four tumour systems were: crispatine, fulvine, heliotrine and its N-oxide, lasiocarpine, monocrotaline, senecionine, spectabiline and supinine. These are all capable of being metabolized in animals to pyrrolic metabolites, though to different extents. High toxicity to the host animals may have prevented some other alkaloids from being given at an effective level. In addition, the 9-chloro derivative of heliotridine (317) showed activity: this is an allylic chloride, itself possessing alkylating reactivity. The tumours affected were: adenocarcinoma 755; sarcoma 180; Walker 256 intramuscular and Walker 256 subcutaneous.

	R¹	R²
(74)	H	H
(413)	—SO—	
(414)	—CHOEt—	
(267)	Ac	Ac

Senecionine and its N-oxide were found to be the principles in Senecio triangularis extracts active against Walker 256 (intramuscular) tumours (Kupchan and Suffness, 1967). Monocrotaline was the constituent of Crotalaria spectabilis extracts found to be active against adenocarcinoma 755 in mice (Kupchan et al., 1964). Chinese workers (Tumour Research Group, 1974, 1975) found that monocrotaline from Crotalaria assamica (see: Group of Crotalaria plant research, 1974) inhibits tumour cells in mice. Wang et al. (1981) studied structure–antitumour relationships for retusine, usaramine, monocrotaline, and some semisynthetic derivatives of monocrotaline, using mouse sarcoma 180, Walker carcinosarcoma 256 and Lewis lung carcinoma 615 as test systems. They found that the sulphite (413) and ethoxymethylene derivative (414) of monocrotaline have more toxicity but less antitumour activity than monocrotaline (74). Di-acetylmonocrotaline (267) has enhanced antitumour activity but not toxicity. Usaramine and the quaternary methiodide of monocrotaline are inactive, but the saturated alkaloid retusine (154) is active. These results appear to show that an unsaturated necine ester is not essential for antitumour activity. Monocrotaline and some derivatives were also tested against sarcoma 180 and other tumours by Huang *et al.* (1980); some were said to be active but with considerable toxicity.

The otonecine ester crosemperine was tested by Sharma and Hebborn (1968) against the Ehrlich ascites tumour in mice: tumour inhibition of borderline sig-nificance was observed after subacute administration at toxic level (five daily doses of 25 mg/kg). Extracts of *Senecio tenuifolius,* which contains senkirkine and senecionine, among other alkaloids (Bhakuni and Gupta, 1982), are active against L-1210 lymphoid leukemia in mice (Dhawan *et al.,* 1977).

2. Indicine N-oxide (INO) (207)

This is the only PA which up to now has undergone clinical trials as an anticancer drug.

(207)   (415)

Extracts of *Heliotropium indicum* were found effective against Walker 256 carcinosarcoma in rats and leukemia 1210 in mice (Kugelman *et al.,* 1976); the active principle was recognised as the *N*-oxide of indicine, an alkaloid previously isolated from this plant (Mattocks *et al.,* 1961). Indicine *N*-oxide exhibited significant antitumour activity in experimental animals, with much lower hepato-toxicity than other PAs.

The compound was investigated in two series of human patients. Courses of INO were given to 29 patients with advanced solid tumours, as daily i.v. infu-sions, each of up to 3 g/m^2 for 5 days (Kovach *et al.,* 1979). There was no objective therapeutic response. The major toxic effect was cumulative bone marrow suppression. Forty percent of the drug was eliminated unchanged in the urine within 24 h; 2% was in the reduced form (indicine base). In a series of 10 patients with advanced acute leukemia, two patients were described as exhibiting a complete remission, and one a partial remission (Letendre *et al.,* 1981). The dosage was similar to that given to the earlier patients, and was well tolerated. Toxic effects were again myelosuppression, and possibly the jaundice and liver failure seen in two patients.

It is not clear whether INO itself or a metabolite from it is the active anti-tumour agent. Powis *et al.* (1979a) studied the metabolism of INO to indicine base. An i.v. dose of INO given to rabbits was eliminated rapidly in the urine, mostly within the first 2 h. Only a small amount of indicine base was excreted; more was formed after oral administration. It was found that INO [like other pyrrolizidine *N*-oxides (Mattocks, 1971c)], is reduced anaerobically to the basic alkaloid mainly by the gut flora, although to a small extent in the liver. Powis *et al.* (1979b) concluded, on the basis of experiments with leukemic mice, that antitumour action is unlikely to result from conversion of INO to indicine base. When given i.v. to mice, indicine itself had much less antitumour action than INO, and INO given orally was ineffective, even though much indicine was formed from it. There was very little reduction of INO to indicine by tumour cells, compared with liver microsomes.

Neither does the antitumour action on INO appear to be due to pyrrolic metabolites. Hepatotoxicity of PAs is associated with the formation of pyrrolic metabolites in the liver (see Chapter 6). INO is probably not metabolized directly to pyrroles: reduction to the basic alkaloid would first be necessary (Mattocks, 1971c). Indicine itself is very water soluble, and its metabolism to pyrroles is low (Powis *et al.*, 1979a; Mattocks, 1972b). Accordingly INO has little effect on the liver: its main toxic effect in the rabbit is to cause bone marrow hypoplasia (Powis *et al.*, 1979a). Heliotrine *N*-oxide is less effective against leukemia in mice than INO in tests which lead to the formation of similar amounts of the respective alkaloid bases (Powis *et al.*, 1979b), even though heliotrine itself is an effective antitumour agent (Culvenor, 1968) and more readily converted to pyrroles (Mattocks, 1972b).

Thus, the mechanism of the antitumour action of INO remains unknown.

A semisynthetic PA *N*-oxide with some structural similarity to INO was prepared in 1982 (Gelbaum *et al.*, 1982). This is 9(2-hydroxy-2-phenylbutyryl)-retronecine *N*-oxide (415): its acid moiety bears some structural similarity to the necic acid of INO (trachelantheic acid). The compound is claimed to be more active than INO in a preliminary screening test against P388 lymphatic leukemia.

It is interesting to speculate whether the unexplained results of Taylor and Taylor (1963), who prolonged the lifespan of tumour-bearing mice by feeding them an extract of comfrey (*Symphytum officinale*), might be due to *N*-oxides of PAs similar to indicine.

3. Pyrrolic Derivatives

When pyrrolic derivatives were found to be cytotoxic metabolites of PAs (see Chapter 6), the possibility was recognised that these alkylating agents might be effective against tumour cells. Tests (Culvenor *et al.*, 1969a) showed that several dehydro-alkaloids are inhibitory to KB cell culture, although their activity is at best only similar to that of some PAs themselves. Thus, dehydromonocrotaline and dehydrosenecionine (ED_{50} 17 and 14 µg/ml, respectively) are more active than the parent alkaloids but dehydroheliotrine and dehydrolasiocarpine are less active (ED_{50} > 100 µg/ml) than heliotrine and lasiocarpine (ED_{50} 15 and 25 µg/ml, respectively), probably because of decomposition under the test conditions. Several pyrrolic alcohols are active, especially 1-hydroxymethyldihydro-5H-pyrrolizine (345) (ED_{50} 1.9 µg/ml). Dehydroheliotridine (272) inhibits the multiplication of Ehrlich ascites cells in mice if added to the cell suspension before injection. In a patent application (Culvenor *et al.*, 1970c), several dihydropyrrolizine alcohols prepared from retronecine or heliotridine were named as potential antitumour, antiviral or immunosuppressive agents. These were dehydroretronecine, dehydroheliotridine and the aldehydes (329) and (416).

(345) R = H   (329) R = OH
(272) R = OH   (416) R = H

Anderson and co-workers (1982, and references therein) have prepared a series of synthetic pyrrolic and dihydropyrrolizine esters for evaluation as tumour inhibitors. The latter compounds are analogues of pyrrolizidine alkaloid metabolites, with substituents to reduce the chemical (alkylating) reactivity and enable them to reach sites such as the cell nucleus where they might react with DNA (Anderson and Corey, 1977). The pyrrolic esters, which have essentially similar chemical properties to the above, bear a family resemblance to pyrrolic derivatives of synthetic pyrrolizidine alkaloid analogues (synthanecine carbamates), first described by Mattocks (1971d).

Antileukaemic activity of a series of dihydropyrrolizine esters was tested against L1210 and P388 cells in mice (Anderson and Corey, 1977); most compounds were active in the second system, and a few in the first. One (417) was 'curative' at 12.5 mg/kg; another (418) was active at doses down to 0.78 mg/kg. Related pyrrole carbamates (419) and (420) were active against human tumour xenografts in nude mice (Anderson, 1982). Further tests on biscarbamates of three pyrroles and four dihydropyrrolizines against eight tumour systems were reported by Anderson *et al.* (1982). These compounds show a different pattern of activity from 'conventional' alkylating agents, in being less effective against the L1210 system.

	R^1	R^2
(417)	Cl	Cl
(418)	H	F

(419) R = OMe
(420) R = nBu

D. General Comments

Tumour inhibiting action is shown by three groups of compounds related to PAs: by some of the alkaloids themselves; by indicine N-oxide (and possibly related compounds); and by pyrrolic alkylating agents. In the last category, those which have lower alkylating reactivity, and are therefore the most stable *in vivo*, are the most effective; this has been exploited in a series of synthetic analogues. It is possible that these pyrroles are further metabolized to proximate antitumour agents (cf. Guengerich and Mitchell, 1980), but in view of their chemical reactivity and their effectiveness in *in vitro* systems it seems more likely that their action is associated with their ability to alkylate cell constituents. It is not clear whether the antitumour activity of pyrrolic alkylating agents, or of PAs which can be metabolized to them, is connected with their powerful antimitotic activity; the latter evidently does not prevent the development of liver tumours in animals given other hepatocarcinogens.

The antitumour action of at least some pyrrolizidines is evidently not due to pyrrolic metabolites. Activity is shown by the saturated PA retusine, which is not susceptible to metabolism to a pyrrole. The antitumour action of indicine N-oxide is not associated with its conversion either to indicine or to a pyrrole. Thus, at least some PAs can inhibit tumours via another mechanism, at present not understood, but probably dependent on the stereochemistry of the molecule and perhaps involving the N-oxide function. It is interesting to speculate whether the antitumour actions of some PAs might be enhanced by metabolism to N-oxides, usually considered a detoxication pathway (Mattocks, 1972a). This could not always be true; for example, heliotrine N-oxide is less active than heliotrine (Culvenor, 1968) or than indicine N-oxide (Powis *et al.*, 1979b).

12

Relationships between Structure, Metabolism and Toxicity

I. INTRODUCTION

Toxic actions of PAs have been described in previous chapters. So also have routes of metabolism and the chemical and biological properties of metabolites. The aim of the present chapter is to show how this information can shed light on molecular mechanisms of PA cytotoxicity.

As a starting point, the following assumptions have been made:

1. PA cytotoxicity is due to metabolites, not the alkaloids themselves.

2. Toxic metabolites are 'pyrrolic' derivatives formed by dehydrogenation of PAs.

3. Toxic effects result from interactions of these active metabolites with tissue constituents.

4. Metabolic activation of PAs in animals occurs chiefly, if not entirely, in the liver: toxic actions in other organs are thus due to metabolites originating in the liver.

5. The toxicity of PA *N*-oxides is due to their first being reduced *in vivo* to the basic alkaloids, which are then activated as above.

It remains possible that other routes of PA activation and toxicity have yet to be discovered; meanwhile the above premises afford a basis for PA toxicity which is consistent with present knowledge.

II. ESSENTIAL FEATURES OF CYTOTOXIC PAs AND METABOLITES

A. Minimum Structural Requirements

Certain minimum structural features are needed for a PA or analogue to be potentially hepatotoxic; other factors, discussed later, determine whether, and to what extent, such toxicity is expressed in a given situation.

The necessary features, illustrated in Fig. 12.1, are as follows.

1. An unsaturated (3-pyrroline) ring. The second (pyrrolidine) ring is not essential, as shown by its absence from the synthetic analogues.

2. One or (preferably) two hydroxyl groups, each attached to the pyrroline ring via one carbon atom.

3. At least one of these hydroxyls is esterified. Monoester PAs are usually much less hepatotoxic than similar diesters; toxicity is still less if the 7-hydroxyl is absent; some macrocyclic diesters have the highest toxicity; synthanecine monoesters are not hepatotoxic.

4. The acid moiety has a branched chain. PA esters with unbranched acids may be hepatotoxic, but only weakly so. Exceptions are some synthetic carbamate and phosphate esters, which can have high toxicity.

B. Features of Pyrrolic Metabolites Necessary for Cytotoxicity

Cytotoxic metabolites possess a pyrrole ring with one or two ester or hydroxyl groups attached to it, each via one carbon atom. Toxic actions of these metabo-

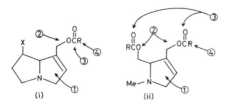

Fig. 12.1.   Structural features of (i) a PA and (ii) a PA analogue essential for hepatotoxicity. X = H or HO or RCOO.

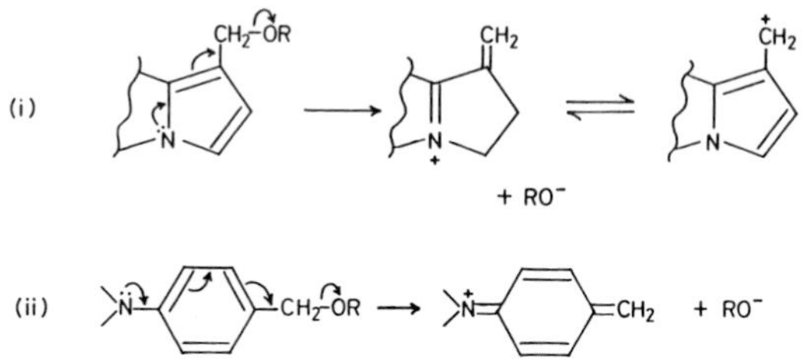

Fig. 12.2.   Activation of (i) a pyrrole, indole or dihydropyrrolizine alcohol (R = H) or ester (R = R'CO), or (ii) dimethylaminobenzyl alcohol (R = H) or its ester (R = R'CO), to an electrophilic (alkylating) agent.

lites are associated with their chemical alkylating reactivity, rather than with a particular molecular structure or stereochemistry, although these latter may play a subsidiary role. The alkylating reactivity is due to the oxygen function being made chemically labile by its relationship with the nitrogen (Fig. 12.2). Esters which are similarly activated by being attached to an indole or dimethylaminobenzyl structure can show the same kind of cytotoxicity as PA metabolites (Mattocks and Driver, 1983). From the above it is apparent that in (405), the 3-ester group is activated but the ester on the 2-sidechain is not; also the pyrrolic metabolite (421) from the non-toxic PA platyphylline cannot act as an alkylating agent.

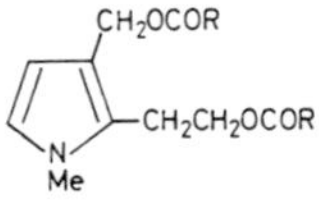

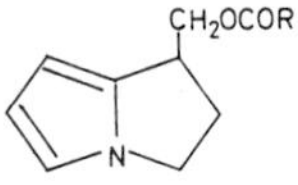

III. METABOLIC ACTIVATION AND DETOXIFICATION

A. Metabolic Routes

   PAs are metabolized chiefly in the liver; major routes are summarized in Fig. 12.3. In many animal species, PAs are activated by dehydrogenation to pyrrolic (dihydropyrrolizine) esters, which are primary toxic metabolites. Hydrolysis of these gives alcohols (dehydronecines) which are secondary toxic metabolites. Major detoxication pathways are ester hydrolysis and N-oxidation: these lead to products which are more water soluble than the parent alkaloids, and can be excreted via urine or bile, either unchanged or as conjugates.

   Most of these pathways are not reversible. However, N-oxides of PAs can be reduced to the basic alkaloids in the gut. This occurs mainly after oral ingestion: such reduction may occur to a minor extent in other tissues, such as liver.

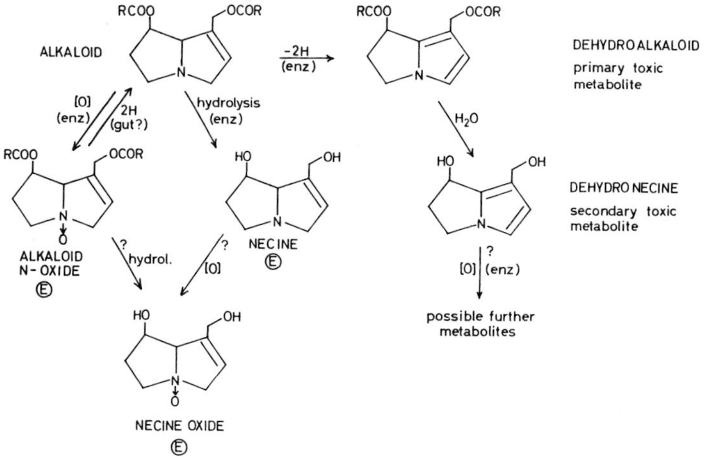

Fig. 12.3. Some metabolic routes important for activation or detoxication of a hepatotoxic PA. Circled E, readily excreted.

Possible additional routes of metabolism might include hydroxylation of the acid moiety (Eastman and Segal, 1982), and the microsomal oxidation of pyrrolic metabolites to further products, perhaps lacking the pyrrole ring; such products, besides being excreted (White, 1977), might contribute to the alkaloid's toxic actions (Guengerich and Mitchell, 1980).

B. Fate of Active Metabolites

Some possible disposal routes of toxic pyrrolic metabolites are illustrated in Fig. 12.4. The primary (pyrrolic ester) metabolite can react rapidly with nucleophilic tissue constituents to give relatively stable products, some of which may be associated with specific cytotoxic actions. Reaction products with soluble nucleophiles could be excreted, whereas products with tissue-bound nucleophilic groups, e.g. in proteins or nucleic acids, may remain in the liver for a considerable time. If it is stable enough, some primary metabolite may escape via the bloodstream to reach and damage other tissues, particularly the lungs. Some reactive pyrrole might polymerize. Some may hydrolyse to the less reactive pyrrolic alcohol. This also can alkylate nucleophiles in the cell, but more slowly and probably more selectively than the ester. It can also become more widely distributed throughout the body, survive longer, and so be capable of damaging tissues remote from the liver.

A further possibility arises from the reversibility of reactions of pyrrolic alkylating agents with certain nucleophiles (e.g. nicotinamide) (Mattocks and Bird, 1983b). Products of such reactions could thus be envisaged as 'reservoir'

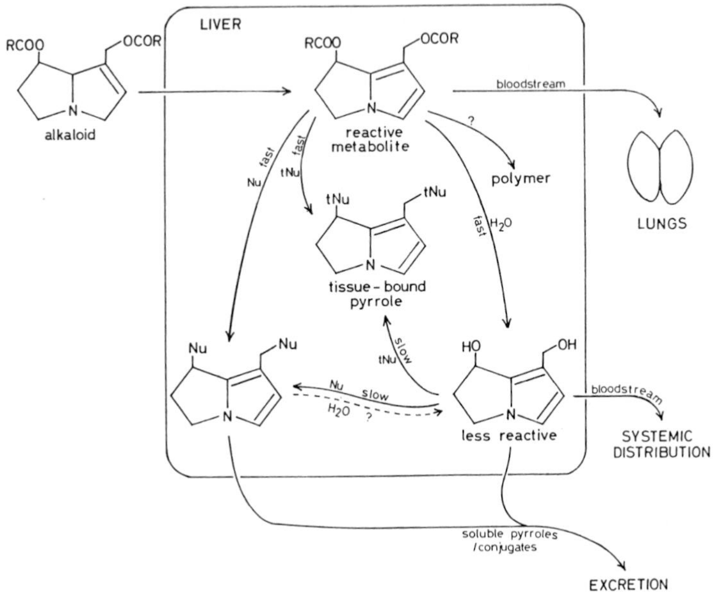

Fig. 12.4. Hypothetical fate of pyrrolic metabolites formed in the liver from a diester PA. Nu, Soluble nucleophile; tNu, tissue-bound nucleophilic group.

compounds, having a longer life than the primary pyrrolic metabolite but capable of being hydrolysed to release the toxic dehydro-necine; this pathway is represented by a broken line in Fig. 12.4.

IV. FACTORS DETERMINING TOXICITY

Toxicity depends firstly on the alkaloid having a structure which is potentially convertible to toxic metabolites, and secondly on the ability of the animal's enzymes to bring about this conversion. The latter has been discussed at length in Chapter 6; the present section deals with activation from the point of view of the alkaloid.

It is assumed that the single primary event from which stem all the toxic actions of PAs considered here, is formation of the dehydro alkaloid. Toxicity is influenced by two sets of factors: those affecting the formation of pyrrolic metabolites, and those affecting the toxicity of these metabolites, once formed. These are discussed below, followed by a consideration of features associated with particular toxic actions.

A. Factors Affecting Formation of Toxic Metabolites

Although the factors below are under separate headings, they are in many ways interdependent.

1. Lipid Solubility

High lipophilicity makes PAs more susceptible to activation by hepatic microsomal enzymes (Mattocks and Bird, 1983a). Conversely, highly water soluble PAs (e.g. indicine) tend to be excreted, have low toxicity, and form relatively low levels of pyrrolic metabolites (Mattocks, 1981a).

2. Base Strength

In general, the more weakly basic PAs tend to be more lipophilic (because a higher proportion is unionised at physiological pH); hence they are often (but not invariably) among the more toxic. For a discussion, see Bull *et al.* (1968).

3. The Acid Moiety

PAs which yield substantial amounts of pyrrolic metabolites are all esters. Unesterified necines give rise to few if any pyrroles. Diesters of retronecine can give up to 15 times more pyrroles in the liver than corresponding monoesters (Mattocks, 1981a), possibly because of their higher lipophilicity as well as greater resistance to hydrolysis.

Most toxic monoester PAs are esterified at the primary 9-hydroxyl; however, 7-monoesters can also be weakly hepatotoxic, as shown by 7-angelylheliotridine (Culvenor *et al.*, 1976b).

The hydrolysis of PAs is a detoxication pathway. Therefore resistance of a PA to hydrolysis enhances its hepatotoxicity. Hydrolytic detoxication *in vivo* is a major reason why PAs given to animals often lead to lower pyrrolic metabolite levels in the liver than are expected from their rates of metabolism to pyrroles in vitro (Mattocks and Bird, 1983a). The allylic primary ester group is especially susceptible to enzymic (as well as to base-catalysed) hydrolysis (Mattocks, 1982b). Hydrolysis is limited if access to the ester groups is restricted by steric hindrance. This can be caused by bulky carbon substituents (Mattocks, 1982b) or by acyl substituents such as acetoxy groups (Mattocks, 1982a) in the acid moiety close to the ester linkage. Hindrance is also enhanced by α,β-unsaturation in the acid moiety (as in tiglate esters), because this restricts the flexibility of the acid chain. Two acid moieties if sufficiently close together may provide mutual hindrance. Macrocyclic diesters have greater rigidity in the acid moiety. Such an alkaloid will be open to hydrolysis if hindering groups are held back from the esters, as in monocrotaline (Bull *et al.*, 1968). However, in other alkaloids steric hindrance due to suitably placed substituents might be increased by such rigidity.

Hydrolysis *in vivo* is also limited if the acid moiety is of a type which inhibits esterase activity, e.g. in PA analogues which are phosphate or carbamate esters (Mattocks, 1971d, 1978b).

In addition to its direct influence on metabolism, the structure of the acid moiety may also favour chemical changes, such as the opening of the second ester linkage in macrocyclic diester PAs (after initial metabolic hydrolysis of the primary ester), through intramolecular transesterification (cf. Bull *et al.*, 1968).

4. The Necine Moiety

The amino alcohol (necine) must have an unsaturated (3-pyrroline) ring: esters of saturated necines either are metabolized to only small amounts of pyrroles (Mattocks and White, 1971a; Mattocks, 1978b), or in a few instances (e.g. platyphylline) form large amounts of non-toxic pyrroles (Mattocks and White, 1971b).

The basic moiety in a toxic PA may either be a 'conventional' necine such as retronecine (183), or the seco-base, otonecine (186). Both types are converted by hepatic microsomal enzymes to similar toxic pyrroles but this process appears to be less efficient for otonecine-based PAs, which must first be N-demethylated (Mattocks and White, 1971a; Culvenor *et al.*, 1971a). Thus, the retronecine ester retrorsine gives about seven times the amount of pyrrole in the liver of young rats, and is about seven times as toxic, as the corresponding otonecine ester hydroxysenkirkine (Schoental, 1970; Mattocks, 1982b).

The type and location of the necine hydroxyls can considerably influence the hydrolytic detoxication of PAs. Hydrolysis is inhibited when two ester groups are close together. This is demonstrated by a comparison of the ditiglates of retronecine, heliotridine and synthanecine A. The first is least hydrolysed *in vivo;* the last, which has two relatively mobile primary esters, is so susceptible to hydrolysis that practically no pyrrolic metabolite is formed from it in rats unless their esterase activity has been inhibited (Mattocks, 1981a).

5. The Total Alkaloid Structure

The molecular structure and conformation of the alkaloid are important in determining the balance between the major routes of metabolism, namely hydrolysis, N-oxidation, and dehydrogenation to a pyrrole, by controlling the accessibility of metabolic sites to the appropriate enzymes. There is evidence to

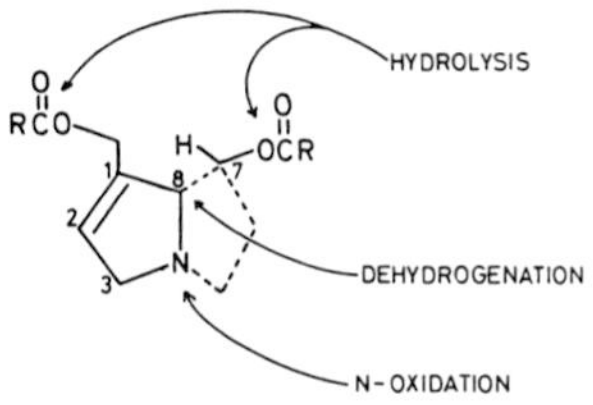

Fig. 12.5. A PA (seen from the α side) showing proposed major sites of metabolic attack. The unsaturated ring is in the plane of the paper.

suggest that pyrroles are formed as a result of microsomal hydroxylation at C8 on the pyrrolizidine nucleus (Mattocks and Bird, 1983a). Thus the most important targets are probably those shown in Fig. 12.5. This model suggests than an optimum structure for conversion to a pyrrole will be a compromise of steric factors which are to some extent mutually incompatible. Thus, for maximum pyrrole formation there should be a minimum of steric hindrance near the 'top' of the necine moiety, between the ester groups (in a diester); but this is inconsistent with maximum hindrance to ester hydrolysis. Moreover, the steric effect of the acid moiety is unlikely to hinder N-oxidation, another microsomal reaction, without causing greater hindrance to oxidation at C8. A comparison of metabolites formed from PAs by rat liver microsomes *in vitro* shows that the ratio of pyrroles to *N*-oxides is highest from macrocyclic diesters and monoesters, and lowest from 'open' diesters which would give greatest hindrance at C8 (Mattocks and Bird, 1983a).

B. Factors Affecting the Toxicity of Pyrrolic Metabolites, Once Formed

1. Chemical Reactivity

The toxic actions of an active metabolite depend on its chemical reactivity and the number and stereochemistry of its reactive groups, which will determine the target sites chiefly affected and thus the specific nature of the toxicity.

Stability and chemical reactivity are of course opposite sides of the same coin. High reactivity is associated with instability, particularly in an aqueous environment. Stability depends on the compound having low enough reactivity to resist decomposition; toxic action depends on its being reactive enough to form bonds with appropriate tissue constituents. The half lives of pyrrolic esters with respect to hydrolysis of alkylation reactions are often only a few minutes or even seconds (Mattocks, 1969a, 1978a). A metabolite with a short half life will tend to react with cell constituents close to its point of formation. An extremely unstable metabolite may lose its activity before it can react with vital cell constituents (such as DNA). On the other hand, a metabolite with low reactivity may be largely dispersed before much of it can react within its tissue of origin. Thus there is probably an optimum level of reactivity for maximum biological action in the liver.

To reach more distant organs a metabolite must first cross the cell membrane and survive contact with blood and vascular tissue. To act within other cells it must penetrate a further cell membrane. Thus the toxic actions of reactive primary metabolites are generally restricted to the liver cells in which they are formed, to lung vascular endothelium, and perhaps to the myocardium (cf. Mattocks and Driver, 1983).

There are three ways in which reactive pyrrolic esters may be rapidly deactivated, independently of enzyme action (see Fig. 12.4).

1. By alkylating nucleophilic tissue constituents such as glutathione to give products which are not harmful and which can be excreted. Thus liver tissue with an increased glutathione (GSH) content is more resistant to PA toxicity (White, 1976).

2. By polymerization. Pyrrolic esters readily polymerize *in vitro* to inert, insoluble products. This has not been demonstrated to occur in the liver of animals, but it appears possible. Pyrrolic alcohols (dehydro-necines) can similarly polymerize: amoebae treated with dehydroretronecine become coated with polymer (unpublished results of M. J. Ord and A. R. Mattocks). A comparison of PAs and analogues shows that compounds whose pyrrolic metabolites have the greatest tendency to polymerize are the least acutely hepatotoxic (Mattocks, 1978b).

3. By hydrolysis. Reaction with water gives the more water soluble pyrrolic alcohols. These are more stable, so they have relatively weak hepatotoxicity but they can penetrate to tissues not accessible to the pyrrolic esters. However, they too have a limited life; thus the DNA-damaging action of dehydroretronecine *in vitro* is more than halved after it has been in aqueous solution for 2 h (Ord *et al.*, 1985).

A few examples will illustrate the variations in toxicity which can be attributed mainly to differences in chemical stability of the primary metabolites. Synthanecine A bisdiethylphosphate is metabolized in rat liver to large amounts of pyrrole, but it causes no acute hepatotoxicity, probably because its labile pyrrolic ester metabolite is hydrolysed before it can react with cell constituents (Driver and Mattocks, 1984); at the opposite extreme, the metabolite from synthanecine A bis-*N,N*-diethylcarbamate is so stable that it survives to cause widespread damage to extrahepatic tissues (Mattocks, 1975). The pyrrole from retrorsine damages the liver of rats but is too unstable for appreciable amounts to reach the lungs, whereas the pyrrole from monocrotaline often reaches and damages the lungs (Mattocks, 1972b).

2. Functionality

Pyrrolic metabolites from PAs may possess either one or two reactive ester groups, or one ester plus one less reactive hydroxyl. Their hydrolysis products can have either one or two hydroxyl groups capable of alkylating action. Thus these pyrroles may be either mono- or bifunctional alkylating agents. Bifunctional pyrroles can cross link macromolecules such as DNA (White and Mattocks, 1972). This might lead to pathological effects not possible with monofunctional pyrroles. However, even if cross linkage is not essential for toxicity, it can enhance toxicity by leading to a stronger attachment between pyrrole and macromolecule, more resistant to repair processes.

Several examples exist of differences between the toxic actions of mono- and bifunctional pyrroles. One is the action of pyrrolic esters on lungs. Many such compounds given i.v. to rats cause acute lung damage, but only those which are bifunctional alkylating agents (including one with an ester and an alcohol function) lead to chronic lung toxicity (Mattocks, 1978b; Mattocks and Driver, 1983). Bifunctionality also enhances the antimitotic effects of PAs and pyrroles: bifunctional pyrrolic alcohols have greater antimitotic action on cultured liver cells, coupled with lower cytotoxicity, than pyrrolic monoalcohols (Mattocks and Legg, 1980). The bis-N-ethylcarbamates of synthanecine A (373) and synthanecine D (394) both cause acute liver necrosis in rats; the first, which can be metabolized to a bifunctional alkylating agent, gives rise to giant hepatocytes but the second, which can only form a monofunctional alkyating agent, does not (Mattocks, 1978b; see Chapter 8, Table 8.2). Nevertheless bifunctionality is not essential for the antimitotic activity of PAs, as shown by the development of giant liver cells in rats given the monofunctional PA supinine (Mattocks, 1978b).

Bifunctional pyrrolic alcohols are much more active than corresponding monoalcohols in inducing sister chromatid exchange in human lymphocytes, but not in their mutagenicity to *Salmonella typhimurium* (Ord *et al.*, 1985).

C. Features Associated with Particular Toxic Actions of PAs

1. Acute Hepatotoxicity

The following evidence indicates that acute liver necrosis is due to primary pyrrolic metabolites, possessing at least one reactive ester group.

1. The liver, where severe necrosis occurs, is the only tissue exposed to high levels of primary PA metabolites.

2. Bound pyrrole metabolite levels in the liver, reached quickly after alkaloid ingestion, correlate well with acute hepatoxicity (Mattocks, 1972b).

3. Pyrrolic alcohols (dehydronecines) are not *acutely* hepatotoxic, even when they are given to animals in larger doses than could be formed from corresponding PAs.

4. Intramesenteric vein injections of pyrrolic esters derived from PAs cause acute liver damage at lower dose levels than the corresponding PAs.

2. Chronic Hepatotoxicity

The antimitotic action on liver, which can lead to chronic development of persistent, giant hepatocytes, is produced both by pyrrolic ester metabolites, such as dehydromonocrotaline (Hsu *et al.*, 1973a), and by pyrrolic alcohols, like dehydroheliotridine and dehydroretronecine (Peterson *et al.*, 1972; Allen and Hsu, 1974). Alkylation by either pyrrolic esters or the corresponding alcohols could lead to similar reaction products, and since the esters are readily hydrolysed it is not possible to tell whether their effects on mitosis are direct or due to the secondary metabolites. The intact dehydro-pyrrolizidine ring is not necessary for this action: synthetic bis-hydroxymethyl pyrroles have similar effects in rats (Mattocks, 1981b). The presence of one alkylating group is sufficient but bifunctional pyrroles are more active; thus it seems that antimitotic action is not due to chemical cross linking.

The pyrrole–tissue interaction alone does not lead to chronic megalocytosis of the liver: a stimulus for cell division is also required. This already exists in very young animals; otherwise it may be provided by the acute liver damage due to primary PA metabolites, or to some other cause.

Thus, chronic PA liver damage results from a combination of two events: (a) the interaction of a pyrrolic metabolite with a cell constituent, as yet unidentified; and (b) sufficient non-specific liver injury to stimulate cell replication.

3. Carcinogenicity

The hepatocarcinogenicity of some PAs, usually manifested after their intermittent chronic administration, is probably due to pyrrolic metabolites. The secondary metabolite dehydroretronecine is a direct acting carcinogen to skin (Shumaker *et al.*, 1976b; Mattocks and Cabral, 1982); at least one synthetic pyrrole ester is even more potently carcinogenic to skin, but pyrrolic esters derived from the natural PAs monocrotaline and retrorsine either are not carcinogenic or are active only with the aid of a tumour promoter (Mattocks and Cabral, 1979, 1982). Thus on present evidence it is not certain which pyrrole species is responsible for PA hepatocarcinogenicity, but it appears that secondary metabolites would be capable of this action.

4. Lung Toxicity

PA lung damage is ascribed to reactive metabolites carried from the liver, especially pyrrolic esters (dehydro-alkaloids) (Butler et al., 1970). Chronic pneumotoxicity is associated particularly with PAs (e.g. monocrotaline) whose dehydro derivatives have two alkylating centres (Mattocks and Driver, 1983), and which are stable enough for a proportion to reach the lungs intact. Pyrrolic alcohols (dehydro-necines) only have weak, chronic pneumotoxicity (cf. Peterson et al., 1983) differing somewhat from the effects produced by PAs themselves (Huxtable et al., 1978) but they may contribute to these effects. Acute lung damage, seen only after massive PA intoxication, is attributed to pyrrolic ester metabolites.

For discussions see Chapter 8, Sections III,C and III,D.

5. Toxicity in Other Tissues

Chronic heart damage (right ventricular hypertrophy) can follow PA-induced lung injury (see Chapter 8); however, pyrrolic esters may have direct action on heart tissue (Mattocks and Driver, 1983). Brain damage (Hooper, 1972) and other neurological effects are secondary results of severe liver injury by PAs. Lung vascular tissue acts as an effective trap for reactive pyrrolic esters, so it is unlikely that much of the already depleted primary metabolite reaching the lungs can escape to other tissues. The more stable pyrrolic alcohols are capable of affecting a range of tissues, particularly in young animals (Peterson et al., 1972; Shumaker et al., 1976a). Enlarged cells occasionally seen in the kidneys and (rarely) the pancreas suggest the antimitotic action characteristic of these of pyrroles. It is likely that the embryotoxicity of heliotrine in rats is due to dehydroheliotridine formed in the maternal liver; this is known to cross the placenta (Peterson and Jago, 1980). Dehydroretronecine can affect the gastric mucosa (Hsu et al., 1973b, 1976), but generally, the amounts of pyrrolic alcohols which are necessary to cause marked pathological effects are greater than those likely to be formed even from severely hepatotoxic doses of PAs. The report of Hsu et al. (1973b) that 'large quantities' of dehydroretronecine are circulating in rats given monocrotaline is misleading: distribution studies in animals given radioactive PAs and analogues, compared with corresponding pyrrolic alcohols, lead to the conclusion that the latter are not major circulating metabolites (see Chapter 6).

V. CONCLUDING REMARKS

Relationships between the toxicity of PAs and their structural and chemical features are complex and are not yet fully understood. However, enough is now known to allow at least partial explanations of some of the toxic properties of individual PAs. It is hoped that this chapter will provide a basis for the better understanding of PA toxicity, for recognition of potentially hazardous compounds, and for planning further experimental work.

References

AASEN, A. J. AND CULVENOR, C. C. J. (1969a). *Aust. J. Chem.* **22,** 2657–2662.

AASEN, A. J. AND CULVENOR, C. C. J. (1969b). *J. Org. Chem.* **34,** 4143–4147.

AASEN, A. J., CULVENOR, C. C. J. AND SMITH, L. W. (1969). *J. Org. Chem.* **34,** 4137–4143.

AASEN, A. J., CULVENOR, C. C. J. AND WILLING, R. I. (1971). *Aust. J. Chem.* **24,** 2575–2580.

ABDULLAEV, U. A., RASHKES, YA. V. AND YUNUSOV, S. YU. (1974a). *Khim. Prir. Soedin.,* 538–539; *Chem. Abstr.* **82,** 16972 (1975).

ABDULLAEV, U. A., RASHKES, YA. V. AND YUNUSOV, S. YU. (1974b). *Khim. Prir. Soedin.,* 620–626; *Chem. Abstr.* **82,** 73270 (1975).

ADAMS, R. AND GIANTURCO, M. (1956a). *J. Am. Chem. Soc.* **78,** 398–400.

ADAMS, R. AND GIANTURCO, M. (1956b). *J. Am. Chem. Soc.* **78,** 1919–1921.

ADAMS, R. AND GIANTURCO, M. (1956c). *J. Am. Chem. Soc.* **78,** 1922–1925.

ADAMS, R. AND GIANTURCO, M. (1956d). *J. Am. Chem. Soc.* **78,** 1926–1928.

ADAMS, R. AND GIANTURCO, M. (1956e). *J. Am. Chem. Soc.* **78,** 4458–4464.

ADAMS, R. AND GIANTURCO, M. (1956f). *J. Am. Chem. Soc.* **78,** 5315–5317.

ADAMS, R. AND GIANTURCO, M. (1957). *J. Am. Chem. Soc.* **79,** 174–177.

ADAMS, R. AND GOVINDACHARI, T. R. (1949a). *J. Am. Chem. Soc.* **71,** 1180–1186.

ADAMS, R. AND GOVINDACHARI, T. R. (1949b). *J. Am. Chem. Soc.* **71,** 1953–1956.

ADAMS, R. AND GOVINDACHARI, T. R. (1949c). *J. Am. Chem. Soc.* **71,** 1956–1960.

ADAMS, R. AND HERZ, W. (1950). *J. Am. Chem. Soc.* **72,** 155–157.

ADAMS R. AND LOOKER, J. H. (1951). *J. Am. Chem. Soc.* **73,** 134–136.

ADAMS, R. AND ROGERS, E. F. (1939). *J. Am. Chem. Soc.* **61,** 2815–2819.

ADAMS, R. AND ROGERS, E. F. (1941). *J. Am. Chem. Soc.* **63,** 537–541.

ADAMS, R. AND VAN DUUREN, B. L. (1952). *J. Am. Chem. Soc.* **74,** 5349–5351.

ADAMS, R. AND VAN DUUREN, B. L. (1953a). *J. Am. Chem. Soc.* **75,** 2377–2379.

ADAMS, R. AND VAN DUUREN, B. L. (1953b). *J. Am. Chem. Soc.* **75,** 4631–4636.

ADAMS, R. AND VAN DUUREN, B. L. (1953c). *J. Am. Chem. Soc.* **75,** 4638–4642.

ADAMS, R. AND VAN DUUREN, B. L. (1954). *J. Am. Chem. Soc.* **76,** 6379–6383.

ADAMS, R., ROGERS, E. F. AND SPRULES, F. J. (1939). *J. Am. Chem. Soc.* **61**, 2819–2821.

ADAMS, R., CARMACK, M. AND ROGERS, E. F. (1942a). *J. Am. Chem. Soc.* **64**, 571–573.

ADAMS, R., HAMLIN, K. E., JELINEK, C. F. AND PHILIPS, R. F. (1942b). *J. Am. Chem. Soc.* **64**, 2760.

ADAMS, R., VAN DUUREN, B. L. AND BRAUN, B. H. (1952). *J. Am. Chem. Soc.* **74**, 5608–5611.

ADAMS, R., GIANTURCO, M. AND VAN DUUREN, B. L. (1956). *J. Am. Chem. Soc.* **78**, 3513–3519.

ADAMS, R., CULVENOR, C. C. J., ROBINSON, C. N. AND STINGL, H. A. (1959). *Aust. J. Chem.* **12**, 706–711.

AFZELIUS, B. A. AND SCHOENTAL, R. (1967). *J. Ultrastruct. Res.* **20**, 328–345.

AKRAMOV, S. T., KIYAMITDINOVA, F. AND YUNUSOV, S. YU. (1961a). *Dokl. Akad. Nauk Uzb. SSR,* 30–32; *Chem. Abstr.* **60**, 16209 (1964).

AKRAMOV, S. T., KIYAMITDINOVA, F. AND YUNUSOV, S. YU. (1961b). *Dokl. Akad. Nauk Uzb SSR* **18**, 35–37; *Chem. Abstr.* **61**, 4700 (1964).

AKRAMOV, S. T. SAMATOV, A. S. AND YUNUSOV, S. YU. (1964). *Dokl. Akad. Nauk Uzb. SSR* **21**, 28.

AKRAMOV, S. T., KIYAMITDINOVA, F. AND YUNUSOV, S. YU. (1965). *Dokl. Akad. Nauk Uzb. SSR* **22**, 35–38; *Chem. Abstr.* **63**, 16770 (1965).

AKRAMOV, S. T. KIYAMITDINOVA, F. AND YUNUSOV, S. YU. (1967). *Khim. Prir. Soedin.,* 288–289; *Chem. Abstr.* **68**, 893 (1968).

AKRAMOV, S. T., SHADMANOV, Z., SAMATOV, A. AND YUNUSOV, S. YU. (1968). *Khim. Prir. Soedin.,* 258; *Chem. Abstr.* **70**, 44831 (1969).

ALDERSON, T. AND CLARK, A. M. (1966). *Nature (London)* **210**, 593–595.

ALDRIDGE, W. N. AND REINER, E. (1972). "Enzyme Inhibitors as Substrates." North-Holland Publ., Amsterdam.

ALEKSEEV, V. S. (1961a). *Farm. Zh. (Kiev)* **16**, 39–44; *Chem. Abstr.* **56**, 13011 (1962).

ALEKSEEV, V. S. (1961b). *Med. Prom-st.SSSR* **15**, 27–29; *Chem. Abstr.* **57**, 7384 (1962).

ALEKSEEV, V. S. (1964). *Chem. Abstr.* **63**, 5943 (1965).

ALEKSEEV, V. S. AND BAN'KOVS'KII (1965). *Farm. Zh. (Kiev)* **20**, 49–54; *Chem. Abstr.* **64**, 9997 (1966).

ALEXANDER, R. S. AND BUTLER, A. R. (1976). *J. Chem. Soc., Perkin Trans. 2,* 696–701.

ALIEVA, SH. A., ABDULLAEV, U. A., TELEZHENETSKAYA, M. V. AND YUNUSOV, S. YU. (1976). *Khim. Prir. Soedin.,* 194–196; *Chem. Abstr.* **85**, 108841 (1976).

ALLEN, J. R. AND CARSTENS, L. A. (1970). *Exp. Mol. Pathol.* **13**, 159–171.

ALLEN, J. R. AND HSU, I. C. (1974). *Proc. Soc. Exp. Biol. Med.* **147**, 546–550.

ALLEN, J. R., CARSTENS, L. A. AND OLSON, B. E. (1967). *Am. J. Pathol.* **50**, 653–667.

ALLEN, J. R., CARSTENS, L. A. AND KATAGIRI, G. J. (1969). *Arch. Pathol.* **87**, 279–289.

ALLEN, J. R., CARSTENS, L. A. AND NORBACK, D. H. (1970a). *Toxicol. Appl. Pharmacol.* **16**, 800–806.

ALLEN, J. R., CARSTENS, L. A., NORBACK, D. H. AND LOH, P. M. (1970b). *Cancer Res.* **30**, 1857–1866.

ALLEN, J. R., CHESNEY, C. F. AND FRAZEE, W. J. (1972). *Toxicol. Appl. Pharmacol.* **23**, 470–479.

ALLEN, J. R., HSU, I. C. AND CARSTENS, L. A. (1975). *Cancer Res.* **35**, 997–1002.

AMES, M. M. AND POWIS, G. (1978). *J. Chromatogr.* **166**, 519–526.

AMIL, H. AND ATES, O. (1971). *Chem. Abstr.* **77**, 72582 (1972).

ANDERSON, C. (1981). *Lancet* **1**, 1424.

ANDERSON, W. K. (1982). *Cancer Res.* **42**, 2168–2170.

ANDERSON, W. K. AND COREY, P. F. (1977). *J. Med. Chem.* **20**, 812–818.

ANDERSON, W. K., CHANG, C.-P., COREY, P. F., HALAT, M. J., JONES, A. N., MCPHERSON, H. L., NEW, J. S. AND RICK, A. C. (1982). *Cancer Treat. Rep.* **66**, 91–97.

ANON. (1949). *Res. Today (Eli Lilly & Co.)* **5**(3), 55–73.

ANON. (1979a). *Br. Med. J.* **1**, 574–575.

ANON. (1979b). *Med. Lett.* **21**, 29–31.

ANON. (1979c). *Br. Med. J.* **1**, 598.

ANON. (1981). "Folk Remedies that Work: Coltsfoot and White Horehound", April 26, p. 47. Sunday Times Magazine, London.

APLIN, R. T., BENN, M. H. AND ROTHSCHILD, M. (1968). *Nature (London)* **219**, 747–748.

ARAYA, O. AND GONZALEZ, S. (1979). *Gac. Vet.* **41**, 743–745.

ARAYA, O., HERNANDEZ, J.R., ESPINOZA, A. E., AND CUBILLOS, V. (1983). *Vet. Hum. Toxical.* **25**, 4–7.

ARMSTRONG, S. J. AND ZUCKERMAN, A. J. (1970). *Nature (London)* **228**, 569–570.

ARMSTRONG, S. J. AND ZUCKERMAN, A. J. (1972). *Br. J. Exp. Pathol.* **53**, 138–144.

ARMSTRONG, S. J., ZUCKERMAN, A. J. AND BIRD, R. G. (1972). *Br. J. Exp. Pathol.* **53**, 145–149.

ARSECULERATNE, S. N., GUNATILAKA, A. A. L. AND PANABOKKE, R. G. (1981). *J. Ethnopharmacol.* **4**, 159–177.

ASADA, Y. AND FUYURA, T. (1982). *Planta Med.* **44**, 182.

ASADA, Y. AND FURUYA, T. (1984a). *Chem. Pharm. Bull.* **32**, 475–482.

ASADA, Y. AND FURUYA, T. (1984b). *Chem. Pharm. Bull.* **32**, 4616–4619.

ASADA, Y., FURUYA, T., SHIRO, M. AND NAKAI, H. (1982a). *Tetrahedron Lett.* **23**, 189–192.

ASADA, Y., FURUYA, T., TAKEUCHI, T. AND OSAWA, Y. (1982b). *Planta Med.* **46**, 125–126.

ASPREY, G. F. AND THORNTON (1955). *West Indian Med. J.* **4**, 145–168.

ATAL, C. K. (1978). *J. Natl. Prod.* **41**, 312–326.

ATAL, C. K. AND SAWHNEY, R. S. (1973). *Indian J. Pharm.* **35**, 1–12.

ATAL, C. K., KAPUR, K. K., CULVENOR, C. C. J. AND SMITH, L. W. (1966a). *Tetrahedron Lett.*, 537–544.

ATAL, C. K., SHARMA, R. K., CULVENOR, C. C. J. AND SMITH. L. W. (1966b). *Aust. J. Chem.* **19**, 2189– 2191.

ATAL, C. K., CULVENOR, C. C. J., SAWHNEY, R. S. AND SMITH, L. W. (1967). *Aust. J. Chem.* **20**, 805–808.

ATAL, C. K., SAWHNEY, R. S., CULVENOR, C. C. J. AND SMITH. L. W. (1968). *Tetrahedron Lett.*, 5605–5608.

ATAL, C. K., CULVENOR, C. C. J., SAWHNEY, R. S. AND SMITH, L. W. (1969). *Aust. J. Chem.* **22**, 1773–1777.

AVANZI, S. (1961). *Cariologia* **14**, 251–261.

AVANZI, S. (1962). *Caryologia* **15**, 351–356.

BALE, N. M. AND CROUT, D. H. G. (1975). *Phytochemistry* **14**, 2617–2622.

BARGER, G. AND BLACKIE, J. J. (1936). *J. Chem. Soc.* 743–745.

BARBOUR, R. H., AND ROBINS, D. J. (1985). *J. Chem. Soc.*, Perkin Trans. **1**, 2475–2478.

BARGER, G. AND BLACKIE, J. J. (1937). *J. Chem. Soc.*, 584–586.

BARGER, G., SESHADRI, T. R., WATT, H. E. AND YABUTA, T. (1935). *J. Chem. Soc.*, 11–15.

BARNES, J. M., MAGEE, P. N. AND SCHOENTAL, R. (1964). *J. Pathol. Bacteriol.* **88**, 521–531.

BARON, R. L., CASTERLINE, J. L., AND ORZEL, R. (1966). *Toxicol. Appl. Pharmacol.* **9**, 6–16.

BARRI, M. E. S. AND ADAM, S. E. I. (1981). *J. Comp. Pathol.* **91**, 621–627.

BATRA, V. AND RAJAGOPALAN, T. R. (1977). *Curr. Sci.* **46**, 141.

BATRA, V., GANDHI, R. N. AND RAJAGOPALAN, T. R. (1975). *Indian J. Chem.* **13**, 989–990.

BATZINGER, R. P., SUH-YEN, L. OU. AND BUEDING, E. (1978). *Cancer Res.* **38**, 4478–4485.

BENN, M., DEGRAVE, J., GNANASUNDERAM, C. AND HUTCHINS, R. (1979). *Experientia* **35**, 731–732.

BERNAYS, E., EDGAR, J. A. AND ROTHSCHILD, M. (1977). *J. Zool.* **182**, 85–87.

BHACCA, N. S. AND SHARMA, R. K. (1968). *Tetrahedron* **24**, 6319–6326.

BHAKUNI, D. S. AND GUPTA, S. (1982). *Planta Med.* **46**, 251.

BHATTACHARYYA, K. J. (1965). *J. Pathol. Bacteriol.* **90**, 151–161.

BICK, Y. A. E. (1970). *Nature (London)* **226**, 1165–1167.

BICK, Y. A. E. AND CULVENOR, C. C. J. (1971). *Cytobios* **3**, 245–255.

BICK, Y. A. E. AND JACKSON, W. D. (1968). *Aust. J. Biol. Sci.* **21**, 469–481.

BICKEL, M. H. (1969). *Pharmacol. Rev.* **21**, 325–355.

BINGLEY, J. B. (1968). *Anal. Chem.* **40**, 1166–1167.

BIRECKA, H. AND CATALFAMO, J. L. (1982). *Phytochemistry* **21**, 2645–2651.

BIRECKA, H., FROHLICH, M. W., HULL, L. AND CHASKES, M. J. (1980). *Phytochemistry* **19**, 421–426.

BIRECKA, H., CATALFAMO, J. L. AND EISEN, R. N. (1981). *Phytochemistry* **20**, 343–344.

BIRECKA, H., FROHLICH, W. AND GLICKMAN, L. M. (1983). *Phytochemistry* **22**, 1167–1171.

BIRNBAUM, G. I. (1974). *J. Am. Chem. Soc.* **96**, 6165–6168.

BIRNBAUM, K. B. (1972). *Acta Crystallogr., Sect. B* **B28**, 2825.

BIRNBAUM, K. B., KLASEK, A., SEDMERA, P., SNATZKE, G., JOHNSON, L. F. AND SANTAVY, F. (1971). *Tetrahedron Lett.*, 3421–3424.

BLACK, D. N. AND JAGO, M. V. (1970). *Biochem. J.* **118**, 347–353.

BLACKIE, J. J. (1937). *Pharm. J.* **138**, 102–104.

BOHLMANN, F., KNOLL, K.-H., ZDERO, C., MAHANTA, P. K., GRENZ, M., SUWITA, A., EHLERS, D., LE VAN, N., ABRAHAN, W.-R. AND NATU, A. A. (1977a). *Phytochemistry* **16**, 965–985.

BOHLMANN, F., ZDERO, C. AND GRENZ, M. (1977b). *Chem. Ber.* **110**, 474–486.

BOHLMANN, F., KLOSE, W. AND NICKISCH, K. (1979). *Tetrahedron Lett.*, 3699–3702.

BOPPRE, M., SEIBT, U. AND WICKLER, W. (1984). *Entomol. Exp. Appl.* **35**, 115–117.

BORCH, R. F. AND HO, B. C. (1977). *J. Org. Chem.* **42**, 1225–1227.

BORKA, L. AND ONSHUUS, I. (1979). *Medd. Nor. Farm. Selsk.* **41**, 165–168.

BOSTWICK, J. L. (1982). *J. Am. Vet. Med. Assoc.* **180**, 386–387.

BOTTOMLEY, W. AND GEISSMAN, T. C. (1964). *Phytochemistry* **3**, 357.

BOUISSOU, R. (1973). *World Health,* (September), 3–17.

BOYD, M. R. (1980). *Crit. Rev. Toxicol.* **7**, 103–176.

BRADBURY, R. B. (1954). *Chem. Ind. (London)* 1022–1023.

BRADBURY, R. B. AND CULVENOR, C. C. J. (1954). *Aust. J. Chem.* **7**, 378–383.

BRADBURY, R. B. AND MASAMUNE, S. (1959). *J. Am. Chem. Soc.* **81**, 5201–5209.

BRADBURY, R. B. AND MOSBAUER, S. (1956). *Chem. Ind. (London)*, 1236–1237.

BRANDANGE, S. AND GRANELLI, I. (1973). *Acta Chem. Scand.* **27**, 1096–1097.

BRANDANGE, S. AND LUNING, B. (1969). *Acta Chem. Scand.* **23**, 1151.

BRAS, G. AND HILL, K. R. (1956). *Lancet* **2**, 161–163.

BRAS, G., JELLIFFE, D. B. AND STUART, K. L. (1954). *Arch. Pathol.* **57**, 285–300.

BRAS, G., BERRY, D. M. AND GYORGY, P. (1957). *Lancet* **2**, 960–962.

BRAS, G., BROOKS, S. E. H. AND WATLER, D. C. (1961). *J. Pathol. Bacteriol.* **82**, 503–512.

BRAUCHLI, J., LUTHY, J., ZWEIFEL, U. AND SCHLATTER, C. (1982). *Experientia* **38**, 1085–1087.

BREDENKAMP, M. W., WIECHERS, A., AND VAN ROOYEN, P. H. (1985). *Tetrahedron Lett.* **26**, 929–932.

BREWSTER, J. H. AND ELIEL, E. L. (1953). *Org. React.* **7**, 99–197.

BRIGGS, L. H., CAMBIE, R. C., CANDY, B. J., O'DONOVAN, G. M., RUSSELL, R. H. AND SEELYE, R. N. (1965). *J. Chem. Soc.* 2492–2498.

BRINK, N. G. (1966). *Mutat. Res.* **3**, 66–72.

BRINK, N. G. (1969). *Mutat. Res.* **8**, 139–146.

BRINK, N. G. (1982). *Mutat. Res.* **104**, 105–111.

BROCH-DUE, A. I. AND AASEN, A. J. (1980). *Acta Chem. Scand., Ser. B* **B34**, 75–77.

BRODERICK, D. J., KRIVAK, B. M. AND DOST, F. N. (1981). *Toxicologist* **1**, 108 (Abstr. 392).

BROWN, C. H. AND SEGALL, H. J. (1982). *Toxicologist* **2**, 23.

BROWN, H. C. AND CAHN, A. (1955). *J. Am. Chem. Soc.* **77**, 1715–1723.

BROWN, K., DEVLIN, J. A. AND ROBINS, D. J. (1983). *J. Chem. Soc., Perkin Trans. 1*, 1819–1824.

BROWN, K., DEVLIN, J. A. AND ROBINS, D. J. (1984). *Phytochemistry* **23**, 457–459.

BROWN, K. S. (1984). *Nature (London)* **309**, 707–709.

BRUEMMERHOFF, S. W. D. AND DE WAAL, H. L. (1961). *J. S. Afr. Chem. Inst.* **14**, 101; *Chem. Abstr.* **56**, 9076 (1962).

BRUNER, L. H., HILLIKER, K. S. AND ROTH, R. A. (1983). *Am. J. Physiol.* **245**, H300–H306.

BUCKMASTER, G. W., CHEEKE, P. R. AND SHULL, L. R. (1976). *J. Anim. Sci.* **43**, 464–473.

BUCKMASTER, G. W., CHEEKE, P. R., ARSCOTT, G. H., DICKINSON, E. O., PIERSON, M. L. AND SHULL, L. R. (1977). *J. Anim. Sci.* **45**, 1322–1325.

BULL, L. B. AND DICK, A. T. (1959). *J. Pathol. Bacteriol.* **78**, 483–502.

BULL, L. B., DICK, A. T. AND MCKENZIE, J. S. (1958). *J. Pathol. Bacteriol.* **75**, 17–25.

BULL, L. B., CULVENOR, C. C. J. AND DICK, A. T. (1968). "The Pyrrolizidine Alkaloids". North-Holland Publ., Amsterdam.

BUNCEL, E., JACKSON, K. G. A. AND JONES, J. K. N. (1965). *Chem. Ind. (London)* 89.

BURGUERA, J. A., EDDS, G. T. AND OSUNA, O. (1983). *Am. J. Vet. Res.* **44**, 1714–1717.

BURNS, J. (1972). *J. Pathol.* **106**, 187–194.

BUTLER, W. H. (1970). *J. Pathol.* **102**, 15–19.

BUTLER, W. H., MATTOCKS, A. R. AND BARNES, J. M. (1970). *J. Pathol.* **100**, 169–175.

BYKOREZ, A. I. (1969). *Chem. Abstr.* **73**, 64504 (1970).

CAMPBELL, J. G. (1956). *Proc. R. Soc. Edinburgh, Sect. B* **66**, 111–129.

CANDRIAN, U., LUTHY, J., GRAF, U. AND SCHLATTER, C. (1984a). *Food Chem. Toxicol.* **22**, 223–225.

CANDRIAN, U., LUTHY, J., SCHMID, P., SCHLATTER, C. AND GALLASZ, E. (1984b). *J. Agric Food Chem.* **32**, 935–937.

CANDRIAN, U., LUTHY, J., AND SCHLATTER, C. (1985). *Chem. Biol. Interact.*, **54**, 57–69.

CARRILLO, L. AND AVIADO, D. M. (1969). *Lab. Invest.* **20**, 243–248.

CATALFAMO, J. L., MARTIN, W. B. AND BIRECKA, H. (1982a). *Phytochemistry* **21**, 2669–2675.

CATALFAMO, J. L., FROHLICH, M. W., MARTIN, W. B. AND BIRECKA, H. (1982b). *Phytochemistry* **21,** 2677–2682.

CAVA, M. P., RAO, M. V., WEISBACH, J. A., RAFFAUF, R. F. AND DOUGLAS, B. (1968). *J. Org. Chem.* **33,** 3570–3573.

CHALMERS, A. H., CULVENOR, C. C. J., AND SMITH, L. W. (1965). *J. Chromatogr.* **20,** 270–277.

CHAMBERLIN, A. R. AND CHUNG, J. Y. L. (1982). *Tetrahedron Lett.* **23,** 2619–2622.

CHAMBERLIN, A. R. AND CHUNG, J. Y. L. (1983). *J. Am. Chem. Soc.* **105,** 3653–3656.

CHEEKE, P. R. AND GARMAN, G. R. (1974). *Nutr. Rep. Int.* **9,** 197–207.

CHEEKE, P. R. AND PIERSON-GOEGER, M. L. (1983). *Toxicol. Lett.* **18,** 343–349.

CHERNOVA, G. P. AND MURAV'EVA, D. A. (1974). *Chem. Abstr.* **84,** 102348 (1976).

CHESNEY, C. F. AND ALLEN, J. R. (1973a). *Toxicol. Appl. Pharmacol.* **26,** 385–392.

CHESNEY, C. F. AND ALLEN, J. R. (1973b). *Am. J. Pathol.* **70,** 489–492.

CHESNEY, C. F. AND ALLEN, J. R. (1973c). *Am. J. Vet. Res.* **34,** 1577–1581.

CHESNEY, C. F., ALLEN, J. R. AND HSU, I. C. (1974a). *Exp. Mol. Pathol.* **20,** 257–268.

CHESNEY, C. F., HSU, I. C. AND ALLEN, J. R. (1974b). *Res. Commun. Chem. Pathol. Pharmacol.* **8,** 567–570.

CHOPRA, R. N., ED. (1933). "Indigenous Drugs of India". The Art Press, Calcutta.

CHRISTIE, G. S. AND LE PAGE, R. N. (1962a). *Biochem. J.* **84,** 25–38.

CHRISTIE, G. S. AND LE PAGE, R. N. (1962b). *Biochem. J.* **84,** 202–212.

CHRISTIE, S. M., KROPMAN, M., LEISEGANG, E. C. AND WARREN, F. L. (1949). *J. Chem. Soc.,* 1700–1702.

CLARK, A. M. (1959). *Nature (London)* **183,** 731–732.

CLARK, A. M. (1960). *Z. Vererbungsl.* **91,** 74–80.

CLARK, A. M. (1976). *Mutat. Res.* **32,** 361–374.

COADY, A. (1973). "Evidence for the exposure of human populations in Ethiopia and elsewhere to liver toxins, including carcinogens, of plant origin." Dissertation submitted to University of Manchester.

COLEMAN, P. C., COUCOURAKIS, E. D. AND PRETORIUS, J. A. (1980). *S. Afr. J. Chem.* **33,** 116–119.

CONSTANTINE, M. F., MEHTA, M. D. AND WARD, R. (1967). *J. Chem. Soc. C,* 397–399.

CONSTANTINESCU, E. AND ALBULESCU, D. (1961). *Farmacia (Bucharest)* **9,** 139–142; *Chem. Abstr.* **56,** 14396 (1962).

COOK, L. M. AND HOLT, A. C. E. (1966). *J. Genet.* **59,** 273–274.

COOK, J. W., DUFFY, E. AND SCHOENTAL, R. (1950). *Br. J. Cancer* **4,** 405–410.

COREY, E. J. AND NICOLAOU, K. C. (1974). *J. Am. Chem. Soc.* **96,** 5614–5616.

COUCOURAKIS, E. D. AND GORDON-GRAY, C. G. (1970). *J. Chem. Soc. C,* 2312–2315.

COUCOURAKIS, E. D., GORDON-GRAY, C. G. AND WHITELEY, C. G. (1972). *J. Chem. Soc., Perkin Trans. 1,* 2339–2343.

CRAIG, A. M. (1979). *In* "Pyrrolizidine Alkaloids" (P. R. Cheeke, ed.), pp. 135–143. Nutr. Res. Inst., Oregon State University, Corvallis.

CRAIG, A. M., MEYER, C., KOLLER, L. D. AND SCHMITZ, J. A. (1978). *Proc. Am. Assoc. Vet. Lab. Diagn.* **21**, 161–177; *Chem. Abstr.* **90**, 198413 (1979).

CRAIG, A. M., SHEGGEBY, G. AND WICKS, C. E. (1984). *Vet. Hum. Toxicol.* **26**, 108–111.

CROUT, D. H. G. (1968a). *Chem. Commun.* 429–430.

CROUT, D. H. G. (1968b). *Phytochemistry* **7**, 1425–1427.

CROUT, D. H. G. (1969). *J. Chem. Soc., C,* 1379–1385.

CROUT, D. H. G. (1972). *J. Chem. Soc., Perkin Trans. 1,* 1602–1607.

CROUT, D. H. G., DAVIES, N. M., SMITH, E. H. AND WHITEHOUSE, D. (1972). *J. Chem. Soc., Perkin Trans. 1,* 671–680.

CROWLEY, H. C. AND CULVENOR, C. C. J. (1955). *Aust. J. Chem.* **8**, 464–465.

CROWLEY, H. C. AND CULVENOR, C. C. J. (1956). *Aust. J. Appl. Sci.* **7**, 359–364.

CROWLEY, H. C. AND CULVENOR, C. C. J. (1959). *Aust. J. Chem.* **12**, 694–705.

CROWLEY, H. C., AND CULVENOR, C. C. J. (1962). *Aust. J. Chem.* **15**, 139–144.

CULVENOR, C. C. J. (1954). *Aust. J. Chem.* **7**, 287–297.

CULVENOR, C. C. J. (1956). *Aust. J. Chem.* **9**, 512–520.

CULVENOR, C. C. J. (1962). *Aust. J. Chem.* **15**, 158.

CULVENOR, C. C. J. (1964). *Aust. J. Chem.* **17**, 233–237.

CULVENOR, C. C. J. (1966). *Tetrahedron Lett.,* 1091–1099.

CULVENOR, C. C. J. (1968). *J. Pharm. Sci.* **57**, 1112–1117.

CULVENOR, C. C. J. (1978). *Bot. Not.* **131**, 473–486.

CULVENOR, C. C. J. (1983). *J. Toxicol. Environ. Health* **11**, 625–635.

CULVENOR, C. C. J. AND DAL BON, R. (1964). *Aust. J. Chem.* **17**, 1296–1300.

CULVENOR, C. C. J. AND EDGAR, J. A. (1972). *Experientia* **28**, 627–628.

CULVENOR, C. C. J. AND GEISSMAN, T. A. (1961a). *J. Am. Chem. Soc.* **83**, 1647–1652.

CULVENOR, C. C. J. AND GEISSMAN, T. A. (1961b). *J. Org. Chem.* **26**, 3045–3050.

CULVERNOR, C. C. J. AND SMITH, L. W. (1954). *Chem. Ind. (London),* 1386.

CULVENOR, C. C. J. AND SMITH, L. W. (1955). *Aust. J. Chem.* **8**, 556–561.

CULVENOR, C. C. J. AND SMITH, L. W. (1957a). *Aust. J. Chem.* **10**, 464–473.

CULVENOR, C. C. J. AND SMITH, L. W. (1957b). *Aust. J. Chem.* **10**, 474–479.

CULVENOR, C. C. J. AND SMITH., L. W. (1958). *Aust. J. Chem.* **11**, 97.

CULVENOR, C. C. J. AND SMITH, L. W. (1961). *Aust. J. Chem.* **14**, 284–294.

CULVENOR, C. C. J. AND SMITH, L. W. (1962). *Aust. J. Chem.* **15**, 121–129.

CULVENOR, C. C. J. AND SMITH, L. W. (1963). *Aust. J. Chem.* **16**, 239–245.

CULVENOR, C. C. J. AND SMITH, L. W. (1966a). *Aust. J. Chem.* **19**, 1955–1964.

CULVENOR, C. C. J. AND SMITH, L. W. (1966b). *Aust. J. Chem.* **19**, 2127–2131.

CULVENOR, C. C. J. AND SMITH, L. W. (1967). *Aust. J. Chem.* **20,** 2499–2503.

CULVENOR, C. C. J. AND SMITH, L. W. (1969). *Tetrahedron Lett.,* 3603–3609.

CULVENOR, C. C. J. AND SMITH, L. W. (1972). *An. Quim.* **68,** 883–892.

CULVENOR, C. C. J. AND WILLETTE, R. E. (1966). *Aust. J. Chem.* **19,** 885–889.

CULVENOR, C. C. J. AND WOODS, W. G. (1965). *Aust. J. Chem.* **18,** 1625–1637.

CULVENOR, C. C. J., DRUMMOND, L. J. AND PRICE, J. R. (1954). *Aust. J. Chem.***7,** 277–286.

CULVENOR, C. C. J., DANN, A. T. AND SMITH. L. W. (1959). *Chem. Ind. (London),* 20–21.

CULVENOR, C. C. J., DANN, A. T. AND DICK, A. T. (1962). *Nature (London)* **195,** 570–573.

CULVENOR, C. C. J., HEFFERNAN, M. L. AND WOODS, W. G. (1965). *Aust. J. Chem.* **18,** 1605–1624.

CULVENOR, C. C. J., O'DONOVAN, G. M. AND SMITH, L. W. (1967a). *Aust. J. Chem.* **20,** 757–768.

CULVENOR, C. C. J., O'DONOVAN, G. M. AND SMITH, L. W. (1967b). *Aust. J. Chem.* **20,** 801–804.

CULVENOR, C. C. J., KORETSKAYA, N. I., SMITH, L. W. AND UTKIN, L. M. (1968). *Aust. J. Chem.* **21,** 1671–1673.

CULVENOR, C. C. J., DOWNING, D. T., EDGAR, J. A. AND JAGO, M. V. (1969a). *Ann. N.Y. Acad. Sci.* **163,** 837–847.

CULVENOR, C. C. J., EDGAR, J. A., SMITH, L. W. AND TWEEDDALE, H. J. (1969b). *Tetrahedron Lett.* **41,** 3599–3602.

CULVENOR, C. C. J., O'DONOVAN, G. M., SAWHNEY, R. S. AND SMITH, L. W. (1970a). *Aust. J. Chem.* **23,** 347–352.

CULVENOR, C. C. J., EDGAR, J. A., SMITH, L. W. AND TWEEDDALE, H. J. (1970b). *Aust. J. Chem.* **23,** 1853–1867.

CULVENOR, C. C. J., EDGAR, J. A., SMITH, L. W. AND TWEEDDALE, H. J. (1970c). *Aust. J. Chem.* **23,** 1869–1879.

CULVENOR, C. C. J., JOHNS, S. R., LAMBERTON, J. A. AND SMITH, L. W. (1970d). *Aust. J. Chem.* **23,** 1279–1282.

CULVENOR, C. C. J., SMITH, L. W., EDGAR, J. A., JAGO, M. V., TWEEDDALE, H. J. AND FRENCH, E. L. (1970e). *Chem. Abstr.* **73,** 55965.

CULVENOR, C. C. J., EDGAR, J. A., SMITH, L. W., JAGO, M. V. AND PETERSON, J. E. (1971a). *Nature (London), New Biol.* **229,** 255–256.

CULVENOR, C. C. J., CROUT, D. H. G., KLYNE, W., MOSE, W. P., RENWICK, J. D. AND SCOPES, P. M. (1971b). *J. Chem. Soc. C,* 3653–3664.

CULVENOR, C. C. J., EDGAR, J. A., FRAHN, J. L., SMITH, L. W., ULUBELEN, A. AND DOGANCA, S. (1975a). *Aust. J. Chem.* **28,** 173–178.

CULVENOR, C. C. J., JOHNS, S. R. AND SMITH, L. W. (1975b). *Aust. J. Chem.* **28,** 2319–2322.

CULVENOR, C. C. J., EDGAR, J. A., SMITH, L. W. AND HIRONO, I. (1976a). *Aust. J. Chem.* **29,** 229–230.

CULVENOR, C. C. J., EDGAR, J. A., JAGO, M. V., OUTTERIDGE, A., PETERSON, J. E. AND SMITH, L. W. (1976b). *Chem.-Biol. Interact.* **12,** 299–324.

CULVENOR, C. C. J., CLARKE, M., EDGAR, J. A., FRAHN, J. L., JAGO, M. V., PETERSON, J. E. AND SMITH, L. W. (1980a). *Experientia* **36,** 377–379.

CULVENOR, C. C. J., EDGAR, J. A., FRAHN, J. L. AND SMITH, L. W. (1980b). *Aust. J. Chem.* **33,** 1105–1113.

CULVENOR, C. C. J., EDGAR, J. A. AND SMITH, L. W. (1981). *J. Agric. Food Chem.* **29,** 958–960.

CULVENOR, C. C. J., JAGO, M. V., PETERSON, J. E., SMITH, L. W., PAYNE, A. L., CAMPBELL, D. G., EDGAR, J. A. AND FRAHN, J. L. (1984). *Aust. J. Agric. Res.* **35,** 293–304.

CURTAIN, C. C. (1975). *Chem.-Biol. Interact.* **10,** 133–139.

CURTAIN, C. C. AND EDGAR, J. A. (1976). *Chem.-Biol. Interact.* **13,** 243–256.

CZYGAN, F.-C. (1983). *Dtsch. Apoth.-Ztg.* **123,** 1779.

DAMIR, H. A., ADAM, S. E. I. AND TARTOUR, G. (1982). *Br. Vet. J.* **138,** 463–472.

DANILOVA, A. V. AND UTKIN, L. M. (1960). *Zh. Obshch. Khim.* **30,** 345; *Chem. Abstr.* **54,** 22698 (1960).

DANISHEFSKY, S., MCKEE, R. AND SINGH, R. K. (1977). *J. Am. Chem. Soc.* **99,** 4783–4788.

DANN, A. T. (1960). *Nature (London)* **186,** 1051.

DANNINGER, T., HAGEMANN, U., SCHMIDT, V. AND SCHONHOFER, P. S. (1983). *Pharm. Ztg.* **128,** 289–303.

DATTA, D. V., KHUROO, M. S., MATTOCKS, A. R., AIKAT, B. K. AND CHHUTTANI, P. N. (1978a). *Postgrad. Med. J.* **54,** 511–515.

DATTA, D. V., KHUROO, M. S., MATTOCKS, A. R., AIKAT, B. K. AND CHHUTTANI, P. N. (1978b). *J. Assoc. Physicians India* **26,** 383–393.

DAVIDSON, J. (1935). *J. Pathol. Bacteriol.* **40,** 285–295.

DEAGEN, J. T. AND DEINZER, M. L. (1977). *Lloydia* **40,** 395–397.

DEINZER, M. L., THOMSON, P. A., BURGETT, D. M. AND ISAACSON, D. L. (1977). *Science* **195,** 497–499.

DEINZER, M. L., THOMPSON, P. A., GRIFFIN, D. AND DICKINSON, E. (1978). *Biomed. Mass Spectrom.* **5,** 175–179.

DEINZER, M. L., ARBOGAST, B. L., BUHLER, D. R. AND CHEEKE, P. R. (1982). *Anal. Chem.* **54,** 1811–1814.

DELAVEAU, P., FERRY, S., BARBAGELATTA, M. AND CASPER, C. (1979). *Ann. Pharm. Fr.* **37,** 13–20.

DEMARLE, A. AND MOULE, Y. (1971). *Int. J. Cancer* **8,** 86–96.

DE PAULA RAMOS, A. L. AND MARQUES, E. K. (1978). *Rev. Bras. Genet.* **1** (4), 279-287.

DEVLIN, J. A. AND ROBINS, D. J. (1981). *Chem. Commun.,* 1272–1274.

DEVLIN, J. A. AND ROBINS, D. J. (1984). *J. Chem. Soc., Perkin Trans.* **1,** 1329–1332.

DEVLIN, J. A., ROBINS, D. J. AND SAKDARAT, S. (1982). *J. Chem. Soc., Perkin Trans.* **1,** 1117–1121.

DE WAAL, H. L. (1941). *Onderstepoort J. Vet. Sci. Anim. Ind.* **16**, 149–166; *Chem Abstr.* **36**, 6308 (1942).

DE WAAL, H. L. AND PRETORIUS, T. P. (1941). *Onderstepoort J. Vet. Sci. Anim. Ind.* **17**, 181; *Chem. Abstr.* **38**, 833 (1944).

DE WAAL, H. L., WIECHERS, A. AND WARREN, F. L. (1963). *J. Chem. Soc.* pp. 953–956.

DHAWAN, B. N., PATNAIK, G. K., RASTOGI, R. P., SINGH, K. K. AND TANDON, J. S. (1977). *Indian J. Exp. Biol.* **15**, 208–219.

DICK, A. T., DANN, A. T., BULL, L. B. AND CULVENOR, C. C. J. (1963). *Nature (London)* **197**, 207–208.

DICKINSON, J. O., COOKE, M. P., KING, R. R. AND MOHAMED, P. A. (1976). *J. Am. Vet. Med. Assoc.* **169**, 1192–1196.

DIMENNA, G. P., KRICK, T. P. AND SEGALL, H. J. (1980). *J. Chromatogr.* **192**, 474–478.

DOWNING, D. T. AND PETERSON, J. E. (1968). *Aust. J. Exp. Biol. Med. Sci.* **46**, 493–502.

DREIFUSS, P. A., BRUMLEY, W. C., SPHON, J. A. AND CARESS, E. A. (1983). *Anal. Chem.* **55**, 1036–1040.

DREWES, S. E. AND PITCHFORD, A. T. (1981). *J. Chem. Soc., Perkin Trans.* **1**, 408–412.

DREWES, S. E., ANTONOWITZ, I., KAYE, P. AND COLEMAN, P. C. (1981). *J. Chem. Soc., Perkin Trans.* **1**, 287–289.

DRIVER, H. E. AND MATTOCKS, A. R. (1984). *Chem.-Biol. Interact.* **51**, 201–218.

D'SILVA, J. B. AND NOTARI, R. E. (1980). *J. Pharm. Sci.* **69**, 471–472.

DYBING, O. AND ERICHSEN, S. (1959). *Acta Pathol. Microbiol. Scand.* **47**, 1–8.

EASTMAN, D. F. AND SEGALL, H. J. (1981). *Toxicol. Lett.* **8**, 217–222.

EASTMAN, D. F. AND SEGALL, H. J. (1982). *Drug Metab. Dispos.* **10**, 696–699.

EASTMAN, D. F., DIMENNA, G. P. AND SEGALL, H. J. (1982). *Drug Metab. Dispos.* **10**, 236–240.

EDGAR, J. A. (1974). *Nature (London)* **248**, 136–137.

EDGAR, J. A. AND CULVENOR, C. C. J. (1974). *Nature (London)* **248**, 614–615.

EDGAR, J. A. AND CULVENOR, C. C. J. (1975). *Experientia* **31**, 393–394.

EDGAR, J. A., CULVENOR, C. C. J. AND SMITH, L. W. (1971). *Experientia* **27**, 761–762.

EDGAR, J. A., COCKRUM, P. A. AND FRAHN, J. L. (1976). *Experientia* **32**, 1535–1537.

EDGAR, J. A., BOPPRE, M. AND SCHNEIDER, D. (1979). *Experientia* **35**, 1447–1448.

EDGAR, J. A., EGGERS, N. J., JONES, A. J. AND RUSSELL, G. B. (1980). *Tetrahedron Lett.* **21**, 2657–2660.

EDWARDS, J. D. AND MATSUMOTO, T. (1967a). *J. Org. Chem.* **32**, 1837–1838.

EDWARDS, J. D. AND MATSUMOTO, T. (1967b). *J. Org. Chem.* **32**, 2561–2563.

EDWARDS, J. D., MATSUMOTO, T. AND HASE, T. (1967a). *J. Org. Chem.* **32**, 244–246.

EDWARDS, J. D., HASE, T. AND ICHIKAWA, N. (1967b). *J. Heterocycl. Chem.* **4**, 487.

EGGERS, N. J. AND GAINSFORD, G. J. (1979). *Cryst. Struct. Commun.* **8**, 597–603.

EISENSTEIN, D., AZARI, J. AND HUXTABLE, R. (1979). *Proc. West. Pharmacol. Soc.* **22**, 193–198.

EVANS, J. V., PENG, A. AND NIELSEN, C. J. (1979). *Biomed. Mass Spectrom.* **6**, 38–43.

EVANS, J. V., DALEY, S. K., MCCLUSKY, G. A. AND NEILSEN, C. J. (1980). *Biomed. Mass Spectrom.* **7**, 65–73.

EVANS, W. C. AND EVANS, E. T. R. (1949). *Nature (London)* **164**, 30–31.

FARNSWORTH, N. R. (1966). *J. Pharm. Sci.* **55**, 225–276.

FERRIS, J. P., GERWE, R. D. AND GAPSKI, G. R. (1967). *J. Am. Chem. Soc.* **89**, 5270–5275.

FERRY, S. (1972). *Ann. Pharm. Fr.* **30**, 145–152; *Chem. Abstr.* **77**, 58770 (1972).

FERRY, S. AND BRAZIER, J. L. (1976). *Ann. Pharm. Fr.* **34**, 133–138; *Chem. Abstr.* **85**, 139754 (1976).

FLITSCH, W. AND RUSSKAMP, P. (1983). *Liebigs Ann. Chem.* 521–528.

FLITSCH, W. AND WERNSMANN, P. (1981). *Tetrahedron Lett.* **22**, 719–722.

FORD, E. J. H., RITCHIE, H. E. AND THORPE, E. (1968). *J. Comp. Pathol.*, **78**, 207–218.

FORSYTH, A. A. (1968). "British Poisonous Plants." H. M. Stationery Office, London.

FOX, D. W., HART, M. C., BERGESON, P. S., JARRETT, P. B., STILLMAN, A. E. AND HUXTABLE, R. J. (1978). *J. Pediatr.* **93**, 980–982.

FRAHN, J. L. (1969). *Aust. J. Chem.* **22**, 1655–1667.

FRAHN, J. L., CULVENOR, C. C. J. AND MILLS, J. A. (1980). *J. Chromatogr.* **195**, 379–383.

FRAYSSINET, C. AND MOULE, Y. (1969). *Nature (London)* **223**, 1269–1270.

FRIDRICHSONS, J., MATHIESON, A. M. AND SUTOR, D. J. (1963). *Acta Crystallogr.* **16**, 1075–1085.

FURUYA, T. AND ARAKI, K. (1968). *Chem. Pharm. Bull.* **16**, 2512–2516.

FURUYA, T. AND HIKICHI, M. (1971). *Phytochemistry* **10**, 2217–2220.

FURUYA, T. AND HIKICHI, M. (1973). *Phytochemistry* **12**, 225.

FURUYA, T., MURAKAMI, K. AND HIKICHI, M. (1971). *Phytochemistry* **10**, 3306–3307.

FURUYA, T., HIKICHI, M. AND IITAKA, Y. (1976). *Chem. Pharm. Bull.* **24**, 1120–1122.

GALLAGHER, C. H. (1968). *Biochem. Pharmacol.* **17**, 533–538.

GALLAGHER, C. H. AND JUDAH, J. E. (1967). *Biochem. Pharmacol.* **16**, 883–895.

GALLAGHER, C. H. AND KOCH, J. H. (1959). *Nature (London)* **183**, 1124–1125.

GANDHI, R. N., RAJAGOPALAN, T. R. AND SESHADRI, T. R. (1966a). *Curr. Sci.* **35**, 121–122.

GANDHI, R. N., RAJAGOPALAN, T. R. AND SESHADRI, T. R. (1966b). *Curr. Sci.* **35**, 514–515.

GANDHI, R. N., RAJAGOPALAN, T. R. AND SESHADRI, T. R. (1968). *Curr. Sci.* **37**, 285–286.

GANEY, P. E., FINK, G. D., AND ROTH, R. A. (1985). *Toxicol. Appl. Pharmacol.* **78**, 55–62.

GARRETT, B. J., CHEEKE, P. R., MIRANDA, C. L., GOEGER, D. E. AND BUHLER, D. R. (1982). *Toxicol. Lett.* **10**, 183–188.

GEISSMAN, T. A. AND WAISS, A. C. (1962). *J. Org. Chem.* **27**, 139–142.

GELBAUM, L. T., GORDON, M. M., MILES, M. AND ZALKOW, L. H. (1982). *J. Org. Chem.* **47**, 2501–2504.

GELLERT, E. AND MATE, C. (1964). *Aust. J. Chem.* **17**, 158.

GENCHEVA, E. (1978). *Farmatsiya (Sofia)* **28**, 21–24; *Chem. Abstr.* **90**, 127593 (1979).

GHARBO, S. A. AND HABIB, A. A. M. (1969). *Lloydia* **32**, 503–508.

GHODSI, F. AND WILL, J. A. (1981). *Am. J. Physiol.* **240**, H149–H155.

GHOSH, M. N. AND SINGH, H. (1974). *Br. J. Pharmacol.* **51**, 503–508.

GILLIS, C. N., HUXTABLE, R. J. AND ROTH, R. A. (1978). *Br. J.Pharmacol.* **63**, 435–443.

GLINSKI, J. A., AND ZALKOW, L. H. (1985). *Tetrahedron Lett.* **26**, 2857–2860.

GLIZIN, V. I. AND SENOV, P. L. (1965). *Farm. Zh. (Kiev)* **20**, 38–40; *Chem. Abstr.* **64**, 8547 (1966).

GLONTI, SH. I. (1956). *Chem. Abstr.* **52**, 12322 (1958).

GOEGER, D. E., CHEEKE, P. R., BUHLER, D. R. AND SCHMITZ, J. A. (1979). *In* "Pyrrolizidine Alkaloids" (P. R. Cheeke, ed.), pp. 77–84. Nutr. Res. Inst., Oregon State University, Corvallis.

GOEGER, D. E., CHEEKE, P. R., SCHMITZ, J. A. AND BUHLER, D. R. (1982a). *Am. J. Vet. Res.* **43**, 252–254.

GOEGER, D. E., CHEEKE, P. R., SCHMITZ, J. A. AND BUHLER, D. R. (1982b). *Am. J. Vet. Res.* **43**, 1631–1633.

GOEGER, D. E., CHEEKE, P. R., RAMSDELL, H. S., NICHOLSON, S. S. AND BUHLER, D. R. (1983). *Toxicol. Lett.* **15**, 19–23.

GONZALEZ, A. G. AND CALERO, A. (1958). *Chem. Ind. (London)* 126.

GOPINATH, C., FORD, E. J. H. AND JONES, R. S. (1972). *J. Pathol.* **107**, 253–263.

GORDON-GRAY, C. G. (1967). *J. Chem. Soc. C,* 781–782.

GORDON-GRAY, C. G. AND WELLS, R. B. (1974). *J. Chem. Soc., Perkin Trans.* **1**, 1556–1561.

GORDON-GRAY, C. G. AND WHITELEY, C. G. (1977). *J. Chem. Soc., Perkin Trans.* **1**, 2040–2046.

GORDON-GRAY, C. G., WELLS, R. B., HALLAK, N., HURSTHOUSE, M. B., NEIDLE, S. AND TOUBE, T. P. (1972). *Tetrahedron Lett.,* 707–710.

GOSS, G. J. (1979). *Environ. Entomol.* **8**, 487–493.

GREEN, C. R. AND CHRISTIE, G. S. (1961). *Br. J. Exp. Pathol.* **42**, 369–378.

GREEN, M. H. L. AND MURIEL, W. J. (1975). *Mutat. Res.* **28**, 331–336.

GREEN, C. E., SEGALL, H. J., AND BYARD, J. L. (1981). *Toxicol. Appl. Pharmacol.* **60**, 176–185.

GROUP OF *CROTALARIA* PLANT RESEARCH (1974). *Chem. Abstr.* **83**, 25052 (1975).

GRUE-SØRENSEN, G. AND SPENSER, I. D. (1981). *J. Am. Chem. Soc.* **103**, 3208–3210.

GRUE-SØRENSEN, G. AND SPENSER, I. D. (1982). *Can. J. Chem.* **60**, 643–662.

GRUE-SØRENSON, G. AND SPENSER, I. D. (1983). *J. Am. Chem. Soc.* **105**, 7401–7404.

GUENGERICH, F. P. (1977). *J. Biol. Chem.* **252**, 3970–3979.

GUENGERICH, F. P. AND MITCHELL, M. B. (1980). *Drug. Metab. Dispos.* **8,** 34–38.

GULICK, B. A., LIU, I. K. M., QUALLS, C. W., GRIBBLE, D. H. AND ROGERS, Q. R. (1980). *Am. J. Vet. Res.* **41,** 1894–1898.

GUPTA, P. S., GUPTA, G. D. AND SHARMA, M. L. (1963). *Br. Med. J.* **1,** 1184–1186.

GUPTA, O. P., ALI, M. M., GHATAK, B. J. AND ATAL, C. K. (1976a). *Indian J. Exp. Biol.* **14,** 34–37.

GUPTA, O. P., SINGH, G. B., GHATAK, B. J. AND ATAL, C. K. (1976b). *Indian J. Exp. Biol.* **14,** 282–284.

GUPTA, O. P., ALI, M. M., GHATAK, B. J. R. AND ATAL, C. K. (1977a). *Indian J. Exp. Biol.* **15,** 220–228.

GUPTA, O. P., ALI, M. M., GHATAK, B. J. R. AND ATAL, C. K. (1977b). *Indian J. Med. Res.* **65,** 436–440.

GUPTA, O. P., SINGH, G. B. AND ATAL, C. K. (1979). *Arzneim.-Forsch.* **29,** 1715–1722.

GUPTA, V. P., HANDOO, S. K. AND SAWHNEY, R. S. (1975a). *Indian J. Pure Appl. Phys.* **13,** 776–779.

GUPTA, V. P., HANDOO, S. K. AND SAWHNEY, R. S. (1975b). *Curr. Sci.* **44,** 451–454.

HABIB, A. A. M. (1974). *Planta Med.* **26,** 279–282.

HABIB, A. A. M. (1975). *Bull. Fac. Sci., Riyad Univ.* **7,** 67–74;*Chem. Abstr.* **85,** 119597 (1976).

HABIB, A. A. M. AND EL-SEBAKHY, N. A. (1978). *Egypt. J. Pharm. Sci.* **19,** 71–76; *Chem. Abstr.* **95,** 30464 (1981).

HABIB, A. A. M., SALEH, M. R. I. AND FARAG, M. (1971). *Lloydia* **34,** 455–456.

HABS, H. (1982). *Dtsch. Apoth.-Ztg.* **122,** 799–804.

HABS, H., HABS, M., MARQUARDT, H., RODER, M., SCHMAHL, D. AND WIEDENFELD, H. (1982). *Arzneim.-Forsch.* **32,** 144–148.

HAKSAR, C. N., SURI, O. P., JAMWAL, R. S. AND ATAL, C. K. (1982). *Indian J. Chem., Sect. B* **21B,** 492–493.

HARRIS, P. N. AND CHEN, K. K. (1970). *Cancer Res.* **30,** 2881–2886.

HARRIS, P. N., ANDERSON, R. C. AND CHEN, K. K. (1942). *J. Pharmacol. Exp. Ther.* **75,** 69–77.

HARRIS, P. N., ROSE, C. L. AND CHEN, K. K. (1957). *Arch. Pathol.* **64,** 152–157.

HART, D. J. AND YANG, T.-K. (1982). *Tetrahedron Lett.,* 2761–2764.

HART, D. J. AND YANG, Y.-K. (1983). *Chem. Commun.,* 135–136.

HAYASHI, K., NATORIGAWA, A. AND MITSUHASHI, H. (1972). *Chem. Pharm. Bull.* **20,** 201–202.

HAYASHI, Y. (1966). *Fed. Proc., Fed. Am. Soc. Exp. Biol.* **25,** 688.

HAYASHI, Y. AND LALICH, J. J. (1967). *Proc. Soc. Exp. Biol. Med.* **124,** 392–396.

HAYASHI, Y. AND LALICH, J. J. (1968). *Toxicol. Appl. Pharmacol.* **12,** 36–43.

HAYASHI, Y., HUSSA, J. F. AND LALICH, J. J. (1967). *Lab. Invest.* **16,** 875–881.

HAYASHI, Y., SHINADA, M. AND KATAYAMA, H. (1977). *Toxicol. Lett.* **1,** 41–44.

HAYASHI, Y., KATO, M. AND OTSUKA, H. (1979). *Toxicol. Lett.* **3**, 151–155.

HAYASHI, Y., KOKUBO, T., TAKAHASHI, M., FURUKAWA, F., OTSUKA, H. AND HASHIMOTO, K. (1984). *Toxicol. Lett.* **21**, 65–71.

HAYES, M. A., ROBERTS, E., JAGO, M. V., SAFE, S. H., FARBER, E. AND CAMERON, R. C. (1984). *J. Toxicol. Environ. Hlth.* **14**, 683–694.

HEATH, D., SHABA, J., WILLIAMS, A., SMITH, P. AND KOMBE, A. (1975). *Thorax* **30**, 399–404.

HENDRICKS, D., SINNHUBER, R. O., HENDERSON, M. C. AND BUHLER, D. R. (1981). *Exp. Mol. Pathol.* **35**, 170–183.

HENDY, R. AND GRASSO, P. (1977). *Chem.-Biol. Interact.* **18**, 309–326.

HENNIG, A. J. (1961). *Lloydia* **24**, 68–70.

HERZ, W., KULANTHAIVEL, P., SUBRAMANIAN, P. S., CULVENOR, C. C. J. AND EDGAR, J. A. (1981). *Experientia* **37**, 683.

HIKICHI, M. AND FURUYA, T. (1974). *Tetrahedron Lett.*, 3657–3660.

HIKICHI, M. AND FURUYA, T. (1976). *Chem. Pharm. Bull.* **24**, 3178–3184.

HIKICHI, M., FURUYA, T. AND IITAKA, Y. (1978). *Tetrahedron Lett.*, 767–770.

HIKICHI, M., ASADA, Y. AND FURUYA, T. (1979). *Tetrahedron Lett.*, 1233–1236.

HIKICHI, M., ASADA, Y. AND FURUYA, T. (1980). *Planta Med., Suppl.*, 1–4.

HILLIKER, K. S. AND ROTH, R. A. (1984). *Br. J. Pharmacol.* **82**, 375–380.

HILLIKER, K. S., BELL, T. G., AND ROTH, R. A. (1982). *Am. J. Physiol.* **242**, H573–H579.

HILLIKER, K. S., GARCIA, C. M. AND ROTH, R. A. (1983a). *Res. Commun. Chem. Pathol. Pharmacol.* **40**, 179–197.

HILLIKER, K. S., DEYO, J. A., BELL, T. G. AND ROTH, R. A. (1983b). *Thromb. Res.* **32**, 325–333.

HILLIKER, K. S., BELL, T. G. AND ROTH, R. A. (1983c). *Thromb. Haemostasis* **50**, 844–847.

HILLIKER, K. S., IMLAY, M. AND ROTH, R. A. (1984). *Biochem. Pharmacol.* **33**, 2690–2692.

HILLS, L. D. (1976). "Comfrey, Past, Present and Future". Faber & Faber, London.

HIRONO, I. (1979). *Gann Monogr. Cancer Res.* **24**, 85–102.

HIRONO, I. (1981). *CRC Crit. Rev. Toxicol.* **8**, 235–277.

HIRONO, I., SHIMIZU, M., FUSHIMI, K., MORI, H. AND KATO, K. (1973). *Gann* **64**, 527–528.

HIRONO, I., MORI, H. AND CULVENOR, C. C. J. (1976). *Gann* **67**, 125–129.

HIRONO, I., MORI, H., YAMADA, K., HIRATA, Y., HAGA, M., TATEMATSU, H. AND KANIE, S. (1977). *J. Natl. Cancer Inst. (U.S.)* **58**, 1155–1157.

HIRONO, I., MORI, H. AND HAGA, M. (1978). *JNCI, J. Natl. Cancer Inst.* **61**, 865–869.

HIRONO, I., HAGA, M., FUJII, M., MATSUURA, S., MATSUBARA, N., NAKAYAMA, M., FURUYA, T., HIKICHI, M., TAKANASHI, H., UCHIDA, E., HOSAKA, S. AND UENO, I. (1979). *J. Natl. Cancer Inst.* **63**, 649–671.

HIRONO, I., UENO, I., AISO, S., YAMAJI, T., AND HAGA, M. (1983). *Cancer Lett.* **20**, 191–198.

HISLOP, A. AND REID, L. (1974). *Br. J. Exp. Pathol.* **55**, 153–163.

HOOPER, P. T. (1972). *Vet. Rec.* **90**, 37–38.

HOOPER, P. T. (1974). *J. Pathol.* **113**, 227–230.

HOOPER, P. T. AND SCANLAN, W. A. (1977). *Aust. Vet. J.* **53**, 109–114.

HOOSON, J. AND GRASSO, P. (1976). *J. Pathol.* **118**, 121–128.

HOQUE, M. S., GHANI, A. AND RASHID, H. (1976). *Bangladesh Pharm. J.* **5**, 13–15; *Chem. Abstr.* **86**, 40191 (1977).

HOSKINS, W. M. AND CROUT, D. H. G. (1977). *J. Chem. Soc., Perkin Trans.* **1**, 538–544.

HRBEK, J., HRUBAN, L., KLASEK, A., KOCHETKOV, N. K., LIKHOSHERSTOV, A. M., SANTAVY, F. AND SNATZE, G. (1972). *Collect. Czech. Chem. Commun.* **37**, 3918–3935.

HSU, I. C. AND ALLEN, J. R. (1975). *J. Labelled Compd.* **11**, 71–76.

HSU, I. C., CHESNEY, C. F. AND ALLEN, J. R. (1973a). *Proc. Soc. Exp. Biol. Med.* **142**, 1133–1136.

HSU, I. C., ALLEN, J. R. AND CHESNEY, C. F. (1973b). *Proc. Soc. Exp. Biol. Med.* **144**, 834–838.

HSU, I. C., SHUMAKER, R. C. AND ALLEN, J. R. (1974). *Chem.-Biol. Interact.* **8**, 163–170.

HSU, I. C., ROBERTSON, K. A. AND ALLEN, J. R. (1976). *Chem.-Biol. Interact.* **12**, 19–28.

HUA, Z., XU, X., WEI, X., TANG, S. AND WU, Y. (1983). *Chem. Abstr.* **100**, 139425 (1984).

HUANG, J. AND MEINWALD, J. (1981). *J. Am. Chem. Soc.* **103**, 861–867.

HUANG, L., WU, K.-M., XUE, Z., CHENG, J.-C., XU, L.-Z.,XU, S.P., AND XI, Y.-G. (1980). *Chem. Abstr.* **95**, 43427 (1981).

HUIZING, H. J. AND MALINGRE, TH. M. (1979a). *J. Chromatogr.* **173**, 187–189.

HUIZING, H. J. AND MALINGRE, TH. M. (1979b). *J. Chromatogr.* **176**, 274–279.

HUIZING, H. J. AND MALINGRE, TH. M. (1981). *J. Chromatogr.* **205**, 218–222.

HUIZING, H. J., DE BOER, F. AND MALINGRE, TH. M. (1980). *J. Chromatogr.* **195**, 407–411.

HUIZING, H. J., DE BOER, R. AND MALINGRE, TH. M. (1981). *J. Chromatogr.* **214**, 257–262.

HURLEY, J. V. AND JAGO, M. V. (1975). *J. Pathol.* **117**, 23–32.

HURLEY, J. V. AND JAGO, M. V. (1976). *Pathology* **8**, 7–20.

HUXTABLE, R. J. (1979a). *Gen. Pharmacol.* **10**, 159–167.

HUXTABLE, R. J. (1979b). *In* "Pyrrolizidine Alkaloids" (P. R. Cheeke, ed.), pp. 43–56. Nutr. Res. Inst., Oregon State University, Corvallis.

HUXTABLE, R. J. (1980a). *Perspect. Biol. Med.* **24**, 1–14.

HUXTABLE, R. J. (1980b). *Trends Pharmacol. Sci.* **1**, 299–303.

HUXTABLE, R. J. AND LAFRANCONI, W. M. (1984). *Thorax* **39**, 159–160.

HUXTABLE, R. J., STILLMAN, E. A. AND CIARAMITARO, D. (1977). *Proc. West. Pharmacol. Soc.* **20**, 455–459.

HUXTABLE, R. J., CIARAMITARO, D. AND EISENSTEIN, D. (1978). *Mol. Pharmacol.* **14,** 1189–1203.

INTERNATIONAL AGENCY FOR RESEARCH IN CANCER (IARC) (1976). *IARC Monogr. Eval. Carcinog. Risk Chem. Man* **10,** 265–342.

INTERNATIONAL AGENCY FOR RESEARCH IN CANCER (IARC) (1983). *IARC Monogr. Eval. Carcinog. Risk Chem. Man* **31,** 207–245.

IRVINE, A. M., FORBES, J. C. AND DRAPER, S. R. (1977). *Weed Res.* **17,** 169–172.

IWASHITA, T., KUSUMI, T. AND KAKISAWA, H. (1979). *Chem. Lett.,* 1337–1340.

IWASHITA, T., KUSUMI, T. AND KAKISAWA, H. (1982). *J. Org. Chem.* **47,** 230–233.

IYER, V. N. AND SZYBALSKI, W. (1964). *Science* **145,** 55–58.

JAGO, M. V. (1969). *Am. J. Pathol.* **56,** 405–421.

JAGO, M. V. (1970). *Aust. J. Exp. Biol. Med. Sci.* **48,** 93–103.

JAGO, M. V. (1971). *J. Pathol.* **105,** 1–11.

JAGO, M. V., LANIGAN, G. W., BINGLEY, J. B., PIERCY, D. W. T., WHITTEM, J. H. AND TITCHEN, D. A. (1969). *J. Pathol.* **98,** 115–128.

JAGO, M. V., EDGAR, J. A., SMITH, L. W. AND CULVENOR, C. C. J. (1970). *Mol. Pharmacol.* **6,** 402–406.

JAMWAL, R. S., SURI, K. A., SURI, O. P., HAKSAR, C. N. AND ATAL, C. K. (1982). *Indian J. Chem. Sect. B* **21B,** 266–267.

JERZMANOWSKA, Z. AND SYKULSKA, Z. (1964). *Diss. Pharm.* **16,** 71–79; *Chem. Abstr.* **61,** 16438 (1964).

JIZBA, J., BUDESINSKY, M., VAREK, T., BOEVA, A., DIMITROVA, K., SANTAVY, F. AND NOVOTNY, L. (1982). *Collect. Czech. Chem. Commun.* **47,** 664–669.

JOHNSON, A. E. (1976). *Am. J. Vet. Res.* **37,** 107–110.

JOHNSON, A. E. (1979). *In* "Pyrrolizidine Alkaloids" (P. R. Cheeke, ed.), pp. 129–134. Nutr. Res. Inst., Oregon State University, Corvallis.

JOHNSON, A. E. (1982). *Am. J. Vet. Res.* **43,** 718–723.

JOHNSON, A. E. AND MOLYNEUX, R. J. (1984). *Am. J. Vet. Res.* **45,** 26–31.

JOHNSON, A. E., MOLYNEUX, R. J. AND MERRILL, G. B. (1985a). *J. Agric. Food Chem.* **33,** 50–55.

JOHNSON, A. E., MOLYNEUX, R. J. AND STUART, L. D. (1985b). *Am. J. Vet. Res.* **46,** 577–582.

JOHNSON, W. D. (1981). *Toxicologist* **1,** 107–108 (Abstr. No. 390).

JOHNSON, W. D., ROBERTSON, K. A., POUNDS, J. G. AND ALLEN, J. R. (1978). *JNCI, J. Natl. Cancer Inst.* **61,** 85–89.

JONES, A. J., CULVENOR, C. C. J. AND SMITH, L. W. (1982). *Aust. J. Chem.* **35,** 1173–1184.

JONES, B. AND GILLIE, O. (1981). "Folk Remedies that Work: Comfrey", March 22, p. 76. Sunday Times Mag., London.

JONES, R. T., DRUMMOND, G. R. AND CHATHAM, R. O. (1981). *Aust. Vet. J.* **57,** 396.

JONES, T. A., BUCKNER, R. C., BURRUS, P. B. AND BUSH, L. P. (1983). *Crop Sci.* **23**, 1135–1140.

JOVCEVA, R. J., BOEVA, A., POTESILOVA, H., KLASEK, A. AND SANTAVY, F. (1978). *Collect. Czech. Chem. Commun.* **43**, 2312–2314.

JUNEJA, T. R., GUPTA, R. L. AND SAMANTA, S. (1984). *Toxicol. Lett.* **21**, 185–189.

KAK, S. N., PURI, S. C. AND KAUL, B. L. (1973). *Cytobiologie* **6**, 481–486.

KAMEJI, R., OTSUKA, H. AND HAYASHI, Y. (1980). *Experientia* **36**, 441–442.

KARCHESY, J. J. AND DEINZER, M. L. (1981). *Heterocycles* **16**, 631–635.

KARCHESY, J. J., DEINZER, M. L. AND GRIFFIN, D. A. (1984a). *J. Agric. Food Chem.* **32**, 1056–1057.

KARCHESY, J. J., DEINZER, M., GRIFFIN, D. AND ROHRER, D. C. (1984b). *Biomed. Mass Spectrom.* **11**, 455–461.

KARIMOV, A., TELEZHENETSKAYA, V., LUTFULLIN, K. L. AND YUNUSOV, S. YU. (1975). *Khim. Prir. Soedin.*, 433–434; *Chem. Abstr.* **84**, 14662 (1976).

KAUL, B. L. AND KAK. S. N. (1974). *Cytobios* **9**, 27–31.

KAY, J. M. AND HEATH, D. (1969). *"Crotalaria spectabilis,* the Pulmonary Hypertension Plant". Thomas, Springfield, Illinois.

KAY, J. M., GILLUND, T. D. AND HEATH, D. (1967). *Am. J. Pathol.* **51**, 1031–1044.

KAY, J. M., CRAWFORD, N. AND HEATH, D. (1968). *Experientia* **24**, 1149–1150.

KAY, J. M., SMITH, P. AND HEATH, D. (1969). *Thorax* **24**, 511–526.

KAY, J. M., SMITH, P., HEATH, D. AND WILL, J. A. (1976). *Cardiovasc. Res.* **10**, 200–205.

KAY, J. M., KEANE, P. M., SUYAMA, K. L. AND GAUTHIER, D. (1982). *Thorax* **37**, 88–96.

KEANE, P. M. AND KAY, J. M. (1984). *Thorax* **39**, 159.

KEANE, P. M., KAY, J. M., SUYAMA, K. L., GAUTHIER, D. AND ANDREW, K. (1982). *Thorax* **37**, 198–204.

KECK, G. E. AND NICKELL, D. G. (1980). *J. Am. Chem. Soc.* **102**, 3632–3634.

KEDZIERSKI, B., AND BUHLER, D. R. (1985). *Toxicol. Lett.* **25**, 115–119.

KERR, J. F. R. (1969). *J. Pathol.* **97**, 557–562.

KHALILOV, D. S. AND TELEZHENETSKAYA, M. V. (1973). *Khim. Prir. Soedin.*, 128–129; *Chem. Abstr.* **78**, 156643 (1973).

KHALILOV, D. S., DAMIROV, I. A. AND TELEZHENETSKAYA, M. V. (1972). *Khim. Prir. Soedin.*, 656; *Chem. Abstr.* **78**, 108214 (1973).

KHAN, H. A. AND ROBINS, D. J. (1981a). *Chem. Commun.*, 146–147.

KAHN, H. A. AND ROBINS, D. J. (1981b). *Chem. Commun.* 554–556.

KHMEL, M. P. (1961). *Farm. Zh. (Kiev)* **16**, 35–39; *Chem. Abstr.* **56**, 13011 (1962).

KIDO, M. (1981). *Fukuoka Igaku Zasshi* **72**, 117–130; *Chem. Abstr.* **95**, 74540 (1981).

KIDO, M., HIROSE, T., TANAKA, K., KUROZUMI, T. AND SHOYAMA, Y. (1981). *Jpn. J. Med.* **20**, 170–177; *Chem. Abstr.* **96**, 117027 (1982).

KIM, H. L. AND JONES, L. P. (1982). *Res. Commun. Chem. Pathol. Pharmacol.* **36**, 341–344.

KING, R. R. AND DICKINSON, J. O. (1979). *In* "Pyrrolizidine Alkaloids" (P. R. Cheeke, ed.), pp. 69–76. Nutr. Res. Inst., Oregon State University, Corvallis.

KIRKLAND, P. D., MOORE, R. E., WALKER, K. H., SEAMAN, J. T. AND DUNN, S. E. (1982). *Aust. Vet. J.* **59,** 64.

KISS, G. AND NEUKOM, H. (1966). *Helv. Chim. Acta* **49,** 989–992.

KIYAMITDINOVA, F., AKRAMOV, S. T. AND YUNUSOV, S. YU. (1967). *Khim. Prir. Soedin.,* 411–412; *Chem. Abstr.* **68,** 75730 (1968).

KIYOOKA, S. AND HASE, T. (1973). *Bull. Chem. Soc. Jpn.* **46,** 3609–3610.

KLASEK, A., VRUBLOVSKY, P. AND SANTAVY, F. (1967). *Collect. Czech. Chem. Commun.* **32,** 2512–2522.

KLASEK, A., SVAROVSKY, V., AHMED, S. S. AND SANTAVY, F. (1968a). *Collect. Czech. Chem. Commun.* **33,** 1738–1743.

KLASEK, A., REICHSTEIN, T. AND SANTAVY, F. (1968b). *Helv. Chim. Acta* **51,** 1088–1095.

KLASEK, A., NEUNER-JEHLE, N. AND SANTAVY, F. (1969). *Collect. Czech. Chem. Commun.* **34,** 1459–1469.

KLASEK, A., SEDMERA, P. AND SANTAVY, F. (1970). *Collect. Czech. Chem. Commun.* **35,** 956–969.

KLASEK, A., SEDMERA, P. AND SANTAVY, F. (1971). *Collect. Czech. Chem. Commun.* **36,** 2205–2215.

KLASEK, A., SEDMERA, P., BOEVA, A. AND SANTAVY, F. (1973a). *Collect. Czech. Chem. Commun.* **38,** 2504–2512.

KLASEK, A., SULA, B. AND SANTAVY, F. (1973b). *Collect. Czech. Chem. Commun.* **38,** 2658–2660.

KLASEK, A., SEDMERA, P. AND SANTAVY, F. (1975a). *Collect. Czech. Chem. Commun.* **40,** 568–573.

KLASEK, A., MNATSAKANYAN, V. A. AND SANTAVY, F. (1975b). *Collect. Czech. Chem. Commun.* **40,** 2524–2528.

KLASEK, A., SEDMERA, P., VOKOUN, J., BOEVA, A., DVORACKOVA, S. AND SANTAVY, F. (1980). *Collect. Czech. Chem. Commun.* **45,** 548–558.

KLEIN, L. L. (1985). *J. Am. Chem. Soc.* **107,** 2573–2574.

KLOSE, W., NICKISCH, K. AND BOHLMANN, F. (1980). *Chem. Ber.* **113,** 2694–2698.

KLOSTERMAN, H. J. AND SMITH, F. (1954). *J. Am. Chem. Soc.* **76,** 1229–1230.

KNIGHT, A. P., KIMBERLING, C. V., STERMITZ, F. R., AND ROBY, M. R. (1984). *J. Am. Vet. Med. Assoc.* **185,** 647–650.

KOCHETKOV, N. K. AND LIKHOSHERSTOV, A. M. (1965). *Adv. Heterocycl. Chem.* **5,** 315–367.

KOCHETKOV, N. K., LIKHOSHERSTOV, A. M. AND KULAKOV, V. N. (1969). *Tetrahedron* **25,** 2313–2323.

KOEKEMOER, M. J. AND WARREN, F. L. (1951). *J. Chem. Soc.* 66–68.

KOEKEMOER, M. J. AND WARREN, F. L. (1955). *J. Chem. Soc.* 63–65.

KOHLMUENZER, S., TOMCZYK, H. AND SAINT-FERMIN, A. (1971). *Diss. Pharm. Pharmacol.* **23**, 419–427; *Chem. Abstr.* **76**, 96972 (1972).

KOLETSKY, A., OYASU, R. AND REDDY, J. K. (1978). *Lab. Invest.* **38**, 352.

KOMPIS, I., SHROTER, H. B., POTESILOVA, H. AND SANTAVY, F. (1960). *Collect. Czech. Chem. Commun.* **25**, 2449–2453.

KONOVALOV, V. S. AND MEN'SHIKOV, G. P. (1945). *J. Gen. Chem. USSR (Engl. Transl.)* **15**, 328; *Chem. Abstr.* **40**, 3760 (1946).

KONOVALOVA, R. A. AND OREKHOV, A. P. (1937). *Bull. Soc. Chim. Fr.* **4**, 2037–2042.

KORNBLUM, N., SELTZER, R. AND HABERFIELD, P. (1963). *J. Am. Chem. Soc.* **85**, 1148–1154.

KOVACH, J. S., AMES, M. M., POWIS, G., MOERTEL, C. G., HAHN, R. G. AND CREAGAN, E. T. (1979). *Cancer Res.* **39**, 4540–4544.

KREHER, R. AND PAWELCZYK, H. (1964). *Angew, Chem., Int. Ed. Engl.* **3**, 510–511.

KROPMAN, M. AND WARREN, F. L. (1949). *J. Chem. Soc.*, 2852–2854.

KRAUS, C., ABEL, G., AND SHIMMER, O. (1985). *Planta Med.* 89–91.

KROPMAN, M. AND WARREN, F. L. (1950). *J. Chem. Soc.* 700–702.

KUGELMAN, M., LIU, W.-C., AXELROD, M., MCBRIDE, T. J. AND RAO, K. V. (1976). *Lloydia* **39**, 125–128.

KUHARA, K., TAKANASHI, H., HIRONO, I., FURUYA, T. AND ASADA, Y. (1980). *Cancer Lett.* **10**, 117–122.

KUMANA, C. R., NG, M., LIN, H. J., KO, W., WU, P.-C. AND TODD, D. (1983). *Lancet* **2**, 1360–1361.

KUMARI, S., KAPUR, K. K. AND ATAL, C. K. (1967). *Curr. Sci.* **35**, 546–547.

KUM-TATT, L. (1960). *J. Pharm. Pharmacol.* **12**, 666–676.

KUPCHAN, S. M. AND SUFFNESS, M. I. (1967). *J. Pharm. Sci.* **56**, 541–543.

KUPCHAN, S. M., DOSKOTCH, R. W. AND VANEVENHOVEN, P. W. (1964). *J. Pharm. Sci.* **53**, 343–345.

KUROZUMI, T., TANAKA, K., KIDO, M. AND SHOYAMA, Y. (1983). *Exp. Mol. Pathol.* **39**, 377–386.

LAFRANCONI, W. M. AND HUXTABLE, R. J. (1981). *Rev. Drug Metab. Drug Interact.* **3**, 271–315.

LAFRANCONI, W. M. AND HUXTABLE, R. J. (1983). *Thorax* **38**, 307–309.

LAFRANCONI, W. M. AND HUXTABLE, R. J. (1984). *Biochem. Pharmacol.* **33**, 2479–2484.

LAFRANCONI, W. M., DUHAMEL, R. C., BRENDEL, K. AND HUXTABLE, R. J. (1984). *Biochem. Pharmacol.* **33**, 191–197.

LAING, M. AND SOMMERVILLE, P. (1972). *Tetrahedron Lett.*, 5183–5186.

LALICH, J. J., JOHNSON, W. D., RACZNIAK, T. J. AND SHUMAKER, R. C. (1977). *Arch. Pathol. Lab. Med.* **101**, 69–73.

LANGLEBEN, D., AND REID, L. M. (1985). *Lab. Invest.* **52**, 298–303.

LANIGAN, G. W. (1970). *Aust. J. Agric. Res.* **21**, 633–639.

LANIGAN, G. W. (1971). *Aust. J. Agric. Res.* **22,** 123–130.

LANIGAN, G. W. (1972). *Aust. J. Agric. Res.* **23,** 1085–1091.

LANIGAN, G. W. AND SMITH, L. W. (1970). *Aust. J. Agric. Res.* **21,** 493–500.

LANIGAN, G. W. AND WHITTEM, J. H. (1970). *Aust. Vet. J.* **46,** 17–21.

LANIGAN, G. W., PAYNE, A. L. AND PETERSON, J. E. (1978). *Aust. J. Agric. Res.* **29,** 1281–1292.

LARIONOV, N. G., KOCHERYA, S. I. AND KRIVUT, B. A. (1980). *Khim.-Farm. Zh.* **14,** 78–81; *Chem. Abstr.* **93,** 31826 (1980).

LARSEN, K. M., ROBY, M. R. AND STERMITZ, F. R. (1984). *J. Natl. Prod.* **47,** 747–748.

LAUTENBERGER, W. J., JONES, E. N. AND MILLER, J. G. (1968). *J. Am. Chem. Soc.* **90,** 1110–1115.

LAWS, L. (1968). *Aust. Vet. J.* **44,** 453.

LEISEGANG, E. C. (1950). *J. S. Afr. Chem. Inst.* **3,** 73–76; *Chem. Abstr.* **46,** 4910 (1952).

LEISEGANG, E. C. AND WARREN, F. L. (1949). *J. Chem. Soc.,* 486–487.

LEISEGANG, E. C. AND WARREN, F. L. (1950). *J. Chem. Soc.,* 702–703.

LEMP. G. (1973). *Planta Med.* **24,** 386–391.

LEONARD, N. J. (1950). *In* "The Alkaloids: Chemistry and Physiology" (R. F. H. Manske and H. L. Holmes, eds.), Vol. 1, pp. 107–164. Academic Press, New York.

LEONARD, N. J. (1960). *In* "The Alkaloids: Chemistry and Physiology" (R. F. H. Manske, ed.), Vol. 6, pp. 35–121. Academic Press, New York.

LEONARD, N. J. AND SATO, T. (1969). *J. Org. Chem.* **34,** 1066–1070.

LETENDRE, L., SMITHSON, W. A., GILCHRIST, G. S., BURGERT, E. O., HOAGLAND, C. H., AMES, M. M., POWIS, G. AND KOVACH, J. S. (1981). *Cancer* **37,** 437–441.

LIANG, X. T. AND ROEDER, E. (1984). *Planta Med.* **50,** 362.

LOCOCK, R. A., BEAL, J. L. AND DOSKOTCH, R. W. (1966). *Lloydia* **29,** 201–205.

LONDAREVA, G. P. AND TIKHOMIROVA, G. B. (1971). *Khim.-Farm. Zh.* **5,** 43–46; *Chem. Abstr.* **75,** 77103 (1971).

LUTHY, J., ZWEIFEL, U., KARLHUBER, B. AND SCHLATTER, C. (1981). *J. Agric. Food Chem.* **29,** 302–305.

LUTHY, J., ZWEIFEL, U., SCHMID, P. AND SCHLATTER, C. (1983a). *Pharm. Acta Helv.* **58,** 98–100.

LUTHY, J., HEIM, T. AND SCHLATTER, C. (1983b). *Toxicol. Lett.* **17,** 283–288.

LUTHY, J., BRAUCHLI, J., ZWEIFEL, U., SCHMID, P. AND SCHLATTER, C. (1984). *Pharm. Acta Helv.* **59,** 242–246.

LYFORD, C. L., VERGARA, G. G. AND MOELLER, D. D. (1976). *Gastroenterology* **70,** 105–108.

MCCOMISH, M., BODEK, I. AND BRANFMAN, A. R. (1980). *J. Pharm. Sci.* **69,** 727–729.

MACDONALD, T. L. AND NARAYANAN, B. A. (1983). *J. Org. Chem.* **48,** 1129–1131.

MCGEE, J. O'D., PATRICK, R. S., WOOD, C. B. AND BLUMGART, L. H. (1976). *J. Clin. Pathol.* **29,** 788–794.

MCGRATH, J. P. M., DUNCAN, J. R. AND MUNNELL, J. F. (1975). *J. Comp. Pathol.* **85,** 185–194.

MACKAY, M. F. AND CULVENOR, C. C. J. (1983). *Acta Crystallogr., Sect. C* **C39,** 1227–1230.

MACKAY, M. F., SADEK, M., CULVENOR, C. C. J. AND SMITH, L. W. (1983). *Acta Crystallogr., Sect. C* **C39,** 1230–1233.

MACKAY, M. F., SADEK, M., CULVENOR, C. C. J. AND SMITH, L. W. (1984). *Acta Crystallogr., Sect. C* **C40,** 473–476.

MCKENZIE, J. S. (1958). *Aust. J. Exp. Biol.* **36,** 11–22.

MACKSAD, A., SCHOENTAL, R. AND COADY, A. (1970). *J. Kuwait Med. Assoc.* **4,** 297–299.

MACLEAN, C. M. U. (1965). *J. Trop. Med. Hyg.* **68,** 237–244.

MCLEAN, E. K. (1970). *Pharmacol. Rev.* **22,** 429–483.

MCLEAN, E. K. (1974). *Isr. J. Med. Sci.* **10,** 436–440.

MCLEAN, E. K. AND MATTOCKS, A. R. (1980). *In* "Toxic Injury of the Liver" (E. Farber and M. M. Fisher, eds.), Part B, pp. 517–539. Dekker, New York.

MCMAHON, R. E. (1966). *J. Pharm. Sci.* **55,** 457–466.

MAHRAN, G., WASSELL, G., EL-MENSHAWI, B., EL-HASSARY, G. AND SAEED, A. (1979). *Acta Pharm. Suec.* **16,** 333–338; *Chem. Abstr.* **92,** 55133 (1980).

MAN'KO, I.V. (1959). *Ukr. Khim. Zh.* **25,** 627–630; *Chem. Abstr.* **54,** 12494 (1960).

MAN'KO, I. V. (1964). *Farm. Zh. (Kiev)* **19,** 22–26; *Chem. Abstr.* **64,** 4125 (1966).

MAN'KO, I. V. (1972). *Rastit. Resur.* **8,** 243–246; *Chem. Abstr.* **77,** 79510 (1972).

MAN'KO I.V. AND BORISYUK, Y. G. (1957). *Ukr. Khim. Zh. (Engl. Transl.)* **23,** 362–366; *Chem. Abstr.* **52,** 2187 (1958).

MAN'KO, I. V. AND KOTOVSKII, B. K. (1970a). *J. Gen. Chem. USSR (Engl. Transl.)* **40,** 2506.

MAN'KO, I. V. AND KOTOVSKII, B. K. (1970b). *J. Gen. Chem. USSR (Engl. Transl.)* **40,** 2519–2520; *Chem. Abstr.* **75,** 1243 (1971).

MAN'KO, I. V. AND MARCHENKO, L. G. (1971). *Khim. Prir. Soedin.,* 537–538; *Chem. Abstr.* **75,** 126598 (1971).

MAN'KO, I. V. AND MARCHENKO, L. G. (1972a). *Khim. Prir. Soedin.,* 655–656; *Chem. Abstr.* **78,** 84611 (1973).

MAN'KO, I. V. AND MARCHENKO, L. G. (1972b). *Khim. Prir. Soedin.,* 812–813; *Chem. Abstr.* **78,** 94812 (1973).

MAN'KO, I.V. AND MARCHENKO, L. G. (1976). *Khim. Prir. Soedin.,* 402–403; *Chem. Abstr.* **85,** 90179 (1976).

MAN'KO, I. V. AND VASIL'KOV, P. N. (1968). *Tr. Leningr. Khim. Farm. Inst.* **26,** 166; *Chem. Abstr.* **73,** 73849 (1970).

MAN'KO, I. V., KOTOVSKII, B. K. AND DENISOV, YU. G. (1970a). *Rastit. Resur.* **6,** 409–411; *Chem. Abstr.* **74,** 61608 (1971).

MAN'KO, I. V., KOTOVSKII, B. K. AND DENISOV, Y. G. (1970b). *Rastit. Resur.* **6,** 582–583; *Chem. Abstr.* **74,** 84023 (1971).

MAN'KO, I. V., MEL'KUMOVA, Z. V. AND MALYSHEVA, U. F. (1972). *Rastit. Resur.* **8,** 538–541; *Chem. Abstr.* **78,** 82085 (1973).

MANSKE, R. F. H. (1931). *Can. J. Res.* **5,** 651–659.

MANSKE, R. F. H. (1936). *Can. J. Res., Sect. B* **14,** 6–11.

MANSKE, R. F. H. (1939). *Can. J. Res., Sect. B* **17,** 1–7.

MARQUEZ, V. C. (1961). *Bol. Soc. Quim. Peru* **27,** 161–172; *Chem. Abstr.* **61,** 15032 (1964).

MARTIN, P. A., THORBURN, M. J., HUTCHINSON, S., BRAS, G. AND MILLER, C. G. (1972). *Br. J. Exp. Pathol.* **53,** 374–380.

MATSUMOTO, T., FUKUI, K. AND EDWARDS, J. D. (1973). *Chem. Lett.,* p. 283.

MATSUMOTO, T., TAKAHASHI, M. AND KASHIHARA, Y. (1979). *Bull. Chem. Soc. Jpn.* **52,** 3329–3336.

MATTOCKS, A. R. (1961). *Nature (London)* **191,** 1281–1282.

MATTOCKS, A. R. (1964a). *J. Chem. Soc.,* 1918–1930.

MATTOCKS, A. R. (1964b). *J. Chem. Soc.,* 1974–1977.

MATTOCKS, A. R. (1967a). *J. Chem. Soc. C,* 329–331.

MATTOCKS, A. R. (1967b). *J. Chromatogr.* **27,** 505–508.

MATTOCKS, A. R. (1967c). *Anal. Chem.* **39,** 443–447.

MATTOCKS, A. R. (1968a). *Nature (London)* **217,** 723–728.

MATTOCKS, A. R. (1968b). *Anal. Chem.* **40,** 1749.

MATTOCKS, A. R. (1968c). *Nature (London)* **219,** 480.

MATTOCKS, A. R. (1968d). *J. Chem. Soc. C,* 235–237.

MATTOCKS, A. R. (1968e). *J. Chem. Soc. C,* 225–226.

MATTOCKS, A. R. (1969a). *J. Chem. Soc. C,* 1155–1162.

MATTOCKS, A. R. (1969b). *J. Chem. Soc. C,* 2698–2700.

MATTOCKS, A. R. (1970). *Nature (London)* **228,** 174–175.

MATTOCKS, A. R. (1971a). *Trop. Sci.* **13,** 65–70.

MATTOCKS, A. R. (1971b). *Xenobiotica* **1,** 451–453.

MATTOCKS, A. R. (1971c). *Xenobiotica* **1,** 563–565.

MATTOCKS, A. R. (1971d). *Nature (London)* **232,** 476–477.

MATTOCKS, A. R. (1972a). *In* "Phytochemical Ecology" (J. B. Harborne, ed.), pp. 179–200. Academic Press, New York.

MATTOCKS, A. R. (1972b). *Chem.-Biol. Interact* **5,** 227–242.

MATTOCKS, A. R. (1973). *Proc. Int. Congr. Pharmacol., 5th, 1972,* 114–123.

MATTOCKS, A. R. (1974). *J. Chem. Soc., Perkin Trans.* **1,** 707–713.

MATTOCKS, A. R. (1975). *Proc. Int. Cancer Congr., 11th, 1974, Excerpta Med. Int. Congr. Ser. No. 350, Vol. 2, pp. 20–24.*

MATTOCKS, A. R. (1977a). J. Chem. Res., Synop., 40–41.

MATTOCKS, A. R. (1977b). *Xenobiotica* **7,** 665–670.

MATTOCKS, A. R. (1978a). *J. Chem. Soc., Perkin Trans.* **1,** 896–905.

MATTOCKS, A. R. (1978b). *In* "Effects of Poisonous Plants on Livestock" (R. F. Keeler, K. R. van Kampen and L. F. James, eds.), pp. 177–187. Academic Press, New York.

MATTOCKS, A. R. (1979). *Toxicol. Lett.* **3**, 79–84.

MATTOCKS, A. R. (1980). *Lancet* **2**, 1136–1137.

MATTOCKS, A. R. (1981a). *Chem.-Biol. Interact.* **35**, 301–310.

MATTOCKS, A. R. (1981b). *Toxicol. Lett.* **8**, 201–205.

MATTOCKS, A. R. (1981c). *Chem. Ind. (London)* 251.

MATTOCKS, A. R. (1982a). *J. Labelled Compd. Radiopharm.* **9**, 479–483.

MATTOCKS, A. R. (1982b). *Toxicol. Lett.* **14**, 111–116.

MATTOCKS, A. R. (1983). *J. Labelled Compd. Radiopharm.* **20**, 285–296.

MATTOCKS, A. R. AND BIRD, I. (1983a). *Chem.-Biol. Interact* **43**, 209–222.

MATTOCKS, A. R. AND BIRD, I. (1983b). *Toxicol. Lett.* **16**, 1–8.

MATTOCKS, A. R. AND CABRAL, J. R. P. (1979). *Tumori* **65**, 289–293.

MATTOCKS, A. R. AND CABRAL, J. R. P. (1982). *Cancer Lett.* **17**, 61–66.

MATTOCKS, A. R. AND DRIVER, H. E. (1983). *Toxicology* **27**, 159–177.

MATTOCKS, A. R. AND LEGG, R. F. (1980). *Chem.-Biol. Interact.* **30**, 325–336.

MATTOCKS, A. R. AND WHITE, I. N. H. (1970). *Anal. Biochem.* **38**, 529–535.

MATTOCKS, A. R. AND WHITE, I. N. H. (1971a). *Chem.-Biol. Interact.* **3**, 383–396.

MATTOCKS, A. R. AND WHITE, I. N. H. (1971b). *Nature (London), New Biol.* **231**, 114–115.

MATTOCKS, A. R. AND WHITE, I. N. H. (1973). *Chem.-Biol. Interact.* **6**, 297–306.

MATTOCKS, A. R. AND WHITE, I. N. H. (1976). *Chem.-Biol. Interact.* **15**, 173–184.

MATTOCKS, A. R., SCHOENTAL, R., CROWLEY, H. C. AND CULVENOR, C. C. J. (1961). *J. Chem. Soc.,* 5400–5403.

MEDVEDEVA, R. G. AND ZOLOTAVINA, Z. M. (1971). *Tr. Inst. Bot., Akad. Nauk. Kaz. SSR* **29**, 181–185; *Chem. Abstr.* **76**, 56583 (1972).

MEINWALD, J., MEINWALD, Y. C. AND MAZZOCCHI, P. H. (1969). *Science* **164**, 1174–1175.

MEINWALD, J., THOMPSON, W. R., EISNER, T. AND OWEN, D. F. (1971). *Tetrahedron Lett.,* 3485–3488.

MEINWALD, J., BORIACK, C., SCHNEIDER, D., BOPPRE, M., WOOD, W. F. AND EISNER, T. (1974). *Experientia* **30**, 721–723.

MEL'KUMOVA, Z. V., TELEZHENETSKAYA, M. V., YUNOSOV, S. YU. AND MAN'KO, I. V. (1974). *Khim. Prir. Soedin.,* 478–480; *Chem. Abstr.* **82**, 54177 (1975).

MEN'SHIKOV, G. P. (1932). *Ber. Dtsch. Chem. Ges. B* **65**, 974–977.

MEN'SHIKOV, G. P. (1946). *J. Gen. Chem. USSR (Engl. Transl.)* **16**, 1311; *Chem. Abstr.* **41**, 3092 (1947).

MEN'SHIKOV, G. P. (1948). *J. Gen. Chem. USSR (Engl. Transl.)* **18**, 1736–1740; *Chem. Abstr.* **43**, 2625 (1949).

MEN'SHIKOV, G. P. AND BORODINA, G. M. (1941). *J. Gen. Chem. (USSR) (Engl. Transl.)* **11**, 209; *Chem. Abstr.* **35**, 7111 (1941).

MEN'SHIKOV, G. P. AND BORODINA, G. M. (1945). *J. Gen. Chem. (USSR)* **15**, 225; *Chem. Abstr.* **40**, 2141 (1946).

MEN'SHIKOV, G. P. AND RUBINSTEIN, W. (1935). *Ber. Dtsch. Chem. Ges. B* **68**, 2039–2044.

MEYRICK, B. O. AND REID, L. M. (1979). *Am. J. Pathol.* **94**, 37–50.

MEYRICK, B. O. AND REID, L. M. (1982). *Am. J. Pathol.* **106**, 84–94.

MEYRICK, B. O., GAMBLE, W. AND REID, L. M. (1980). *Am. J. Physiol.* **239**, H692–H702.

MILLER, W. C., RICE, D. L., KREUSEL, R. G. AND BEDROSSIAN, C. W. M. (1978). *J. Appl. Physiol.* **45**, 962–965.

MIRANDA, C. L., CHEEKE, P. R. AND BUHLER, D. R. (1980a). *Res. Commun. Chem. Pathol. Pharmacol.* **29**, 573–587.

MIRANDA, C. L., CHEEKE, P. R. AND BUHLER, D. R. (1980b). *Biochem. Pharmacol.* **29**, 2645–2649.

MIRANDA, C. L., CARPENTER, H. M., CHEEKE, P. R. AND BUHLER, D. R. (1981a). *Chem.-Biol. Interact.* **37**, 95–107.

MIRANDA, C. L., CHEEKE, P. R., GOEGER, D. E. AND BUHLER, D. R. (1981b). *Toxicol. Lett.* **8**, 343–347.

MIRANDA, C. L., REED, R. L., CHEEKE, P. R. AND BUHLER, D. R. (1981c). *Toxicol. Appl. Pharmacol.* **59**, 424–430.

MIRANDA, C. L., HENDERSON, M. C. AND BUHLER, D. R. (1981d). *Toxicol. Appl. Pharmacol.* **60**, 418–423.

MIRANDA, C. L., HENDERSON, M. C., REED, R. L., SCHMITZ, J. A. AND BUHLER, D. R. (1982a). *J. Toxicol. Environ. Health* **9**, 359–366.

MIRANDA, C. L., HENDERSON, M. C., BUHLER, D. R. AND SCHMITZ, J. A. (1982b). *J. Toxicol. Environ. Health* **9**, 933–939.

MIRANDA, C. L., BUHLER, D. R., RAMSDELL, H. S., CHEEKE, P. R. AND SCHMITZ, J. A. (1982c). *Toxicol. Lett.* **10**, 177–182.

MODY, N. V., SAWHNEY, R. S., RAJINDA, S. AND PELLETIER, S. W. (1979). *J. Nat. Prod.* **42**, 417–420.

MOHABBAT, O., SRIVASTAVA, R. N., YOUNOS, M. S., SEDIQ, G. G., MERZAD, A. A. AND ARAM, G. N. (1976). *Lancet* **2**, 269–271.

MOHANRAJ, S. AND HERZ, W. (1982). *J. Nat. Prod.* **45**, 328–336.

MOHANRAJ, S., SUBRAMANIAN, R. S., CULVENOR, C. C. J., EDGAR, J. A., FRAHN, J. L., SMITH, L. W. AND COCKRUM, P. A. (1978). *Chem. Commun.*, 423–424.

MOHANRAJ, S., KULANTHAIVEL, P., SUBRAMANIAN, P. S. AND HERZ, W. (1981). *Phytochemistry* **20**, 1991–1995.

MOHANRAJ, S., SUBRAMANIAN, P. S. AND HERZ, W. (1982). *Phytochemistry* **21**, 1775–1779.

MOLTENI, A., WARD, W. F., TS'AO, C.-H., PORT, C. D. AND SOLLIDAY, N. H. (1984). *Proc. Soc. Exp. Biol. Med.* **176**, 88–94.

MOLYNEUX, R. J., AND JOHNSON, A. E. (1984). *J. Nat. Prod.* **47,** 1030–1032.

MOLYNEUX, R. J. AND ROITMAN, J. N. (1980). *J. Chromatogr.* **195,** 412–415.

MOLYNEUX, R. J., JOHNSON, A. E., ROITMAN, J. N. AND BENSON, M. E. (1979). *J. Agric. Food Chem.* **27,** 494–499.

MOLYNEUX, R. J., ROITMAN, J. N., BENSON, M. AND LUNDIN, R. E. (1982). *Phytochemistry* **21,** 439–443.

MORI, H., SUGIE, S., YOSHIMI, N., ASADA, Y., FURUYA, T., AND WILLIAMS, G. (1985). *Cancer Res.* **45,** 3125–3129.

MORTIMER, P. H. AND WHITE, E. P. (1967). *Nature (London)* **214,** 1255–1256.

MORTIMER, P. H. AND WHITE, E. P. (1975). *Proc. N.Z. Weed Pest Control Conf.* **28,** 88–91; *Chem. Abstr.* **84,** 70126 (1976).

MOTIDOME, M. AND FERREIRA, P. C. (1966). *Chem. Abstr.* **67,** 79670 (1967).

MUNIER, R. (1953). *Bull. Soc. Chim. Biol.* **35,** 1225.

MUNIER, R., MACHEBOEUF, M. AND CHERRIER, N. (1952). *Bull. Soc. Chim. Biol.* **34,** 204.

MUNOZ QUEVEDO, F. C. (1976). *Rev. Colomb. Cienc. Quim. Farm.* **3,** 45–64; *Chem. Abstr.* **86,** 27671 (1977).

NARASAKA, K. AND UCHIMARU, T. (1982). *Chem. Lett.,* 57–58.

NARASAKA, K., SAKAKURA, T., UCHIMARU, T., MORIMOTO, K. AND MUKAIYAMA, T. (1982). *Chem. Lett.,* 455–458.

NARASAKA, K., SAKAKURA, T., UCHIMARU, T. AND GUEDIN-VUONG, D. (1984). *J. Am. Chem. Soc.* **106,** 2954–2961.

NEUNER-JEHLE, N., NESVABDA, H. AND SPITELLER, G. (1965). *Monatsh. Chem.* **96,** 321–338.

NEWBERNE, P. M. (1968). *Cancer Res.* **28,** 2327–2337.

NEWBERNE, P. M. AND ROGERS, A. E. (1973). *Plant Foods Man* **1,** 23–31.

NEWBERNE, P. M., WILSON, R. AND ROGERS, A. E. (1971). *Toxicol. Appl. Pharmacol.* **18,** 387–397.

NEWBERNE, P. M., WING CHEUNG MICHAL CHAN, AND ROGERS, A. E. (1974). *Toxicol. Appl. Pharmacol.* **28,** 200–208.

NGHIA, T. N., SEDMERA, P., KLASEK, A., BOEVA, A., DRJANOVSKA, L., DOLEJS, L. AND SANTAVY, F. (1976). *Collect. Czech. Chem. Commun.* **41,** 2952–2963.

NIWA, H., ISHIWATA, H. AND YAMADA, K. (1983a). *J. Chromatogr.* **257,** 146–150.

NIWA, H., ISHIWATA, H., KURODA, A. AND YAMADA, K. (1983b). *Chem. Lett.,* 789–790.

NIWA, H., KURODA, A. AND YAMADA, K. (1983c). *Chem. Lett.,* 125–126.

NIWA, H., UOSAKI, Y. AND YAMADA, K. (1983d). *Tetrahedron Lett.* **24,** 5731–5732.

NOSSIN, P. P. M. AND SPECKAMP, W. N. (1979). *Tetrahedron Lett.,* 4411–4414.

NOVELLI, A. (1958). *Chem. Abstr.* **53,** 3606 (1959).

NOWACKI, E. AND BYERRUM, R. U. (1962). *Life Sci.* **1,** 157.

OHNUMA, T., NAGASAKI, M., TABE, M. AND BAN, Y. (1983). *Tetrahedron Lett.* **24,** 4253–4256.

OHSAWA, T., IHARA, M., FUKUMOTO, K. AND KAMETANI, T. (1983). *J. Org. Chem.* **48,** 3644–3648.

OHTSUBO, K., ITO, Y., SAITO, M., FURUYA, T. AND HIKICHI, M. (1977). *Experientia* **33,** 498–499.

O'KELLY, J. AND SARGEANT, K. (1961). *J. Chem. Soc.,* 484.

OMAR, M., DEFEO, J. AND YOUNGKEN, H. W. (1983). *J. Nat. Prod.* **46,** 153–156.

ORD, M. J., HERBERT, A. AND MATTOCKS, A. R. (1985). *Mutat. Res.* **149,** 485–493.

OREKHOV, A., KONOVALOVA, R. A. AND TIDEBEL, V. (1935). *Ber. Dtsch. Chem. Ges.* **68,** 1886.

PANDEY, V. B., SINGH, J. P., RAO, Y. V. AND ACHARYA, S. B. (1982). *Planta Med.* **45,** 229–233.

PANDEY, V. B., SINGH, J. P., MATTOCKS, A. R. AND BAILEY, E. (1983). *Planta Med.* **49,** 254.

PANIZO, F. M. AND RODRIGUEZ, B. (1974). *An. Quim.* **70,** 1043–1048; *Chem. Abstr.* **83,** 179345 (1975).

PASS, D. A., HOGG, G. G., RUSSELL, R. G., EDGAR, J. A., TENCE, I. M. AND RIKARD-BELL, L. (1979). *Aust. Vet. J.* **55,** 284–288.

PASTEWKA, V., WIEDENFELD, H. AND ROEDER, E. (1980). *Arch. Pharm. (Weinheim, Ger.)* **313,** 785–790.

PEDERSEN, E. (1970). *Dan. Tidsskr. Farm.* **44,** 287–291; *Chem. Abstr.* **74,** 72780 (1971).

PEDERSEN, E. (1975a). *Arch. Pharm. Chem., Sci. Ed.* **3,** 55–64.

PEDERSEN, E. (1975b). *Phytochemistry* **14,** 2086–2087.

PEDERSEN, E. AND LARSEN, E. (1970). *Org. Mass Spectrom.* **4,** 249–256.

PERCY, J. J. AND PIERCE, A. E. (1971). *Immunology* **21,** 273–280.

PEREZ-SALAZAR, A. (1978). *An. Quim.* **74,** 196–198.

PEREZ-SALAZAR, A., CANO, F. H., FAYOS, J., MARTINEZ-CARRERA, S. AND GARCIA-BLANCO, S. (1977). *Acta Crystallogr., Sect. B* **B33,** 3525–3527.

PEREZ-SALAZAR, A., CANO, F. H. AND GARCIA-BLANCO, S. (1978). *Cryst. Struct. Commun.* **7,** 105–109.

PERSAUD, T. V. N. AND HOYTE, D. A. N. (1974). *Exp. Pathol.* **9,** 59–63.

PERSAUD, T. V. N., PUTZKE, H. P., TESSMANN, D. AND BIENENGRAEBER, A. (1970). *Acta Histochem.* **37,** 369–378.

PETERSON, J. E. (1965). *J. Pathol. Bacteriol* **89,** 153–171.

PETERSON, J. E. AND CULVENOR, C. C. J. (1983). *In* "Handbook of Natural Toxins" (R. F. Keeler and A. T. Tu, eds.), Vol. 1, pp. 637–671. Dekker, New York.

PETERSON, J. E. AND JAGO, M. V. (1980). *J. Pathol.* **131,** 339–355.

PETERSON, J. E. AND JAGO, M. V. (1984). *Aust. J. Agric. Res.* **35,** 305–315.

PETERSON, J. E., SAMUEL, A. AND JAGO, M. V. (1972). *J. Pathol.* **107,** 175–189.

PETERSON, J. E., JAGO, M. V., REDDY, J. K. AND JARRETT, R. G. (1983). *JNCI, J. Natl. Cancer Inst.* **70,** 381–386.

PETRY, T. W., BOWDEN, G. T., HUXTABLE, R. J. AND SIPES, I. G. (1984). *Cancer Res.* **44,** 1505–1509.

PHILLIPSON, J. D. (1971). *Xenobiotica* **1,** 419–447.

PHILLIPSON, J. D. AND HANDA, S. S. (1978). *J. Nat. Prod.* **41,** 385–431.

PICKARD, P. L. AND IDDINGS, F. A. (1959). *Anal. Chem.* **31,** 1228–1230.

PIERSON, M. L., CHEEKE, P. R. AND DICKINSON, E. O. (1977). *Res. Commun. Chem. Pathol. Pharmacol.* **16,** 561–564.

PILBEAM, D. J., LYON-JOYCE, A. J. AND BELL, E. A. (1983). *J. Nat. Prod.* **46,** 601–605.

PIMENOV, M. G., YAKHONTOVA, L. D., PAKALNE, D. AND SAPUNOVA, L. A. (1975). *Rastit. Resur.* **11,** 72–77; *Chem. Abstr.* **82,** 152218 (1975).

PINNICK, H. W. AND CHANG, Y.-H. (1978). *J. Org. Chem.* **43,** 4662–4663.

PIPER, J. R., KARI, P. AND SHEALY, Y. F. (1981). *J. Labelled Compd. Radiopharm.* **18,** 1579–1591.

PIZZORNO, M. T. AND ALBONICO, S. M. (1974). *J. Org. Chem.* **39,** 731.

PLESCIA, S., DAIDONE, G. AND SPRIO, V. (1976). *Phytochemistry* **15,** 2026.

PLESTINA, R. AND STONER, H. B. (1972). *J. Pathol.* **106,** 235–249.

PLESTINA, R., STONER, H. B., JONES, G., BUTLER, W. H. AND MATTOCKS, A. R. (1977). *J. Pathol.* **121,** 9–18.

PLISKE, T. E., EDGAR, J. A. AND CULVENOR, C. C. J. (1976). *J. Chem. Ecol.* **2,** 255–262.

POHLENZ, J., LUTHY, J., MINDER, H. P. AND BIVETTI, A. (1980). *Schweiz. Arch. Tierheilkd.* **122,** 183–193.

POMEROY, A. R. AND RAPER, C. (1971a). *Br. J. Pharmacol.* **41,** 683–690.

POMEROY, A. R. AND RAPER, C. (1971b). *Eur. J. Pharmacol.* **14,** 374–383.

POOL, B. L. (1982). *Toxicology* **24,** 351–355.

PORTER, L. A. AND GEISSMAN, T. A. (1962). *J. Org. Chem.* **27,** 4132–4134.

POWIS, G. AND WINCENTSEN, L. (1980). *Biochem. Pharmacol.* **29,** 347–351.

POWIS, G., AMES, M. M. AND KOVACH, J. S. (1979a). *Cancer Res.* **39,** 3564–3570.

POWIS, G., AMES, M. M. AND KOVACH, J. S. (1979b). *Res. Commun. Chem. Pathol. Pharmacol.* **24,** 559–569.

POWIS, G., SVINGEN, B. A. AND DEGRAW, C. (1982). *Biochem. Pharmacol.* **31,** 293–299.

PRETORIUS, T. P. (1949). *Onderstepoort J. Vet. Sci. Anim. Ind.* **22,** 297–300; *Chem. Abstr.* **44,** 3217 (1950).

PURI, S. C., SAWHNEY, R. S. AND ATAL, C. K. (1973). *Experientia* **29,** 390.

PURI, S. C., SAWHNEY, R. S. AND ATAL, C. K. (1974). *J. Indian Chem. Soc.* **51,** 628.

PUTZKE, H. P. AND PERSAUD, T. V. (1976). *Exp. Pathol.* **12,** 329–335.

QUALLS, C. W. (1980). *Diss. Abstr. Int. B* **41,** 2080–2081.

QUALLS, C. W. AND SEGALL, H. J. (1978). *J. Chromatogr.* **150,** 202–206.

RACZNIAK, T. J., SHUMAKER, R. C., ALLEN, J. R., WILL, J. A. AND LALICH, J. J. (1979). *Respiration* **37,** 252–260.

RAJAGOPALAN, T. R. AND BATRA, V. (1977a). *Indian J. Chem., Sect. B* **15B**, 455–457.

RAJAGOPALAN, T. R. AND BATRA, V. (1977b). *Indian J. Chem., Sect. B.* **15B**, 494.

RAMSDELL, H. S. AND BUHLER, D. R. (1979). *In* "Pyrrolizidine Alkaloids" (P. R. Cheeke, ed.), pp. 19–22. Nutr. Res. Inst., Oregon State University, Corvallis.

RAMSDELL, H. S. AND BUHLER, D. R. (1981a). *J. Chromatogr.* **210**, 154–158.

RAMSDELL, H. S. AND BUHLER, D. R. (1981b). *Toxicologist* **1**, 90 (Abstr. No. 326).

RANA, J. AND ROBINS, D. J. (1983). *Chem. Commun.*, 1222–1224.

RANA, J. AND ROBINS, D. J. (1984). *Chem. Commun.*, 517–519.

RAO, M. S. AND REDDY, J. K. (1978). *Br. J. Cancer* **37**, 289–293.

RAO, M. S., JAGO, M. V. AND REDDY, J. K. (1983). *Hum. Toxicol.* **2**, 15–26.

RAO, P. G., SURI, O. P., SAWHNEY, R. S. AND ATAL, C. K. (1974). *Indian J. Pharm.* **36**, 163.

RAO, P. G., SAWHNEY, R. S. AND ATAL, C. K. (1975a). *Indian J. Chem.* **13**, 835–836.

RAO, P. G., SAWHNEY, R. S. AND ATAL, C. K. (1975b). *Indian J. Chem.* **13**, 870–871.

RAO, P. G., SAWHNEY, R. S. AND ATAL, C. K. (1975c). *Experientia* **31**, 878.

RAO, P. G., SAWHNEY, R. S., GUPTA, O. P. AND ATAL, C. K. (1975d). *Indian J. Pharm.* **37**, 127–129.

RATNOFF, O. D. AND MIRICK, G. S. (1949). *Bull. Johns Hopkins Hosp.* **84**, 507–525.

REDDY, J. K. AND SVOBODA, D. (1972). *Arch. Pathol.* **93**, 55–60.

REDDY, J. K., HARRIS, G. AND SVOBODA, D. (1968). *Nature (London)* **217**, 659–661.

REDDY, J. K., CHIGA, M. AND SVOBODA, D. (1969). *Lab. Invest.* **20**, 405–411.

REDDY, J. K., RAO, M. S. AND JAGO, M. V. (1976). *Int. J. Cancer* **17**, 621–625.

RED'KO, A. L. (1956). *Chem. Abstr.* **53**, 20695 (1959).

RESCH, J. F., ROSBERGER, D. F., MEINWALD, J. AND APPLING, J. W. (1982). *J. Nat. Prod.* **45**, 358–362.

RESCH, J. F., GOLDSTEIN, S. A. AND MEINWALD, J. (1983). *Planta Med.* **47**, 255.

RETIEF, G. P. (1962). *J. S. Afr. Vet. Med. Assoc.* **33**, 405–407.

REYES, Q. A., RODRIQUEZ, M. O. AND SILVA, O. M. (1982). *Bol. Soc. Chil. Quim.* **27**, 305–306; *Chem. Abstr.* **96**, 196575 (1982).

RICHARDSON, M. F. AND WARREN, F. L. (1943). *J. Chem. Soc.*, 452–454.

RIDKER, P. M., OHKUMA, S., MCDERMOTT, W. V., TREY, C. AND HUXTABLE, R. J. (1985). *Gastroenterology* **88**, 1050–1054.

RIMINGTON, C., KROL, S. AND TOOTH, B. (1956). *Scand. J. Clin. Lab. Invest.* **8**, 251.

ROBERTSON, K. A. (1982). *Cancer Res.* **42**, 8–14.

ROBERTSON, K. A., SEYMOUR, J. L., HSIA, M.-T. AND ALLEN, J. R. (1977). *Cancer Res.* **37**, 3141–3144.

ROBINS, D. J. (1979). *Adv. Heterocycl. Chem.* **24**, 247–291.

ROBINS, D. J. (1982a). *Fortschr. Chem. Org. Naturst.* **41**, 115–203.

ROBINS, D. J. (1982b). *Chem. Commun.*, 1289–1290.

ROBINS, D. J. AND CROUT, D. H. G. (1969). *J. Chem. Soc.* C, 1386–1391.

ROBINS, D. J. AND SAKDARAT, S. (1979a). *Chem. Commun.*, 1181–1182.

ROBINS, D. J. AND SAKDARAT, S. (1979b). *J. Chem. Soc., Perkin Trans.* 1, 1734–1735.

ROBINS, D. J. AND SAKDARAT, S. (1980). *Chem. Commun.*, 282–283.

ROBINS, D. J. AND SAKDARAT, S. (1981). *J. Chem. Soc., Perkin Trans.* 1, 909–913.

ROBINS, D. J. AND SWEENEY, J. R. (1981). *J. Chem. Soc., Perkin Trans.* 1, 3083–3086.

ROBY, M. R., AND STERMITZ, F. R. (1984). *J. Nat. Prod.* 47, 846–853.

RODRIGUEZ, F. D. AND GONZALEZ, A. G. (1969). *An. Quim.* 65, 307.

RODRIGUEZ, F. D. AND GONZALEZ, A. G. (1971). *Farm. Nueva* 36, 810–812; *Chem. Abstr.* 76, 83572 (1972).

RODRIGUEZ, F. D., GONZALEZ, A. G. AND MENDEZ, A. M. (1971). *Farm. Nueva* 36, 803–810; *Chem. Abstr.* 76, 83573 (1972).

ROEDER, E. AND WIEDENFELD, H. (1977). *Phytochemistry* 16, 1462–1463.

ROEDER, E., WIEDENFELD, H. AND PASTEWKA, U. (1979). *Planta Med.* 37, 131–136.

ROEDER, E., WIEDENFELD, H. AND FRISSE, M. (1980a). *Phytochemistry* 19, 1275–1277.

ROEDER, E., WIEDENFELD, H. AND FRISSE, M. (1980b). *Arch. Pharm.* (Weinheim, Ger.) 313, 803–806.

ROEDER, E., WIEDENFELD, H. AND STENGL, P. (1980c). *Planta Med., Suppl.*, pp. 182–184.

ROEDER, E., WIEDENFELD, H. AND JOST, E. J. (1981). *Planta Med.* 43, 99–102.

ROEDER, E., WIEDENFELD, H. AND STENGL, P. (1982a). *Arch. Pharm.* (Weinheim, Ger.) 315, 87–89.

ROEDER, E., WIEDENFELD, H. AND JOST, E. J. (1982b). *Planta Med.* 44, 182–183.

ROEDER, E., WIEDENFELD, H. AND HOENIG, A. (1983). *Planta Med.* 49, 57–59.

ROEDER, E., WIEDENFELD, H. AND BRIZ-KIRSTGEN, R. (1984a). *Phytochemistry* 23, 1761–1763.

ROEDER, E., WIEDENFELD, H. AND SCHRAUT, R. (1984b). *Phytochemistry* 23, 2125–2126.

ROGERS, A. E. AND NEWBERNE, P. M. (1971). *Toxicol. Appl. Pharmacol.* 18, 356–366.

ROGERS, Q. R., KNIGHT, H. D. AND GULICK, B. A. (1979). *In* "Pyrrolizidine Alkaloids" (P. R. Cheeke, ed.), pp. 145–147. Nutr. Res. Inst., Oregon State University, Corvallis.

ROITMAN, J. N. (1981). *Lancet* 1, 944.

ROITMAN, J. N. (1983a). *Aust. J. Chem.* 36, 769–778.

ROITMAN, J. N. (1983b). *Aust. J. Chem.* 36, 1203–1213.

ROITMAN, J. N., MOLYNEUX, R. J. AND JOHNSON, A. E. (1979). *In* "Pyrrolizidine Alkaloids" (P. R. Cheeke, ed.), pp. 23–33. Nutr. Res. Inst., Oregon State University, Corvallis.

ROSBERGER, D. F., RESCH, J. F. AND MEINWALD, J. (1981). *Mitt. Geb. Lebensmittelunters. Hyg.* 72, 432–436; *Chem. Abstr.* 98, 105865 (1983).

ROSE, C. L., HARRIS, P. N. AND CHEN, K. K. (1959). *J. Pharmacol. Exp. Ther.* 126, 179–184.

ROSE, E. F. (1972). *S. Afr. Med. J.* **46,** 1039–1043.

ROSILES, M. R. AND PAASCH, L. H. (1982). *Veterinaria (Mexico City)* **13,** 151–153; *Vet. Bull.,* p. 959 (Abstr. No. 6690) (1983).

ROSS, A. J. (1977). *J. Agric. Sci.* **89,** 101–105.

ROSS, A. J. AND TUCKER, J. W. (1977). *J. Agric. Sci.* **89,** 95–99.

ROTH, R. A. (1981). *Toxicol. Appl. Pharmacol.* **57,** 69–78.

ROTH, R. A., DOTZLAF, L. A., BARANYI, B., KUO, C.-H. AND HOOK, J. B.(1981). *Toxicol. Appl. Pharmacol.* **60,** 193–203.

ROTHSCHILD, M., APLIN, R. T., COCKRUM, P. A., EDGAR, J. A., FAIRWEATHER, P. AND LEES, R. (1979). *Biol. J. Linn. Soc.* **12,** 305–326.

RUEGER, H. AND BENN, M. (1982). *Heterocycles* **19,** 1677–1680.

RUEGER, H. AND BENN, M. H. (1983a). *Can. J. Chem.* **61,** 2526–2529.

RUEGER, H. AND BENN, M. (1983b). *Heterocycles* **20,** 1331–1334.

RYSKIEWICZ, E. E. AND SILVERSTEIN, R. M. (1954). *J. Am. Chem. Soc.* **76,** 5802.

SAMUEL, A. AND JAGO, M. V. (1975). *Chem.-Biol. Interact.* **10,** 185–197.

SANTAVY, F., SULA, B. AND MANIS, V. (1962). *Collect. Czech. Chem. Commun.* **27,** 1666–1671.

SAPIRO, M. L. (1949). *Onderstepoort J. Vet. Sci.* **22,** 291–295; *Chem. Abstr.* **44,** 3217 (1950).

SAPUNOVA, L. C. AND BAN'KOVSKII, A. I. (1968). *Khim. Prir. Soedin.,* 389; *Chem. Abstr.* **70,** 109120 (1969).

SAWHNEY, R. S. AND ATAL, C. K. (1966). *Indian J. Pharm.* **28,** 274–275.

SAWHNEY, R. S. AND ATAL, C. K. (1968). *J. Indian Chem. Soc.* **45,** 1052–1053.

SAWHNEY, R. S. AND ATAL, C. K. (1970). *J. Indian Chem. Soc.* **47,** 667–668.

SAWHNEY, R. S. AND ATAL, C. K. (1972). *Planta Med.* **21,** 435–437.

SAWHNEY, R. S. AND ATAL, C. K. (1973). *Indian J. Chem.* **11,** 88–89.

SAWHNEY, R. S., GIROTRA, R. N., ATAL, C. K., CULVENOR, C. C. J. AND SMITH, L. W. (1967). *Indian J. Chem.* **5,** 655–656.

SAWHNEY, R. S., ATAL, C. K., CULVENOR, C. C. J. AND SMITH, L. W. (1974). *Aust. J. Chem.* **27,** 1805–1808.

SCHLOSSER, F. D. AND WARREN, F. L. (1965). *J. Chem. Soc.,* 5707–5710.

SCHMID, W. (1975). *Mutat. Res.* **31,** 9–15.

SCHNEIDER, D., BOPPRE, M., SCHNEIDER, H., THOMPSON, W. R., BORIACK, C. J., PETTY, R. L. AND MEINWALD, J. (1975). *J. Comp. Physiol.* **97,** 245–256.

SCHNEIDER, D., BOPPRE, M., ZWEIG, J., HORSLEY, S. B., BELL, T. W., MEINWALD, J., HANSEN, K. AND DIEHL, E. W. (1982). *Science* **215,** 1264–1265.

SCHOENTAL, R. (1959). *J. Pathol. Bacteriol.* **77,** 485–495.

SCHOENTAL, R. (1960). *J. Chem. Soc.,* 2375.

SCHOENTAL, R. (1963). *Aust. J. Chem.* **16,** 233–238.

SCHOENTAL, R. (1966). *J. Pathol. Bacteriol.* **91,** 629–631.

SCHOENTAL, R. (1968a). *Cancer Res.* **28**, 2237–2246.

SCHOENTAL, R. (1968b). *Isr. J. Med. Sci.* **4**, 1133–1145.

SCHOENTAL, R. (1970). *Nature (London)* **227**, 401–402.

SCHOENTAL, R. (1972). *Nature (London)* **238**, 106–107.

SCHOENTAL, R. (1975a). *Cancer Res.* **35**, 2020–2024.

SCHOENTAL, R. (1975b). *Biochem. Soc. Trans.* **3**, 292–294.

SCHOENTAL, R. AND BENSTED, J. P. M. (1963). *Br. J. Cancer* **17**, 242–251.

SCHOENTAL, R. AND CAVANAGH, J. B. (1972). *J. Natl. Cancer Inst. (U.S.)* **49**, 665–671.

SCHOENTAL, R. AND COADY, A. (1968). *East Afr. Med. J.* **45**, 577–580.

SCHOENTAL, R. AND HEAD, M. A. (1955). *Br. J. Cancer* **9**, 299–237.

SCHOENTAL, R. AND HEAD, M. A. (1957). *Br. J. Cancer* **11**, 535–544.

SCHOENTAL, R. AND MAGEE, P. N. (1957). *J. Pathol. Bacteriol.* **74**, 305–319.

SCHOENTAL, R. AND MAGEE, P. N. (1959). *J. Pathol. Bacteriol.* **78**, 471–482.

SCHOENTAL, R. AND MATTOCKS, A. R. (1960). *Nature (London)* **185**, 842–843.

SCHOENTAL, R. AND PULLINGER, B. D. (1972). *East Afr. Med. J.* **49**, 436–439.

SCHOENTAL, R., HEAD, M. A. AND PEACOCK, P. R. (1954). *Br. J. Cancer* **8**, 458–465.

SCHOENTAL, R., FOWLER, M. E. AND COADY, A. (1970). *Cancer Res.* **30**, 2127–2131.

SCHROTER, H. B. AND SANTAVY, F. (1960). *Collect. Czech. Chem. Commun.* **25**, 472–482.

SEAMAN, W. AND ALLEN, E. (1951). *Anal. Chem.* **23**, 592–594.

SEDMERA, P., KLASEK, A., DUFFIELD, A. M. AND SANTAVY, F. (1972). *Collect. Czech. Chem. Commun.* **37**, 4112–4119.

SEGALL, H. J. (1978). *Toxicol. Lett.* **1**, 279–284.

SEGALL, H. J. (1979a). *J. Liq. Chromatogr.* **2**, 429–436.

SEGALL, H. J. (1979b). *J. Liq. Chromatogr.* **2**, 1319–1323.

SEGALL, H. J. AND DALLAS, J. L. (1983). *Phytochemistry* **22**, 1271–1273.

SEGALL, H. J. AND KRICK, T. P. (1979). *Toxicol. Lett.* **4**, 193–198.

SEGALL, H. J. AND MOLYNEUX, R. J. (1978). *Res. Commun. Chem. Pathol. Pharmacol.* **19**, 545–548.

SEGALL, H. J., BROWN, C. H. AND PAIGE, D. F. (1983). *J. Labelled Compd. Radiopharm.* **20**, 671–689.

SEGALL, H. J., DALLAS, J. L. AND HADDON, W. F. (1984). *Drug. Metab. Dispos.* **12**, 68–71.

SELZER, G. AND PARKER, R. G. F. (1951). *Am. J. Pathol.* **27**, 885–907.

SETHI, M. L. AND ATAL, C. K. (1964). *Planta Med.* **12**, 173–176.

SHA, S.-Y. AND TSENG, C.-Y. (1980). *Yao Hsueh Tung Pao* **15**, 4–6; *Chem. Abstr.* **94**, 90416h (1981).

SHARMA, M. L., SINGH, G. B. AND GHATAK, B. J. R. (1967). *Indian J. Exp. Biol.* **5**, 149.

SHARMA, R. K. AND HEBBORN, P. (1968). *J. Med. Chem.* **11**, 620–621.

SHARMA, R. K., KHAJURIA, G. S. AND ATAL, C. K. (1965). *J. Chromatogr.* **19**, 433–434.

SHEVELEVA, G. P., PLEKHANOVA, N. V. AND SARGAZAKOV, D. S. (1969). *Chem. Abstr.* **76**, 23067 (1972).

SHONO, T., MATSUMURA, Y., UCHIDA, K., TSUBATA, K. AND MAKINO, A. (1984). *J. Org. Chem.* **49**, 300–304.

SHULL, L. R., BUCKMASTER, G. W. AND CHEEKE, P. R. (1976a). *J. Anim. Sci.* **43**, 1024–1027.

SHULL, L. R., BUCKMASTER, G. W. AND CHEEKE, P. R. (1976b). *J. Anim. Sci.* **43**, 1247–1253.

SHULL, L. R., BUCKMASTER, G. W. AND CHEEKE, P. R. (1977). *Res. Commun. Chem. Pathol. Pharmacol.* **17**, 337–340.

SHUMAKER, R. C., HSU, I. C. AND ALLEN, J. R. (1976a). *J. Pathol.* **119**, 21–28.

SHUMAKER, R. C., ROBERTSON, K. A., HSU, I. C. AND ALLEN, J. R. (1976b). *J. Natl. Cancer Inst. (U.S.)* **56**, 787–789.

SHUMAKER, R. C., SEYMOUR, J. L. AND ALLEN, J. R. (1976c). *Res. Commun. Chem. Pathol. Pharmacol.* **14**, 53–61.

SHUMAKER, R. C., RACZNIAK, T. J., JOHNSON, W. D. AND ALLEN, J. R. (1977). *Proc. Soc. Exp. Biol. Med.* **154**, 57–59.

SIDDIQI, M. A., SURI, K. A., SURI, O. P. AND ATAL, C. K. (1978a). *Phytochemistry* **17**, 2049–2050.

SIDDIQI, M. A., SURI, K. A., SURI, O. P. AND ATAL, C. K. (1978b). *Indian J. Chem., Sect. B* **16B**, 1132–1133.

SIDDIQI M. A., SURI, K. A., SURI, O. P. AND ATAL, C. K. (1978c). *Phytochemistry* **17**, 2143–2144.

SIDDIQI, M. A., SURI, K. A., SURI, O. P. AND ATAL, C. K. (1979a). *Indian J. Pharm.* **41**, 129–130.

SIDDIQI, M. A., SURI, K. A., SURI, O. P. AND ATAL, C. K. (1979b). *Phytochemistry* **18**, 1413–1415.

SIMANEK, V., KLASEK, A. AND SANTAVY, F. (1969). *Collect. Czech. Chem. Commun.* **34**, 1832–1836.

SINGH, G. B., SHARMA, M. L. AND GHATAK, B. L. R. (1969). *Indian J. Exp. Biol.* **7**, 144–147.

SIPPEL, W. L. (1964). *Ann. N.Y. Acad. Sci.* **111**, 562–570.

SMITH, L. W. AND CULVENOR, C. C. J. (1981). *J. Nat. Prod.* **44**, 121–152.

SMITH, L. W. AND CULVENOR, C. C. J. (1984). *Phytochemistry* **23**, 457–459.

SMITH, P. AND HEATH, D. (1978). *J. Pathol.* **124**, 177–183.

SNEHALATA, S. AND GHOSH, M. N. (1968). *Indian J. Med. Res.* **56**, 1386–1390.

SRUNGBOONMEE, S. AND MASKASAME, C. (1981). *Sattawaphaet San* **32**, 91–107; *Chem. Abstr.* **96**, 137560 (1982).

STENMARK, K. R., MORGANROTH, M. L., REMIGIO, L. K., VOELKEL, N. F., MURPHY, R. C., HENSON, P. M., MATHIAS, M. M., AND REEVES, J. T. (1985). *Am. J. Physiol.* **248**, H859–H866.

STERMITZ, F. R. AND ADAMOVICS, J. S. (1977). *Phytochemistry* **16,** 500.

STERMITZ, F. R. AND SUESS, T. R. (1978). *Phytochemistry* **17,** 2142.

STERNBACH, D. D., ABOOD, L. G. AND HOSS, W. (1974). *Life Sci.* **14,** 1847–1856.

STILLMAN, A. S., HUXTABLE, R., CONSROE, P., KOHNEN, P. AND SMITH, S. (1977). *Gastroenterology* **73,** 349–352.

STOECKLI-EVANS, H. (1979a). *Acta Crystallogr., Sect. B* **B35,** 231–234.

STOECKLI-EVANS, H. (1979b). *Acta Crystallogr., Sect. B* **B35,** 2798–2800.

STOECKLI-EVANS, H. (1982). *Acta Crystallogr., Sect. B.* **B38,** 1614–1617.

STOECKLI-EVANS, H. AND CROUT, D. H. G. (1976). *Helv. Chim. Acta* **59,** 2168–2178.

STOECKLI-EVANS, H. AND ROBINS, D. J. (1983). *Helv. Chim. Acta* **66,** 1376–1380.

STOYEL, C. AND CLARK, A. M. (1980). *Mutat. Res.* **74,** 393–398.

STYLES, J. (1977). *Br. J. Cancer* **36,** 558–563.

STYLES, J., ASHBY, J. AND MATTOCKS, A. R. (1980). *Carcinogenesis (N.Y.)* **1,** 161–164.

SUBRAMANIAN, S. S. AND NAGARAJAN, S. (1967). *Indian J. Pharm.* **29,** 311–312.

SUBRAMANIAN, S. S., NAGARAJAN, S. AND GHOSH, M. N. (1968). *Indian J. Pharm.* **30,** 153–155.

SUBRAMANIAN, P. S., MOHANRAJ, S., COCKRUM, P. A., CULVENOR, C. C. J., EDGAR, J. A., FRAHN, J. L. AND SMITH, L. W. (1980). *Aust. J. Chem.* **33,** 1357–1363.

SUGITA, T., HYERS, T. M., DAUBER, I. M., WAGNER, W. W., MCMURTRY, I. F. AND REEVES, J. T. (1983a). *J. Appl. Physiol.* **54,** 371–374.

SUGITA, T., STENMARK, K. R., WAGNER, W. W., HENSON, P. M., HENSON, J. E., HYERS, T. M. AND REEVES, J. T. (1983b). *Exp. Lung Res.* **5,** 201–215.

SULLIVAN, G. (1981). *Vet. Hum. Toxicol.* **23,** 6–7.

SULLMAN, S. F. AND ZUCKERMAN, A. J. (1969). *Br. J. Exp. Pathol.* **50,** 361–370.

SUN, P. S., HSIA, M.-T. S., CHU, F. S. AND ALLEN, J. R. (1977). *Food Cosmet. Toxicol.* **15,** 419–422.

SUNDARESON, A. E. (1942). *J. Pathol. Bacteriol.* **54,** 289–298.

SURI, K. A., SAWHNEY, R. S. AND ATAL, C. K. (1975a). *Indian J. Pharm.* **37,** 36–38.

SURI, K. A., SAWHNEY, R. S. AND ATAL, C. K. (1975b). *Indian J. Pharm.* **37,** 69–70.

SURI, K. A., SAWHNEY, R. S. AND ATAL, C. K. (1975c). *Indian J. Pharm.* **37,** 96–97.

SURI, K. A., SAWHNEY, R. S. AND ATAL, C. K. (1976a). *Indian J. Chem., Sect. B* **14B,** 471.

SURI, K. A., SAWHNEY, R. S., GUPTA, O. P. AND ATAL, C. K. (1976b). *Indian J. Pharm.* **38,** 23–25.

SURI, K. A., SURI, O. P., DHAR, K. L. AND ATAL, C. K. (1978). *Indian J. Chem., Sect. B* **16B,** 78.

SURI, O. P. AND ATAL, C. K. (1967). *Curr. Sci.* **36,** 614–615.

SURI, O. P., SAWHNEY, R. S. AND ATAL, C. K. (1975). *Indian J. Chem.* **13,** 505–506.

SURI, O. P., SAWHNEY, R. S., BHATIA, M. S. AND ATAL, C. K. (1976). *Phytochemistry* **15,** 1061–1063.

SURI, O. P., JAMWAL, R. S., KHAJURIA, R. K., ATAL, C. K. AND HAKSAR, C. N. (1982). *Planta Med.* **44**, 181.

SUSSMAN, J. L. AND WODAK, S. J. (1973). *Acta Crystallogr.* **29**, 2918–2926.

SUZUKI, H. AND TWAROG, B. M. (1982). *Am. J. Physiol.* **242**, H907–H915.

SVOBODA, M. D. AND REDDY, J. K. (1972). *Cancer Res.* **32**, 908–912.

SVOBODA, M. D. AND REDDY, J. K. (1974). *J. Natl. Cancer Inst. (U.S.)* **53**, 1415–1418.

SVOBODA, M. D. AND SOGA, J. (1966). *Am. J. Pathol.* **48**, 347–373.

SWAIN, C. G. AND SCOTT, C. B. (1953). *J. Am. Chem. Soc.* **75**, 141–147.

SWICK, R. A., CHEEKE, P. R. AND BUHLER, D. R. (1979). *In* "Pyrrolizidine Alkaloids" (P. R. Cheeke, ed.), pp. 115–123. Nutr. Res. Inst., Oregon State University, Corvallis.

SWICK, R. A., CHEEKE, P. R., GOEGER, D. E. AND BUHLER, D. R. (1982a). *J. Anim. Sci.* **55**, 1411–1416.

SWICK, R. A., CHEEKE, P. R., PATTON, N. M. AND BUHLER, D. R. (1982b). *J. Anim. Sci.* **55**, 1417–1424.

SWICK, R. A., CHEEKE, P. R. AND BUHLER, D. R. (1982c). *J. Anim. Sci.* **55**, 1425–1430.

SWICK, R. A., CHEEKE, P. R., MIRANDA, C. L. AND BUHLER, D. R. (1982d). *J.Toxicol. Environ. Health* **10**, 757–768.

SWICK, R. A., MIRANDA, C. L., CHEEKE, R. P. AND BUHLER, D. R. (1982e). *Proc., Annu. Meet.—Am. Soc. Anim. Sci., West. Sect.* **33**, 67–70.

SWICK, R. A., CHEEKE, P. R., RAMSDELL, H. S. AND BUHLER, D. R. (1983a). *J. Anim. Sci.* **56**, 645–651.

SWICK, R. A., MIRANDA, C. L., CHEEKE, P. R. AND BUHLER, D. R. (1983b). *J. Anim. Sci.* **56**, 887–894.

SYDNES, L. K., SKATTEBOL, L., CHAPLEO, C. B., LEPPARD, D. G., SVANHOLT, K. L. AND DREIDING, A. S. (1975). *Helv. Chim. Acta* **58**, 2061–2073.

SYKULSKA, Z. (1962). *Acta Pol. Pharm.* **19**, 183–184; *Chem. Abstr.* **59**, 2876 (1963).

TAKANASHI, H., UMEDA, M. AND HIRONO, I. (1980). *Mutat. Res.* **78**, 67–77.

TAKEOKA, O., ANGEVINE, D. M. AND LALICH, J. J. (1962). *Am. J. Pathol.* **40**, 545–554.

TANDON, B. N., TANDON, R. K., TANDON, H. D., NARNDRANATHAN, M. AND JOSHI, Y. K. (1976). *Lancet* **2**, 271–272.

TANDON, H. D., TANDON, B. N. AND MATTOCKS, A. R. (1978). *Am. J. Gastroenterol.* **70**, 607–613.

TANDON, R. K., TANDON, B. N., TANDON, H. D., BHATIA, M. L., BHARGAVA, S., LAL, P. AND ARORA, R. R. (1976). *Gut* **17**, 849–855.

TASHKHODZHAEV, B., TELEZHENETSKAYA, M. V. AND YUNOSOV, S. YU. (1979a). *Khim. Prir. Soedin.*, pp. 363–367; *Chem. Abstr.* **92**, 111199 (1980).

TASHKODZHAEV, B., YAGUDAEV, M. R. AND YUNOSOV, S. YU. (1979b). *Khim. Prir. Soedin.*, pp. 368–373; *Chem. Abstr.* **92**, 111194 (1980).

TATSUTA, K., TAKAHASHI, H., AMERIYA, Y. AND KINOSHITA, M. (1983). *J. Am. Chem. Soc.* **105**, 4096–4097.

TAYLOR, A. AND TAYLOR, N. C. (1963). *Proc. Soc. Exp. Biol. Med.* **114,** 772–774.

TERAO, Y., IMAI, N., ACHIWA, K. AND SEKIWA, M. (1982). *Chem. Pharm. Bull.* **30,** 3167–3717.

THORPE, E. AND FORD, E. J. H. (1968). *J. Comp. Pathol.* **78,** 195–205.

TITTEL, G., HINZ, H. AND WAGNER, H. (1979). *Planta Med.* **37,** 1–8.

TODD, L., MULLEN, M., OLLEY, P. M. AND RABINOVITCH, M. (1985). *Pediatric Res.* **19,** 731–737.

TRIVEDI, B. AND SANTAVY, F. (1963). *Collect. Czech. Chem. Commun.* **28,** 3455.

TSUDA, Y. AND MARION, L. (1963). *Can. J. Chem.* **41,** 1919–1923.

TSUGI, T. (1979). *Nihon Univ. J. Med.* **21,** 221–234; *Chem. Abstr.* **93,** 5459 (1980).

TUCHWEBER, B., KOVACS, K., JAGO, M. V. AND BEAULIEU, T. (1974). *Res. Commun. Chem. Pathol. Pharmacol.* **7,** 459–480.

TUCKER, A., BRYANT, S. E., FROST, H. H. AND MIGALLY, N. (1983). *Can. J. Physiol. Pharmacol.* **61,** 356–362.

TUFARIELLO, J. J. AND LEE, G. E. (1980). *J. Am. Chem. Soc.* **102,** 373–374.

TUFARIELLO, J. J. AND TETTE, J. P. (1971). *Chem. Commun.,* pp. 469–470.

TUFARIELLO, J. J. AND TETTE, J. P. (1975). *J. Org. Chem.* **40,** 3866–3869.

TUMOR RESEARCH GROUP (CHINA) (1974). *Chem. Abstr.* **80,** 66666.

TUMOR RESEARCH GROUP (CHINA) (1975). *Chem. Abstr.* **82,** 11176.

TURNER, J. H. AND LALICH, J. J. (1965). *Arch. Pathol.* **79,** 409–418.

UCHIYAMA, T., HAYASHI, H., HAYASHI, Y., TSUGI, T., EGUCHI, T. AND OKAYASU, M. (1982). *Nihon Univ. J. Med.* **24,** 123–132; *Chem. Abstr.* **97,** 86443 (1982).

ULUBELEN, A. AND DOGANCA, S. (1971). *Phytochemistry* **10,** 441–442.

ULUBELEN, A. AND OCAL, F. (1977). *Phytochemistry* **16,** 499–500.

UMEDA, M. AND SAITO, M. (1971). *Acta Pathol. Jpn.* **21,** 507–514.

UNGAR, H., SULLMAN, S. F. AND ZUCKERMAN, A. J. (1976). *Br. J. Exp. Pathol.* **57,** 157–164.

VALDIVIA, E., SONNAD, J., HAYASHI, Y., AND LALICH, J. J. (1967a). *Angiology* **18,** 378–383.

VALDIVIA, E., LALICH, J. J., HAYASHI, Y. AND SONNAD, J. (1967b). *Arch. Pathol.* **84,** 64–76.

VAN DER WATT, J. J. AND PURCHASE, I. F. H. (1970). *Br. J. Exp. Pathol.* **51,** 183–190.

VAN DER WATT, J. J., PURCHASE, I. F. H. AND TUSTIN, R. C. (1972). *J. Pathol.* **107,** 279–287.

VDOVIKO, E. A., POKHMELKINA, S. A., PETRENKO, V. V. AND CHERNENKA, N. I. (1977). *Chem. Abstr.* **88,** 126258a (1978).

VDOVIKO, E. A., PRYAKHIN, O. R. AND POKHMELKINA, S. A. (1979). *Khim. Prir. Soedin.,* pp. 674–676. *Chem. Abstr.* **95,** 30458j (1981); cf. *Chem. Abstr.* **92,** 203668 (1980).

VEDEJS, E. AND LARSEN, S. D. (1984). *J. Am. Chem. Soc.* **106,** 3030–3032.

VEDEJS, E. AND MARTINEZ, G. R. (1980). *J. Am. Chem. Soc.* **102,** 7993–7994.

VEDEJS, E., LARSEN, S., AND WEST, F. G. (1985). *J. Org. Chem.* **50,** 2170–2174.

VENZANO, A. J. AND VOTTERO, D. A. J. (1982). *Rev. Med. Vet. (Buenos Aires)* **63,** 426–438. *Vet. Bull.,* p. 961 (Abstr. No. 6705) (1983).

VILLA-TREVINO, S. AND LEAVER, D. D. (1968). *Biochem. J.* **109,** 87–91.

VISCONTINI, M. AND GILLHOF-SCHAUFELBERGER, H. (1971). *Helv. Chim. Acta* **54,** 449–456.

WADE, A., ED. (1977). "Martindale. The Extra Pharmacopoeia", 27th ed., p. 1810. Pharmaceutical Press, London.

WAGENVOORT, C. A., DINGEMANS, K. P. AND LOTGERING, G. G. (1974a). *Thorax* **29,** 511–521.

WAGENVOORT, C. A., WAGENVOORT, N. AND DIJK, H. J. (1974b). *Thorax* **29,** 522–529.

WAGNER, H., NEIDHARDT, U. AND TITTEL, G. (1981). *Planta Med.* **41,** 232–239.

WALDI, D., SCHNACKERZ, K. AND MUNTER, F. (1961). *J. Chromatogr.* **6,** 61–73.

WALI, B. K. AND HANDA, K. L. (1964). *Curr. Sci.* **33,** 585.

WALKER, K. H. AND KIRKLAND, P. D. (1981). *Aust. Vet. J.* **57,** 1–7.

WANG, S.-D. (1980a). *Chem. Abstr.* **92,** 164133.

WANG, S.-D. (1980b). *Chem. Abstr.* **92,** 181444.

WANG, S.-D. AND HU, N. (1981). *Sci. Sin. (Engl. Ed.)* **24,** 1536–1544; *Chem. Abstr.* **96,** 44186 (1982).

WANG, X., HAN, G., CUI, J., PAN, J., LI, C. AND ZHENG, G. (1981). *Chem. Abstr.* **97,** 33110 (1982).

WARREN, F. L. (1955). *Fortschr. Chem. Org. Naturst.* **12,** 198–269.

WARREN, F. L. (1966). *Fortschr. Chem. Org. Naturst.* **24,** 329–406.

WARREN, F. L., KROPMAN, M., ADAMS, R., GOVINDACHARI, T. R. AND LOOKER, J. H. (1950). *J. Am. Chem. Soc.* **72,** 1421–1422.

WATT, J. M. AND BREYER-BRANDWIJK, M. G., EDS. (1962). "The Medicinal and Poisonous plants of Southern & Eastern Africa". Livingstone, Edinburgh & London.

WEHNER, F. C., THIEL, P. G. AND VAN RENSBURG, S. J. (1979). *Mutat. Res.* **66,** 187–190.

WEI, Y., PENG, Q., WANG, Z., LUO, Q., LUO, F., WU, L., ZHAO, J. AND REN, W. (1982). *Zhongcaoyao* **13,** 435–436; *Chem. Abstr.* **98,** 113583 (1983).

WHITE, E. P. (1969). *N.Z. J. Sci.* **12,** 165.

WHITE, I. N. H. (1976). *Chem.-Biol. Interact.* **13,** 333–342.

WHITE, I. N. H. (1977). *Chem.-Biol. Interact.* **16,** 169–180.

WHITE, I. N. H. AND MATTOCKS, A. R. (1971). *Xenobiotica* **1,** 503–505.

WHITE, I. N. H. AND MATTOCKS, A. R. (1972). *Biochem. J.* **128,** 291–297.

WHITE, I. N. H. AND MATTOCKS, A. R. (1976). *Chem.-Biol. Interact.* **15,** 185–198.

WHITE, I. N. H., MATTOCKS, A. R. AND BUTLER, W. H. (1973). *Chem.-Biol. Interact.* **6,** 207–218.

WHITE, R. D., KRUMPERMAN, P. H., CHEEKE, P. R. AND BUHLER, D. R. (1983). *Toxicol. Lett.* **15**, 25–31.

WHITE, R. D., SWICK, R. A. AND CHEEKE, P. R. (1984a). *Am. J. Vet. Res.* **45**, 159–161.

WHITE, R. D., KRUMPERMAN, P. H., CHEEKE, P. R., DEINZER, M. L. AND BUHLER, D. R. (1984b). *J. Anim. Sci.* **58**, 1245–1254.

WICKRAMANAYAKE, P. P., ARBOGAST, B. L., BUHLER, D. R., DEINZER, M. L., AND BURLINGAME, A. L. (1985). *J. Am. Chem. Soc.* **107**, 2485–2488.

WIEDENFELD, H. (1982). *Phytochemistry* **21**, 2767–2768.

WIEDENFELD, H. AND ROEDER, E. (1979). *Phytochemistry* **18**, 1083–1084.

WIEDENFELD, H., PASTEWKA, U., STENGL, P. AND ROEDER, E. (1981). *Planta Med.* **41**, 124–128.

WIEDENFELD, H., KNOCH, F., ROEDER, E. AND APPEL, R. (1984). *Arch. Pharm. (Weinheim)* **317**, 97–102.

WILL, J. A. (1981). *Am. J. Physiol.* **241**, H894.

WILLETTE, R. E. AND CAMMARATO, L. V. (1972). *J. Pharm. Sci.* **61**, 122.

WILLETTE, R. E. AND DRISCOLL, R. C. (1972). *J. Med. Chem.* **15**, 110–112.

WILLIAMS, A. O. AND SCHOENTAL, R. (1970). *Trop. Geogr. Med.* **22**, 201–210.

WILLIAMS, A. O., EDINGTON, G. M. AND OBAKPONOVWE, P. C. (1967). *Br. J. Cancer* **21**, 474–482.

WILLIAMS, D. E., MIRANDA, C. L. AND BUHLER, D. R. (1983). *Biochem. Pharmacol.* **32**, 2443–2447.

WILLIAMS, G. M., MORI, H., HIRONO, I. AND NAGAO, M. (1980). *Mutat. Res.* **79**, 1–5.

WILLIAMS, H. (1960). *Chem. Ind. (London)*, 900.

WILLMOT, F. C. AND ROBERTSON, G. W. (1920). *Lancet* **2**, 848–849.

WODAK, S. J. (1975). *Acta Crystallogr., Sect. B* **B31**, 569–573.

WOLF, H. U. (1983). *Dtsch. Apoth.-Ztg.* **123**, 2116–2117.

WONG, R. Y. AND ROITMAN, J. N. (1984). *Acta Crystallogr., Sect. C* **C40**, 163–166.

WUNDERLICH, J. A. (1962). *Chem. Ind. (London)*, 2089–2090.

WUNDERLICH, J. A. (1967). *Acta Crystallogr.* **23**, 846–855.

YADAR, V. K., RUEGER, H. AND BEHN, M. (1984). *Heterocycles* **2**, 2735–2738.

YAKHONTOVA, L. D., PIMENOV, M. G. AND SAPUNOVA, L. A. (1976). *Khim. Prir. Soedin.*, 122–123; *Chem. Abstr.* **85**, 59575 (1976).

YAMADA, K., TATEMATSU, H., SUZUKI, M., HIRATA, Y., HAGA, M. AND HIRONO, I. (1976a). *Chem. Lett.*, 461–464.

YAMADA, K., TATEMATSU, H., HIRATA, Y., HAGA, M. AND HIRONO, I. (1976b). *Chem. Lett.*, 1123–1126.

YAMADA, K., TATEMATSU, H., UNNO, R. AND HIRATA, Y. (1978). *Tetrahedron Lett.*, 4543–4546.

YAMANAKA, H., NAGAO, M., SUGIMURA, T., FURUYA, T., SHIRAI, A. AND MATSUSHIMA, T. (1979). *Mutat. Res.* **68**, 211–216.

YATES, S. G. AND TOOKEY, H. L. (1965). *Aust. J. Chem.* **18,** 53–60.

YUNUSOV, S. YU. AND PLEKHANOVA, N. V. (1953). *Dokl. Akad. Nauk. Uzb. SSR,* (4), 28; *Chem. Abstr.* **51,** 1539 (1957).

YUNUSOV, S. YU. AND PLEKHANOVA, N. V. (1959). *Zh. Obshch. Khim.* **29,** 677; *Chem. Abstr.* **54,** 1580 (1960).

ZALKOW, L. H., GELBAUM, L. AND KEINAN, E. (1978). *Phytochemistry* **17,** 172.

ZALKOW, L. H., BONETTI, S., GELBAUM, L., GORDON, M. M., PATIL, B. B., SHANI, A. AND VAN DERVEER, I. D. (1979). *J. Nat. Prod.* **42,** 603–614.

ZALKOW, L. H., GLINSKI, J. A., GELBAUM, L. T., FLEISCHMANN, T. J., MCGOWAN, L. S., AND GORDON, M. M. (1985). *J. Med. Chem.* **28,** 687–694.

ZAPESOCHNAYA, G. G., FEDYUNINA, N. A., LARIONOV, N. G. AND BAN'KOVSKII, A. I. (1973). *Khim.-Farm. Zh.* **7,** 37–38; *Chem. Abstr.* **79,** 96898 (1973).

ZIMMER, H., LANKIN, D. C. AND HORGAN, S. W. (1971). *Chem. Rev.* **71,** 229–246.

Index

A

Abortion, 257
ACE, *see* Angiotensin converting enzyme
Acetic anhydride, conversion of *N*-oxides to
  pyrroles, 56, 57, 82, 84, 140, 142
Acetyl derivatives of PAs, 79, 134
Acetyl group, steric effect, 177, 178, 321
Acetyl heliotrine, toxicity, 221
Acetyl lycopsamine, countercurrent purifica-
  tion, 48
Acetylaminofluorene (AAF), *see* *N*-2-
  Fluorenylacetamide
Acetylation
  of PAs, 78, 79
  of pyrrolic alcohols, 84
  toxicity increase, 221
Acid catalysis
  of alkylation reactions, 145–147, 187
  of polymerization, 150, 155
Acid moiety, influence on PA metabolism,
  321, 322
Acidity, stomach tissue, 187
Actinomycin D, action on cell cycle, 289
Activation
  metabolic, 168, 316
  of mutagens, 275, 278
Active metabolites, 318–320
Acute hepatotoxicity, metabolites responsible,
  325, 326
Acyl derivatives, of PAs, 79, 134
Acyl substituents, causing steric hindrance,
  321
Additive effect of carcinogens, 271
Adenine, alkylation of, 154, 155
Adenocarcinoma 755, 311

Adenoma
  liver cell, 295, 297, 299, 300, 306, 307
  pulmonary, 294
Adrenal glands, haemorrhage, 211
Afghanistan, outbreaks of poisoning, 260
Aflatoxin B1, 214
  hepatocarcinogenicity, 305, 310
Aflatoxins, in diet, 271
Age
  effect on metabolism, 165, 178
  effect on susceptibility to PAs, 196, 202, 203
Albumin, alkylation, 150
Alcohols
  oxidation to aldehydes, 87, 88
  pyrrolic, preparation, 83
Aldehyde
  derivatives, of necines, 137
  metabolite, 171
  from pyrrolic alcohol, 88
  reduction, 138
Alkyl-oxygen fission, 134
Alkylating
  agents
    mutagenic, 275
    pneumotoxicity, 241–243
  reactivity, 56, 150, 151, 242, 243
    relationship to antitumour activity, 314,
      315
  role in toxicity, 318, 323, 324
Alkylation
  of alcohols, 147
  by allylic ester PAs, 134
  of amines, 148
  of DNA, 279
  of enols and phenols, 149, 150
  equilibrium, 154